Smart Materials for Advanced Environmental Applications

RSC Smart Materials

Series Editors:
Professor Hans-Jörg Schneider, *Saarland University, Germany*
Professor Mohsen Shahinpoor, *University of Maine, USA*

Titles in this Series:
1: Janus Particle Synthesis, Self-Assembly and Applications
2: Smart Materials for Drug Delivery: Volume 1
3: Smart Materials for Drug Delivery: Volume 2
4: Materials Design Inspired by Nature
5: Responsive Photonic Nanostructures: Smart Nanoscale Optical Materials
6: Magnetorheology: Advances and Applications
7: Functional Nanometer-Sized Clusters of Transition Metals: Synthesis, Properties and Applications
8: Mechanochromic Fluorescent Materials: Phenomena, Materials and Applications
9: Cell Surface Engineering: Fabrication of Functional Nanoshells
10: Biointerfaces: Where Material Meets Biology
11: Semiconductor Nanowires: From Next-Generation Electronics to Sustainable Energy
12: Supramolecular Materials for Opto-Electronics
13: Photocured Materials
14: Chemoresponsive Materials: Stimulation by Chemical and Biological Signals
15: Functional Metallosupramolecular Materials
16: Bio-Synthetic Hybrid Materials and Bionanoparticles: A Biological Chemical Approach Towards Material Science
17: Ionic Polymer Metal Composites (IPMCs): Smart Multi-Functional Materials and Artificial Muscles Volume 1
18: Ionic Polymer Metal Composites (IPMCs): Smart Multi-Functional Materials and Artificial Muscles Volume 2
19: Conducting Polymers: Bioinspired Intelligent Materials and Devices
20: Smart Materials for Advanced Environmental Applications

How to obtain future titles on publication:
A standing order plan is available for this series. A standing order will bring delivery of each new volume immediately on publication.

For further information please contact:
Book Sales Department, Royal Society of Chemistry, Thomas Graham House, Science Park, Milton Road, Cambridge, CB4 0WF, UK
Telephone: +44 (0)1223 420066, Fax: +44 (0)1223 420247
Email: booksales@rsc.org
Visit our website at www.rsc.org/books

Smart Materials for Advanced Environmental Applications

Edited by

Peng Wang
King Abdullah University of Science and Technology, Thuwal, Saudi Arabia
Email: peng.wang@kaust.edu.sa

ROYAL SOCIETY OF CHEMISTRY

THE QUEEN'S AWARDS FOR ENTERPRISE: INTERNATIONAL TRADE 2013

RSC Smart Materials No. 20

Print ISBN: 978-1-78262-108-9
PDF eISBN: 978-1-78262-219-2
EPUB eISBN: 978-1-78262-820-0
ISSN: 2046-0066

A catalogue record for this book is available from the British Library

Published by the Royal Society of Chemistry,
Thomas Graham House, Science Park, Milton Road,
Cambridge CB4 0WF, UK

Registered Charity Number 207890

For further information see our web site at www.rsc.org

Printed in the United Kingdom by CPI Group (UK) Ltd, Croydon, CR0 4YY, UK

Preface

Water pollution and water scarcity are among the most severe grand environmental challenges facing mankind nowadays. With rapid population growth, steadily improving living standards, fast industrialization and modernization of the developing countries, these challenges will persist, if not worsen, in the years to come. Conventional water treatment technologies, including adsorption, chemical treatment, membrane-based filtration, biological treatment, *etc.*, with no doubt have made critical contributions in sustaining human society in the past century. However, the ever-increasing demand for safe and clean water by the ever-growing human population has gradually pushed these conventional technologies to their limits over the past 100 years or so. Therefore, it is now a popular perception that the solutions to existing and future environmental problems highly hinge on further developments in materials science.

The concept of smart materials, since its inception in 1990s, has extended its presence in a variety of applications and has led to development of many new technologies. Smart materials are conventionally defined as materials that are designed to have one or more properties which can be significantly changed in a controlled fashion in response to external stimuli, such as stress, temperature, moisture, pH, electric fields or magnetic fields. Piezoelectric materials and shape-memory materials were the early types of smart materials. However, in essence, the response mechanism of all smart materials lies in the change in molecular movement in response to external stimuli, which then brings about the macroscopic property change of the materials. Following this line of thought, the origin of all smart materials is biomimicry as nature is the ultimate builder of machines. Thus, smart materials can also be generically defined as materials which are astute or "operating as if by intelligence", and the term "smart materials" can thus be used broadly

RSC Smart Materials No. 20
Smart Materials for Advanced Environmental Applications
Edited by Peng Wang

Published by the Royal Society of Chemistry, www.rsc.org

to refer to materials that are designed and fabricated from bioinspirations, mimicking nature's procedures, structures or strategies.

The concept of smartness in materials design involves thinking outside the box and integrates multiple synergistic and advanced functions into one single material or device, which helps create new applications, broaden existing applications, enhance performances, and elongate the lifetime of the materials. Therefore, the development of smart materials represents the future of materials design and fabrication. It is not a surprise that smart materials have also set foot in the environmental field, and have shown great promise in coming up with novel and next-generation solutions to the large environmental challenges. In return, the interaction between environmental science and materials science has also promoted the development of new smart materials, and a series of novel smart materials and their application in environmental areas have been put forward and explored.

However, this book is not intended to be exhaustive about all areas of smart material applications to advanced environmental problem solving. Rather, it focuses on some selected topics, especially bioinspired smart interfacial materials, which I believe are novel and inspirational to conventional thinking. As you will see, the topics of the book chapters are truly multidisciplinary. They span from an introduction to smart materials and their applications (Chapter 1, Zhang and Professor Wang), smart draw solutions in forward osmosis (Chapter 2, Chen and Professor Lu), superwetting materials for oil–water separation (Chapter 3, Gao and Professor Jin), responsive particle-stabilized emulsions (Chapter 4, Kwok and Professor Ngai), self-healing materials (Chapter 5, Professor Zeng *et al.*), bioinspired fog collection (Chapter 6, Zhang and Professor Wang), nature-inspired "slippery" liquid-infused surfaces (Chapter 7, Professor Zacharia) to challenges and opportunities of superhydrophobic/superamphiphobic coatings in real applications (Chapter 8, Paven, Mammen and Professor Vollmer). The contributors are all established researchers who are in their early or mid career. In my opinion, they are the ones to watch in the years to come in this emerging field of smart materials and their applications to environmental problem solving.

I hope this book will provide an inspiration for readers who are interested in smart materials and who are passionate at further exploring smart materials to make contributions to the solutions to our grand environmental challenges.

Peng Wang
King Abdullah University of Science and Technology
Thuwal, Saudi Arabia

Contents

RSC Smart Materials No. 20
Smart Materials for Advanced Environmental Applications
Edited by Peng Wang

Published by the Royal Society of Chemistry, www.rsc.org

CHAPTER 1

Introduction

LIANBIN ZHANG[a] AND PENG WANG*[a]

[a]Water Desalination and Reuse Center (WDRC), Division of Biological and Environmental Sciences and Engineering, King Abdullah University of Science and Technology (KAUST), Thuwal, Saudi Arabia
*E-mail: Peng.Wang@kaust.edu.sa

With the development of modern industry and modern economies, environmental problems, especially water pollution and water scarcity, have become the most serious global challenges.[1–4] In dealing with these challenges, various kinds of functionalized materials and devices are purposefully developed, fabricated, and utilized.[5–10] For instance, to clean up spilled oil from an oil leakage accident, absorbent materials have been widely used to selectively remove the oil from the aqueous system.[9,10] Shown in Figure 1.1 are examples of absorptive pads used in the Kalamazoo River oil spill (Michigan, US, 2010) and the Huangdao oil pipeline explosion incident (Qingdao, China, 2013). These absorptive pads are generally made of recycled cellulose or polypropylene, and have highly porous structures.[10] Upon the oil removal, the pads, which have the absorbed oil, are collected and disposed of by direct burning. In other areas, such as wastewater treatment and seawater desalination, membrane-based separation processes are essential, which necessitate specifically functionalized membranes, such as nanofiltration (NF), ultrafiltration (UF), and reverse osmosis (RO) membranes.[11–16] Therefore, it is now a popular perception that the solutions to existing and future environmental problems highly rely on developments in material sciences.

RSC Smart Materials No. 20
Smart Materials for Advanced Environmental Applications
Edited by Peng Wang

Published by the Royal Society of Chemistry, www.rsc.org

The concept of smart materials, since its inception in 1990s,[17,18] has emerged as a novel design concept, which promise a variety of new technologies, extending from shape-memory materials, artificial muscles, nanoscale motors, and biosensors, to new drug-delivery devices, *etc.* (Figure 1.2a).[19–29] Smart materials are conventionally defined as materials that are designed to have one or more properties which can be significantly changed in a controlled fashion in response to external stimuli, such as stress, temperature, moisture, pH, electric fields, or magnetic fields.[30–33] Piezoelectric materials[34–38] and shape-memory materials[17–20] (including shape-memory alloys and shape-memory polymers) are early types of smart materials. Piezoelectric materials produce electrical charges when subjected to dynamic strain (Figure 1.3a). The piezoelectric effect is understood as the linear electromechanical interaction between the mechanical and the electrical state in crystalline materials with no inversion symmetry.[34] Piezoelectric materials have been used to capture and store vibrational energy that can later be used to

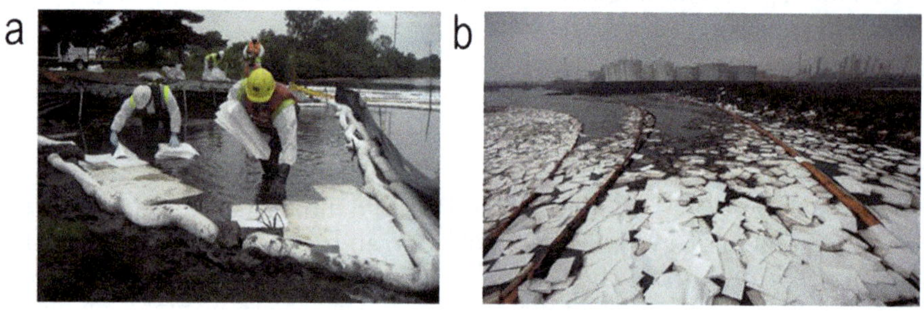

Figure 1.1 Oil spill absorbents used in oil leakage incidents. (a) Sorbent booms, sweeps, and snares were used to help aid in the massive cleanup efforts after the Kalamazoo River oil spill, Michigan, US, July 28, 2010. (Picture from http://www.chemtexinc.com/whats-new-at-chemtex.html.) (b) Oil fences and pads to control leaked oil after a Sinopec Corp oil pipeline explosion, Huangdao, Qingdao, China, November 24, 2013. (Picture from http://www.citylab.com/politics/2013/11/scenes-destruction-chinas-deadly-pipeline-explosion/7694/).

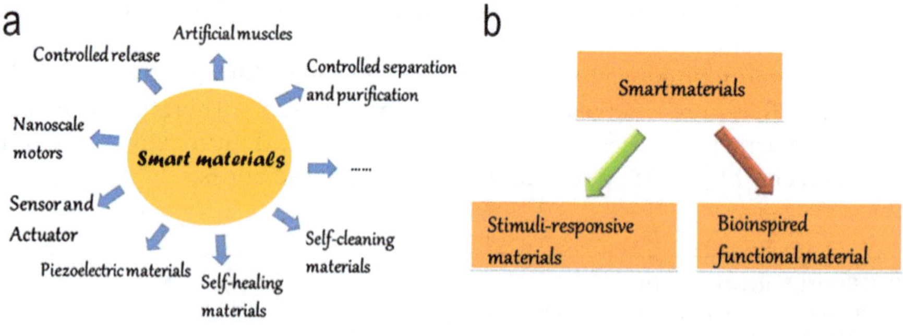

Figure 1.2 (a) Various applications of smart materials. (b) Classification of the smart materials.

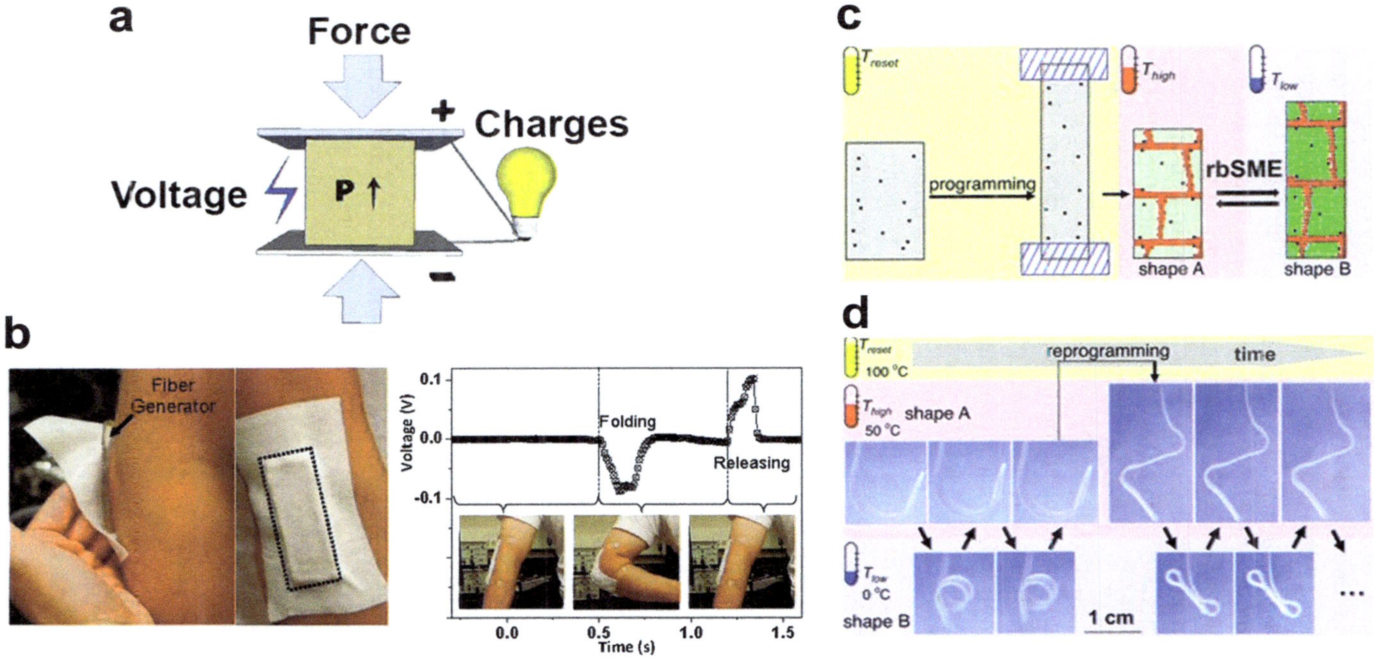

Figure 1.3 (a) A schematic illustration of the piezoelectric effect. (b) A hybrid piezoelectric structure for wearable nanogenerators attached on a human arm. (Reprinted with permission from M. Lee, C.-Y. Chen, S. Wang, S. N. Cha, Y. J. Park, J. M. Kim, L.-J. Chou and Z. L. Wang, A Hybrid Piezoelectric Structure for Wearable Nanogenerators, *Adv. Mater.*, 2012, **24**, 1759. Copyright © [2012] John Wiley and Sons.) (c) Scheme of the bidirectional shape-memory free-standing copolymer network: after deformation at T_{reset} the skeleton domains (red), which determine the shape shifting geometry, are crystallized by cooling (programming). The rbSME is triggered by the reversible crystallization and melting of oriented actuator domains (green). (d) Photograph series showing bidirectional shape memory of a polymer ribbon. The bowed shape was obtained after programming by deformation in a helix-like shape at T_{reset}, cooling to T_{low} and subsequent heating to T_{high}. The bidirectional shape memory occurred as a reversible shift between shape A (bow) at T_{high} and shape B (helix) at T_{low}. The sample was reprogrammed by T_{reset} into an open shape (new shape A), which could be shifted reversibly to a folded shape (new shape B). (Reprinted with permission from M. Behl, K. Kratz, J. Zotzmann, U. Nöchel and A. Lendlein, Reversible Bidirectional Shape-Memory Polymers, *Adv. Mater.*, 2013, **25**, 4466–4469. Copyright © [2013] John Wiley and Sons.)

power up small devices. Shown in Figure 1.3b is a hybrid-fiber nanogenerator based on the piezoelectric effect, which comprises zinc oxide nanowires (NWs) and a poly(vinylidene fluoride) (PVDF) polymer around a conducting fiber.[38] The zinc oxide NWs serve as a piezoelectric-potential generator and also as an additive to enhance the surface contact area, which guides the formation of a uniform layer of piezoelectric polymer (*i.e.*, PVDF) around the fiber during the dip-coating process. By elongating or bending the hybrid fiber, mechanical energy is converted into electricity owing to constructive piezoelectric-potential generation from the two components (*i.e.*, PVDF and ZnO NWs). The unique structure of the hybrid fiber may inspire future research in wearable energy-harvesting technology.

Shape-memory materials are featured by their ability to recover their original shape from a significant and seemingly plastic deformation when a particular stimulus is applied, which is known as the shape-memory effect.[17–20,39–41] Shape-memory materials have been utilized in many fields, from aerospace engineering (*e.g.*, in deployable structures and morphing wings) to medical devices (*e.g.*, in stents and filters). Figure 1.3c and d show an example of a reversible bidirectional shape-memory free-standing copolymer network with two types of crystallizable domains that is capable of a fully reversible bidirectional shape-memory effect.[41] This copolymer network is made of a multiphase copolyester urethane and consists of one set of crystallizable domain that determines the shape-shifting geometry, while the other domain of the network provides the thermally controlled actuation capability. This reversible bidirectional shape-memory technology can be inspirational to the design of many smart devices such as self-sufficient grippers and reversible fastening devices.

In essence, the response mechanism of smart materials lies in the change in molecular movement in response to external stimuli, which brings about the macroscopic property change of the materials. Following this line of thought, the origin of all smart materials is biomimicking, as nature is the ultimate builder of machines.[30] Therefore, in the comprehensive view, 'smart materials' describe materials which are astute or 'operating as if by intelligence', and they can also widely refer to materials that are fabricated from bioinspirations, mimicking nature's procedures, structure, or strategy (Figure 1.2b). The simplest of organisms in nature can utilize atomic-scale engineering to build systems from fundamental chemistry into tissue with functional gradation to alter properties, integrated with electrical, chemical, and mechanical capabilities.[42] Many fascinating biological processes are systems that exhibit motor functions that perform mechanical transformations triggered by a fuel input. Different triggering stimuli, such as adenosine triphosphate (ATP) fuel, a pH gradient, and light signals, are common activators of biological motors.[22] For instance, the myosin–actin couple represents an ATP-driven motor that is translated to macroscopic mechanical functions of muscles.[43,44]

Inspired by biomotors, researchers have artificially fabricated smart molecular motor or nanoscale motor (nanomotor) systems, and further developed artificial muscles and robots.[21–23,43–50] Ingenious supramolecular architectures that enable the signal-triggered mechanical translocation of molecular units have been developed. Figure 1.4a shows a linear vectorial translocation of a molecular unit on a molecular wire.[22] The system consists of a crown ether threaded onto a wire that includes an ammonium group and a bipyridinium group as binding sites and two bulky groups that act as stoppers at the ends of the wire. The crown ether is stabilized on the ammonium site in the form of structure (**I**). Deprotonation of the ammonium unit (chemical stimulus) results in the shuttling of the crown ether to the less-favored bipyridinium unit to form (**II**), whereas acidification of the system, and protonation of the amine site, reshuttles the crown ether ring to the protonated site to re-form (**I**). Beside this molecular-level motor, nanomotors with macroscopic movement have been developed after Whitesides and coworkers first reported self-propelled movement of centimeter-sized plates,[51] capable of converting chemical energy into autonomous movement.[21,23] Several propelling mechanisms have been employed for the realization of the movement of the nanomotors, including bubble recoil, bubble implosion, Brownian ratchet, thermal gradient, interfacial tension, bipolar electrophoresis, and self-diffusiophoresis.[52–57] Figure 1.4b shows an example of the autonomous movement of a nanomotor. The nanomotor is prepared by selectively entrapping catalytically active platinum nanoparticles within polymer stomatocytes nanocavities and subsequently using catalysis as a driving force for movement.[58] Hydrogen peroxide can access the inner stomatocyte cavity, where it is decomposed by the active catalyst (the entrapped platinum nanoparticles) into oxygen and water. This generates a rapid discharge, which induces thrust and directional movement.

Another type of important smart materials is the self-healing material.[59–63] The inspiration for self-healing materials also comes from biological systems, which have the ability to heal after being wounded. Self-healing materials can be polymers, metals, ceramics, and their composites. When damaged through thermal, mechanical, ballistic, or other means they have the ability to heal and restore the material to its original set of properties. This is a very valuable characteristic to design into a material since it effectively expands the lifetime use of the product and has desirable economic and human safety attributes. White and coworkers pioneered a structural polymeric material with the ability to autonomously heal cracks, as illustrated in Figure 1.5a.[64] This material incorporates a microencapsulated healing agent that is released upon crack intrusion. Polymerization of the healing agent is then triggered by contact with an embedded catalyst, bonding the crack faces. Their fracture experiments have revealed as much as 75% recovery in toughness.

Human skin has a unique structure that enables minor damage to be healed effectively. The outer epidermal layer of skin is composed of multiple sublayers that work in concert to continually rebuild the surface of the skin, whereas the underlying dermal layer supplies the epidermis with nutrient-laden blood and regulates temperature.[65] A cut in the skin triggers blood flow from the capillary network in the dermal layer to the wound site, rapidly

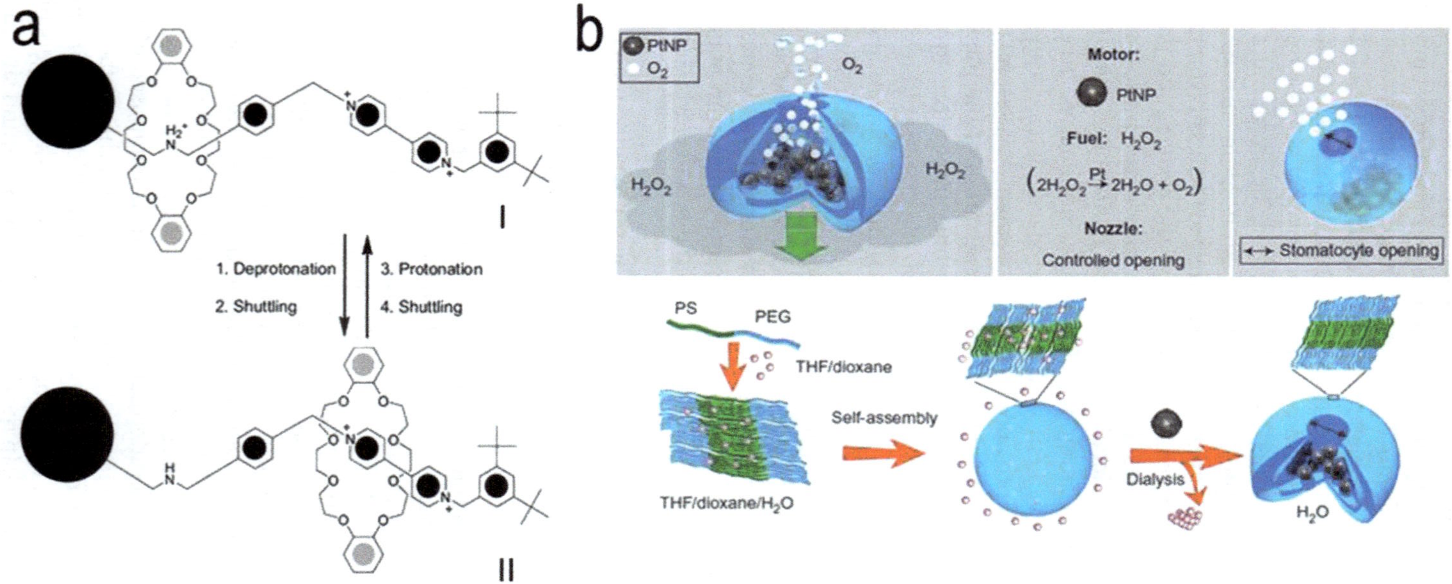

Figure 1.4 (a) Schematic diagram of pH-induced shuttling of a crown ether interlocked in a rotaxane configuration between two distinct binding sites. (Reprinted with permission from I. Willner, B. Basnar and B. Willner, From Molecular Machines to Microscale Motility of Objects: Application as "Smart Materials", Sensors, and Nanodevices, *Adv. Funct. Mater.*, 2007, **17**, 702. Copyright © [2007] John Wiley and Sons.) (b) Design of the supramolecular nanomotor. The autonomous movement of artificial stomatocytes (blue) is made possible by entrapping catalytically active nanoparticles (grey, here Pt nanoparticles) through control of the opening and adding H_2O_2. (Reprinted with permission from Macmillan Publishers Ltd., Nature Publishing Group: *Nat. Chem.*, 2012, **4**, 268–274 copyright (2012).)

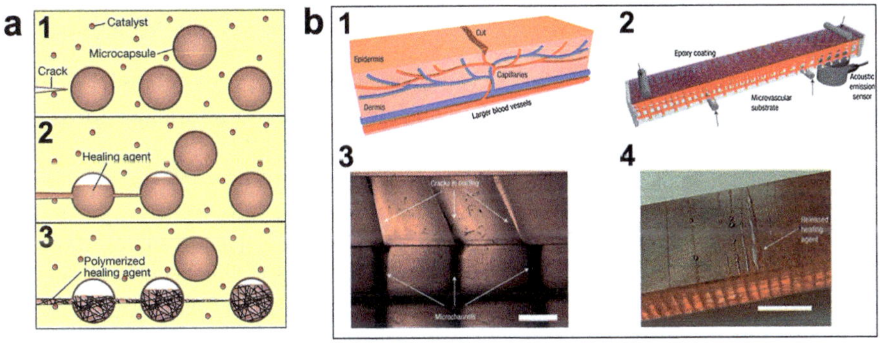

Figure 1.5 (a) Concept of structural self-healing materials. A microencapsulated healing agent is embedded in a structural composite matrix containing a catalyst capable of polymerizing the healing agent. (1) Cracks form in the matrix wherever damage occurs; (2) the crack ruptures the microcapsules, releasing the healing agent into the crack plane through capillary action; (3) the healing agent contacts the catalyst, triggering polymerization that bonds the crack faces closed. (Reprinted with permission from Macmillan Publishers Ltd., Nature Publishing Group: *Nature*, 2001, **409**, 794, copyright (2001).) (b) Schematic diagram of the skin-inspired self-healing structure composed of a microvascular substrate and a brittle epoxy coating containing embedded catalyst in a four-point bending configuration monitored with an acoustic-emission sensor. (1) Schematic diagram of a capillary network in the dermis layer of skin with a cut in the epidermis layer. (2) Self-healing structure composed of a microvascular substrate and a brittle epoxy coating. (3) High-magnification cross-sectional image of the coating showing that cracks, which initiate at the surface, propagate towards the microchannel openings at the interface (scale bar = 0.5 mm). (4) Optical image of a self-healing structure after cracks are formed in the coating, revealing the presence of excess healing fluid on the coating surface (scale bar = 5 mm). (Reprinted with permission from Macmillan Publishers Ltd., Nature Publishing Group: *Nat. Mater.*, 2007, **6**, 581, copyright (2007).)

forming a clot that serves as a matrix through which cells and growth factors migrate as healing ensues. Owing to the vascular nature of this supply system, minor damage to the same area can be healed repeatedly. By mimicking the human skin, White *et al.* reported the use of microvascular networks for autonomic healing of structural damage by exploring a microvascular coating–substrate architecture (Figure 1.5b).[66] To this end, an epoxy coating is deposited on a ductile substrate that contains a pervasive three-dimensional (3D) microvascular network. Solid catalyst particles are incorporated within the coating and the network is filled with a liquid healing agent. After damage occurs in the coating, healing agent wicks from the microchannels into the crack(s) through capillary action, and interacts with the catalyst particles in the coating to initiate polymerization, rebonding the crack faces autonomously. After a sufficient time period, the cracks are healed and the structural integrity of the coating is restored. Clearly, with this smart self-healing ability, materials can have longer lifetimes and lower production and maintenance costs.

Bioinspired interfacial material is another important type of smart materials.[67–73] A famous example of the bioinspired interfacial materials is lotus-leaf inspired superhydrophobic self-cleaning material.[74,75] Since Jiang *et al.* first revealed that the combination of spatial micro- and nanometer-scale hierarchical structures and proper chemical composition results in the superhydrophobic self-cleaning effect of lotus leaf,[75] numerous efforts have been devoted to this area and many artificial superhydrophobic surfaces have been developed, which allowed for a series of applications, such as self-cleaning surfaces, marine coatings, anti-adhesive coatings, microfluidic channels with reduced flow resistance, and so forth.[67–76] Furthermore, in combination with the responsive polymer materials, smart responsive surfaces were developed.[77–80] Jiang *et al.* first demonstrated reversible switching between superhydrophilicity and superhydrophobicity.[77] By grafting temperature-responsive polymer of poly(*N*-isopropylacrylamide) (PNIPAM) onto a rough substrate, thermo-responsive switching between superhydrophilicity and superhydrophobicity was realized. As shown in Figure 1.6a, with increasing temperature the water contact angle on such surface increases. This effect can be explained by the competition between intermolecular and intramolecular hydrogen bonding below and above the lower critical solution temperature (LCST). At temperatures below the LCST the predominantly intermolecular hydrogen bonding contributes to the hydrophilicity of PNIPAM films whereas at temperatures above the LCST increased intramolecular hydrogen bonding results in hydrophobic films (Figure 1.6b). Later on, Jiang *et al.* further demonstrated that the PNIPAM surface could effectively control the platelet adhesion *in vitro*.[78] They grafted the polymer onto silicon nanowire arrays *via* surface-initiated atom transfer radical polymerization and found that the as-prepared surface showed largely reduced platelet adhesion *in vitro* both below and above the LCST of PNIPAM (~32 °C), while a smooth PNIPAM surface exhibited anti-adhesion to platelets only below the LCST (Figure 1.6c). Contact angle and adhesive force measurements on oil droplets in water demonstrated that the nanoscale topography kept a relatively high ratio of water content on the as-prepared surface and played a key role in largely reducing the adhesion of platelets (Figure 1.6d). The results have potential in the applications of PNIPAM in the fields of biomaterials and biomedicine under human physiological temperature and provide a new strategy for fabricating new blood-compatible materials.

From the above examples, we can see that the concept of smartness in material design rationally integrates multiple synergistic and advanced functions into one single material or device, which helps broaden the application and elongate the lifetime of the materials. Therefore, the development of smart materials represents the future of material design and fabrication. Therefore, it is not a surprise that smart materials have also set foot in the environmental community, and have shown great promises in coming up with novel and next generation solutions to the grand environmental challenges. In turn, the interaction between environmental science with materials

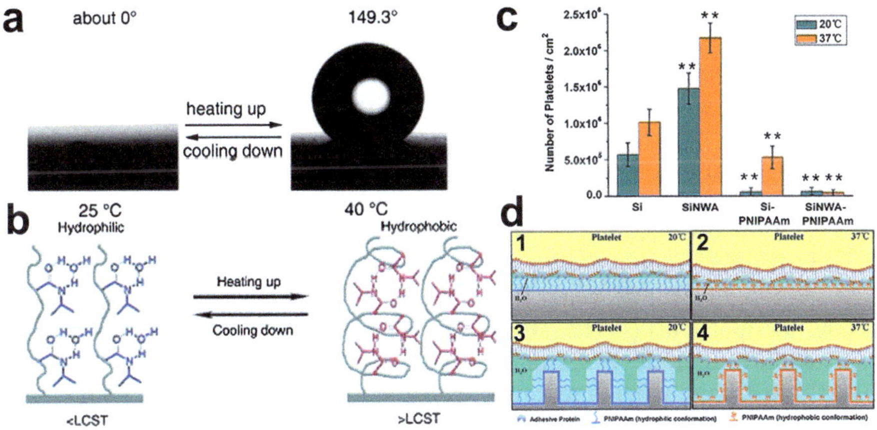

Figure 1.6 (a) Water drop profile for thermally responsive switching between superhydrophilicity and superhydrophobicity of a PNIPAM-modified rough surface, at 25 °C and 40 °C. (b) Diagram of reversible formation of intermolecular hydrogen bonding between PNIPAM chains and water molecules (left) and intramolecular hydrogen bonding between C=O and N–H groups in PNIPAM chains (right) below and above the LCST, which is considered to be the molecular mechanism of the thermally responsive wettability of a PNIPAM thin film. (Reprinted with permission from T. Sun, G. Wang, L. Feng, B. Liu, Y. Ma, L. Jiang and D. Zhu, Reversible Switching Between Superhydrophilicity and Superhydrophobicity, *Angew. Chem., Int. Ed.*, 2004, **43**, 357. Copyright © [2004] John Wiley and Sons.) (c) Statistical histogram of the number of adhered platelets on different surfaces at 20 °C (blue bars) and 37 °C (orange bars): Si, pristine smooth silicon wafer; SiNWA, silicon nanowire array; Si–PNIPAM, PNIPAM grafted onto Si; SiNWA–PNIPAM, PNIPAM grafted onto SiNWAc. (d) Hypothetical platelet adhesion mechanisms for different surfaces. (1) On the Si–PNIPAM surface at 20 °C, extended polymer chains form a hydration layer that prevents adhesive proteins of platelets from adsorbing. (2) On the Si–PNIPAM surface at 37 °C, the polymer chains crouch to a hydrophobic state, enabling adhesive proteins to adsorb. (3) On the SiNWA–PNIPAM surface at 20 °C, water is trapped into the interstices of the nanowire arrays, causing a higher hydration ratio of the surface and lower protein adhesion. (4) On the SiNWA–PNIPAM surface at 37 °C, the water-trapping effect changes the surface wettability from a hydrophobic state to a hydrophilic state, thus largely reducing the adsorption of adhesive proteins. (Reprinted with permission from L. Chen, M. Liu, H. Bai, P. Chen, F. Xia, D. Han and L. Jiang, *J. Am. Chem. Soc.*, 2009, **131**, 10467–10472. Copyright (2009) American Chemical Society.)

science has also promoted the development of smart materials, and a series of novel smart materials and their applications in environmental area have been explored. For instance, forward osmosis (FO) is a promising membrane technology for seawater desalination and wastewater treatment at low energy cost.[15,81–83] FO is a naturally occurring osmosis-driven process, and consists of two key components: (i) a semipermeable membrane that allows the permeation of water while rejecting salts and other unwanted elements; and (ii) a draw solution that is able to generate sufficiently high osmotic pressure to draw water from the feed solution across the membrane. In an applicable FO desalination process, the draw solute should be recycled and regenerated in economical and efficient ways. The use of smart draw agents with facile recovery methods can effectively lower the energy consumption of the process at a great potential.[84–90] To this end, Lu *et al.* reported a new design of thermoresponsive magnetic nanoparticles (TMNPs) to significantly improve their osmolality for seawater desalination (Figure 1.7).[90] In this work, magnetic nanoparticles (MTPs, *i.e.*, Fe_3O_4) are functionalized with a copolymer, poly(sodium styrene-4-sulfonate)-*co*-poly(*N*-isopropylacrylamide) (PSSS-PNIPAM). This copolymer integrates two functions: poly(sodium styrene-4-sulfonate) (PSSS) as a polyelectrolyte can dissociate to give a large number of ions in solution and to provide high osmolality, whereas thermoresponsive PNIPAM facilitates draw-solute regeneration *via* particle aggregation at enhanced temperatures. The MNPs functionalized with PSSS-PNIPAM successfully draw water from seawater. In addition, regeneration of the MNP-based draw solution can be achieved with magnetic separation assisted by mild heating.

In the field of oil-spill cleanup, smart materials also show tremendous advantages over conventional methods. A bioinspired smart membrane with superwetting behavior can easily realize gravity-driven oil–water separation, which is of great importance to facilitate the oil-spill cleanup, contributing to much reduced response time and operation cost.[91–94] For instance, inspired by the superhydrophobic lotus leaf, Jiang and coworkers in 2004 first prepared superhydrophobic and superoleophilic mesh for gravity-driven oil–water separation, by depositing low-surface-energy material of polytetrafluoroethylene (PTFE) on stainless steel mesh.[95] The coated mesh showed a water contact angle greater than 150°, and diesel contact angle of ~0°, and thus can be used effectively for the separation of oil and water, allowing oil to pass though while retaining water. Later, researchers gradually realized that although this superhydrophobic and superoleophilic material is effective for the separation of an oil–water mixture, this ‘oil-removing’ type of material can be easily fouled or even blocked by oils because of their intrinsic oleophilic property. The adhered oils, especially high-viscosity oils, seriously affect the separation efficiency after usage for a limited period of time. In addition, oils adhered or absorbed are hard to remove, which results in secondary pollution during the post-treatment process as well as a waste of both oil and oleophilic materials. To solve this problem, inspired by the oil-repellent capability of fish scale (Figure 1.8a),[96] Jiang and coworkers recently fabricated a novel superhydrophilic and underwater superoleophobic hydrogel-coated mesh (Figure 1.8b).[97] This novel ‘water-removing’ type of material has completely opposite wettability to traditional hydrophobic and oleophilic materials, and thus overcomes the easy-fouling and recycling problem because, in essence: (a) it allows water to pass, which effectively avoids or reduces the possibility of the membrane being clogged by the viscous oil;

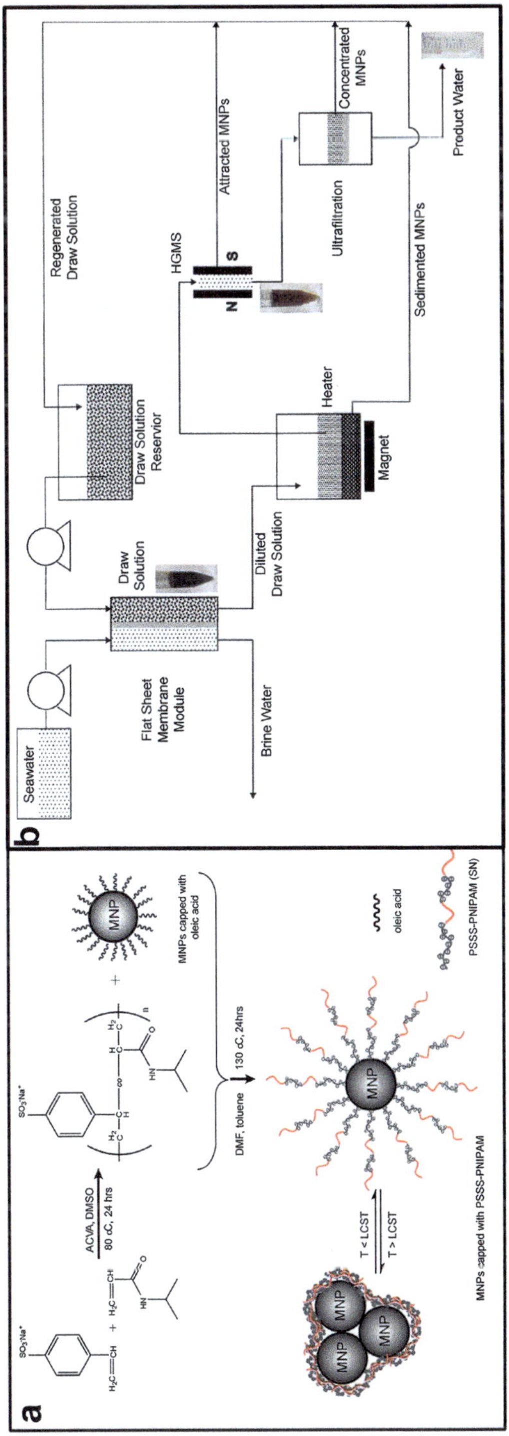

Figure 1.7 (a) Synthesis of MNPs functionalized with the thermoresponsive copolymer PSSS-PNIPAM. (b) Schematic illustration of the FO process by using smart draw solution of thermoresponsive magnetic nanoparticles and the regeneration of the draw solute. (Reprinted with permission from Q. Zhao, N. Chen, D. Zhao and X. Lu, *ACS Appl. Mater. Interfaces*, 2013, **5**, 11453. Copyright (2013) American Chemical Society.)

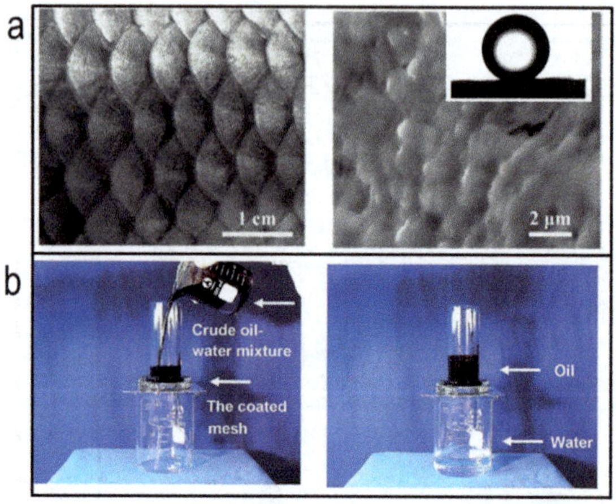

Figure 1.8 (a) Surface structures of fish scale. Inset in (a): shape of an oil droplet on fish scales in water, showing the superoleophobicity of the fish scales. (Reprinted with permission from, M. Liu, S. Wang, Z. Wei, Y. Song and L. Jiang, Bioinspired Design of a Superoleophobic and Low Adhesive Water/Solid Interface, *Adv. Mater.*, 2009, **21**, 665. Copyright © [2009] John Wiley and Sons.) (b) Oil/water separation of the hydrogel-coated mesh. The coated mesh was fixed between two glass tubes, the mixture of crude oil and water was put into the upper glass tube. Water selectively permeated through the coated mesh, while the oil was repelled and kept in the upper glass tube. (Reprinted with permission from, Z. Xue, S. Wang, L. Lin, L. Chen, M. Liu, L. Feng and L. Jiang, A Novel Superhydrophilic and Underwater Superoleophobic Hydrogel-Coated Mesh for Oil/Water Separation, *Adv. Mater.*, 2011, **23**, 4270. Copyright © [2011] John Wiley and Sons.)

(2) it prevents the formation of the water barrier between the membranes and the oil phase, which would otherwise occur with the conventional hydrophobic and oleophilic separation materials due to the fact that water is generally heavier than oil phase and it thus prevents the contact between oil and separation membranes.[97–100] They have shown that this novel material can selectively separate water from oil/water mixtures such as vegetable oil, gasoline, diesel, and even crude oil/water mixtures effectively (more than 99%) and without any extra power.

In view of the diversity of oil–water mixtures and the complexity of oil-spill incidents, an on-demand or highly controllable oil–water separation is highly desirable, *i.e.*, the separation of the oil–water mixture can be controlled so to let either oil or water pass through in one system. Such a controllable separation entails a smart surface with controlled oil wettability, or more desirably a smart surface that switches its oil wettability in response to external stimuli in aqueous media.[101–105] Wang and coworkers for the first time demonstrated a smart surface with switchable superoleophilicity and superoleophobicity in aqueous media for controllable oil–water separation, as illustrated in Figure 1.9.[101] To obtain a smart surface

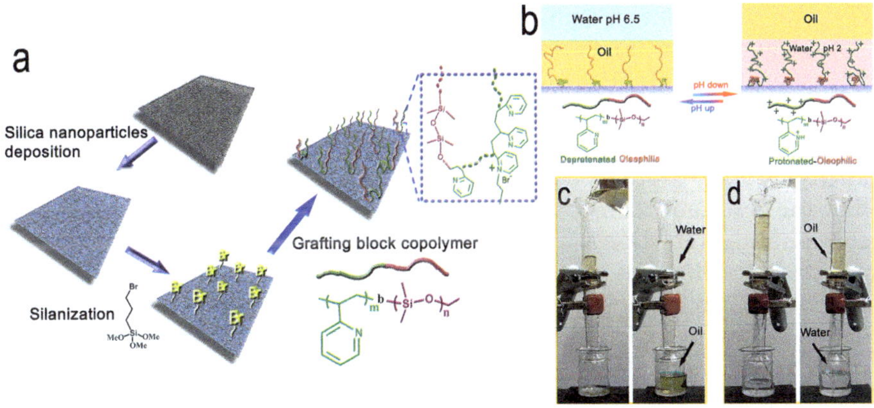

Figure 1.9 (a) Schematic showing the preparation strategy for a surface with switchable superoleophilicity and superoleophobicity on a non-woven textile. (b) Schematic diagrams for the switchable oil wettability of the P2VP-*b*-PDMS grafted textile. (c) Gasoline selectively passed through the textile, whereas the water remained in the upper glass tube when pH of water is 6.5. (d) Here the functionalized textile was first wetted with acidic water (pH 2.0). Water selectively passed through the textile, whereas gasoline remained in the upper glass tube. (Reprinted with permission from Nature Publishing Group: *NPG Asia Mater.*, 2012, **4**, e8, copyright (2012).)

with switchable oil wettability in aqueous media, especially between superoleophobicity and superoleophilicity, the chemistry on the surface should be delicately designed such that it comprises both hydrophilic and oleophilic/hydrophobic characteristics, with either characteristic becoming dominantly exposed over the other in response to environmental conditions. By employing a block copolymer-grafting strategy that uses a block copolymer comprising pH-responsive poly(2-vinylpyridine) and oleophilic/hydrophobic polydimethylsiloxane blocks (*i.e.*, P2VP-*b*-PDMS) to functionalize inexpensive and easily available materials, including non-woven textiles and polyurethane sponges, surfaces with switchable superoleophilicity and superoleophobicity in aqueous media were prepared. The P2VP block on the grafted block copolymer can alter its wettability and its conformation *via* protonation and deprotonation in response to the pH of the aqueous media, which in turn provides controllable and switchable access of oil by the oleophilic PDMS block on the surface, thus realizing a smart surface with switchable oil wettability in aqueous media. With such a functionalized surface, highly controllable oil–water separation was realized. It is anticipated that such a surface with controlled oil wettability would offer great promise in the design and fabrication of intelligent materials for advanced applications.

Besides the superwetting membranes, bioinspired nanomotors are used for the removal of oil from water. Wang *et al.* demonstrated the use of artificial nanomotors for effective interaction, capture, transport, and removal of oil droplets from an aqueous medium. The catalytic microtubular Au/Ni/PEDOT/Pt motors were fabricated by electrochemical deposition and further functionalized with alkanethiols to form a hydrophobic monolayer (SAM) on the outer gold surface of the microtube (Figure 1.10(A)).[106] The strong interactions between the alkanethiol chains and oil from the solution enable micromotors to capture oil droplets and transport them. The authors showed that the resultant SAM-coated Au/Ni/PEDOT/Pt microsubmarine displayed continuous interaction with large oil droplets and was capable of loading and transporting multiple small oil droplets (Figure 1.10(B) and (C)). Such a bioinspired nanomotor has great potential in the autonomous removal of emulsified oil droplets, which is a very demanding task for conventional oil spill cleanup methods.

From these examples, it is clear that smart materials have not only provided effective strategies for solving environmental problems, but have also exhibited unprecedented advantages over traditional materials by integration of multifunction and/or processes into one advanced device/material. In the rest of this book, we will present a broad collection of bioinspired smart materials and systems that are used in environmental problem solving. The topics of these chapters span from bioinspired fog collection, self-healing materials, responsive particle-stabilized emulsions, smart draw solution in forward osmosis, slippery coatings, insightful analysis of problems and opportunities of hydrophobic surface application in real condition, to superwetting materials for oil–water separation. We hope this book will provide an inspiration for readers to further explore smart materials to solve environmental problems.

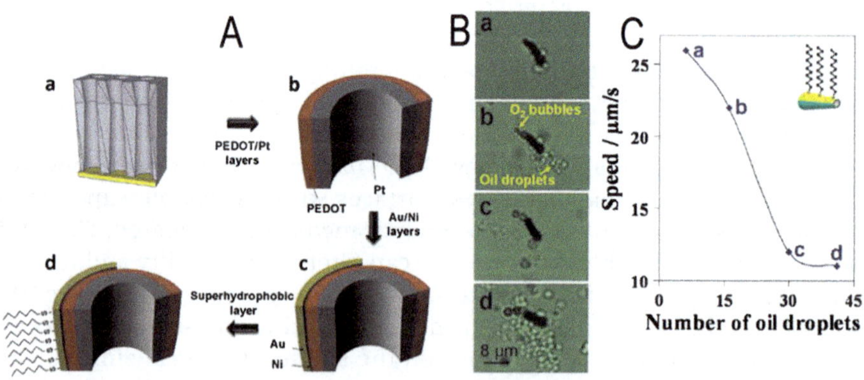

Figure 1.10 (A) Fabrication of the SAM-Au/Ni/PEDOT/Pt micromotors for oil droplet removal. (B) Dodecanethiol (C12-SAM)-modified microsubmarine carrying floating olive oil droplets. (C) Dependence of the micromotors speed upon the number of cargos (olive oil droplets). Inset: cartoon of the dodecanethiol-modified micromotor. (Reprinted with permission from M. Guix, J. Orozco, M. García, W. Gao, S. Sattayasamitsathit, A. Merkoçi, A. Escarpa and J Wang, *ACS Nano*, 2012, **6**, 4445. Copyright (2012) American Chemical Society.)

References

1. W. Barnaby, *Nature*, 2009, **459**, 163.
2. J. C. Padowski and S. M. Gorelick, *Environ. Res. Lett.*, 2014, **9**, 104004.
3. Z. W. Kundzewicz and P. Kowalczak, *Nature*, 2009, **459**, 31.
4. Q. Schiermeier, *Nature*, 2014, **505**, 10.
5. K. Yoon, B. S. Hsiao and B. Chu, *J. Mater. Chem.*, 2008, **18**, 5326.
6. X. Hu, G. Li and J. C. Yu, *Langmuir*, 2010, **26**, 3031.
7. M. Padaki, R. S. Murali, M. S. Abdullah, N. Misdan, A. Moslehyani, M. A. Kassim, N. Hilal and A. F. Ismail, *Desalination*, 2015, **357**, 197.
8. B. Wang, W. Liang, Z. Guo and W. Liu, *Chem. Soc. Rev.*, 2015, **44**, 336.
9. H. M. Choi and R. M. Cloud, *Environ. Sci. Technol.*, 1992, **26**, 772.
10. M. O. Adebajo, R. L. Frost, J. T. Kloprogge, O. Carmody and S. Kokot, *J. Porous Mater.*, 2003, **10**, 159.
11. R. J. Petersen, *J. Membr. Sci.*, 1993, **83**, 81.
12. C. Fritzmann, J. Löwenberg, T. Wintgens and T. Melin, *Desalination*, 2007, **216**, 1.
13. M. A. Shannon, P. W. Bohn, M. Elimelech, J. G. Georgiadis, B. J. Mariñas and A. M. Mayes, *Nature*, 2008, **452**, 301.
14. B. Van Der Bruggen, C. Vandecasteele, T. Van Gestel, W. Doyen and R. Leysen, *Environ. Prog.*, 2003, **22**, 46.
15. T. Y. Cath, A. E. Childress and M. Elimelech, *J. Membr. Sci.*, 2006, **281**, 70.
16. Q. Ge, J. Su, G. L. Amy and T.-S. Chung, *Water Res.*, 2012, **46**, 1318.
17. K. Kuribayashi, *Sens. Actuators, A*, 1989, **20**, 57.
18. J. A. Walker and K. J. Gabriel, *Sens. Actuators, A*, 1990, **21**, 243.
19. A. Lendlein and S. Kelch, *Angew. Chem., Int. Ed.*, 2002, **41**, 2034.
20. J. Hu, Y. Zhu, H. Huang and J. Lu, *Prog. Polym. Sci.*, 2012, **37**, 1720.
21. M. Liu, T. Zentgraf, Y. Liu, G. Bartal and X. Zhang, *Nat. Nanotechnol.*, 2010, **5**, 570.
22. I. Willner, B. Basnar and B. Willner, *Adv. Funct. Mater.*, 2007, **17**, 702.
23. J. Wang and K. M. Manesh, *Small*, 2010, **6**, 338.
24. C. Alvarez-Lorenzo and A. Concheiro, *Chem. Commun.*, 2014, **50**, 7743.
25. P. Yang, S. Gai and J. Lin, *Chem. Soc. Rev.*, 2012, **41**, 3679.
26. M. Colilla and M. Vallet-Regí, in *Chemoresponsive Materials: Stimulation by Chemical and Biological Signals*, ed. H.-J. Schneider, Royal Society of Chemistry, Cambridge, 2015.
27. B. A. Cornell, V. L. B. BraachMaksvytis, L. G. King, P. D. J. Osman, B. Raguse, L. Wieczorek and R. J. Pace, *Nature*, 1997, **387**, 580.
28. I. I. Slowing, B. G. Trewyn, S. Giri and V. S. Y. Lin, *Adv. Funct. Mater.*, 2007, **17**, 1225.
29. M. Gerard, A. Chaubey and B. D. Malhotra, *Biosens. Bioelectron.*, 2002, **17**, 345.
30. L. Jiang and L. Feng, *Bioinspired Intelligent Nanostructured Interfacial Materials*, World Scientific Publishing Company, Singapore, 2010.
31. M. V. Gandhi and B. S. Thompson, *Smart Materials and Structures*, Chapman & Hall, London, UK, 1992.

32. M. A. C. Stuart, W. T. S. Huck, J. Genzer, M. Muller, C. Ober, M. Stamm, G. B. Sukhorukov, I. Szleifer, V. V. Tsukruk, M. Urban, F. Winnik, S. Zauscher, I. Luzinov and S. Minko, *Nat. Mater.*, 2010, **9**, 101.
33. D. Roy, J. N. Cambre and B. S. Sumerlin, *Prog. Polym. Sci.*, 2010, **35**, 278.
34. G. Gautschi, *Piezoelectric Sensorics: Force, Strain, Pressure, Acceleration and Acoustic Emission Sensors, Materials and Amplifiers*, Springer Science & Business Media, Berlin, Germany, 2002.
35. Z. L. Wang, *J. Phys.: Condens. Matter*, 2004, **16**, R829.
36. U. Ozgur, Y. I. Alivov, C. Liu, A. Teke, M. A. Reshchikov, S. Dogan, V. Avrutin, S. J. Cho and H. Morkoc, *J. Appl. Phys.*, 2005, **98**, 041301.
37. Z. L. Wang and J. H. Song, *Science*, 2006, **312**, 242.
38. M. Lee, C.-Y. Chen, S. Wang, S. N. Cha, Y. J. Park, J. M. Kim, L.-J. Chou and Z. L. Wang, *Adv. Mater.*, 2012, **24**, 1759.
39. K. Otsuka and C. M. Wayman, *Shape Memory Materials*, Cambridge University Press, Cambridge, New York, 1999.
40. W. M. Huang, Z. Ding, C. C. Wang, J. Wei, Y. Zhao and H. Purnawali, *Mater. Today*, 2010, **13**, 54.
41. M. Behl, K. Kratz, J. Zotzmann, U. Nöchel and A. Lendlein, *Adv. Mater.*, 2013, **25**, 4466.
42. E. Garcia, *Smart Mater. Struct.*, 2013, **22**, 010201.
43. K. Kitamura, M. Tokunaga, A. H. Iwane and T. Yanagida, *Nature*, 1999, **397**, 129.
44. I. Dobbie, M. Linari, G. Piazzesi, M. Reconditi, N. Koubassova, M. A. Ferenczi, V. Lombardi and M. Irving, *Nature*, 1998, **396**, 383.
45. J.-P. Collin, J.-M. Kern, L. Raehm and J.-P. Sauvage, in *Molecular Switches*, ed. B. L. Feringa, Wiley-VCH, Weinheim, Germany, 2001, ch. 8, pp. 249–280.
46. V. Balzani, A. Credi and M. Venturi, *Molecular Devices and Machines: A Journey into the Nano World*, Wiley-VCH, Weinheim, Germany, 2003.
47. V. Balzani, M. Gomex-Lopez and J. F. Stoddart, *Acc. Chem. Res.*, 1998, **31**, 405.
48. J.-P. Sauvage, *Acc. Chem. Res.*, 1998, **31**, 611.
49. R. F. Shepherd, A. A. Stokes, J. Freake, J. Barber, P. W. Snyder, A. D. Mazzeo, L. Cademartiri, S. A. Morin and G. M. Whitesides, *Angew. Chem.*, 2013, **125**, 2964.
50. R. F. Shepherd, F. Ilievski, W. Choi, S. A. Morin, A. A. Stokes, A. D. Mazzeo, X. Chen, M. Wang and G. M. Whitesides, *Proc. Natl. Acad. Sci. U. S. A.*, 2011, **108**, 20400.
51. R. Ismagilov, A. Schwartz, N. Bowden and G. M. Whitesides, *Angew. Chem., Int. Ed.*, 2002, **41**, 652.
52. W. F. Paxton, K. C. Kistler, C. C. Olmeda, A. Sen, S. K. St. Angelo, Y. Cao, T. E. Mallouk, P. E. Lammert and V. H. Crespi, *J. Am. Chem. Soc.*, 2004, **126**, 13424.
53. J. Wu, S. Balasubramanian, D. Kagan, K. M. Manesh, S. Campuzano and J. Wang, *Nat. Commun.*, 2010, **1**, 1.

54. K. M. Manesh, M. Cardona, R. Yuan, M. Clark, D. Kagan, S. Balasubramanian and J. Wang, *ACS Nano*, 2010, **4**, 1799.
55. M. Pumera, *Nanoscale*, 2010, **2**, 1643.
56. Y. Mei, A. A. Solovev, S. Sanchez and O. G. Schmidt, *Chem. Soc. Rev.*, 2011, **40**, 2109.
57. J. R. Howse, R. A. L. Jones, A. J. Ryan, T. Gough, R. Vafabakhsh and R. Golestanian, *Phys. Rev. Lett.*, 2007, **99**, 048102.
58. D. A. Wilson, R. J. M. Nolte and J. C. M. van Hest, *Nat. Chem.*, 2012, **4**, 268.
59. S. K. Ghosh, *Self-healing Materials: Fundamentals, Design Strategies, and Applications*, Wiley-VCH, Weinheim, Germany, 2008.
60. D. Y. Wu, S. Meure and D. Solomon, *Prog. Polym. Sci.*, 2008, **33**, 479.
61. R. P. Wool, *Soft Matter*, 2008, **4**, 400.
62. E. B. Murphy and F. Wudl, *Prog. Polym. Sci.*, 2010, **35**, 223.
63. D. G. Shchukin and H. Mohwald, *Small*, 2007, **3**, 926.
64. S. R. White, N. R. Sottos, P. H. Geubelle, J. S. Moore, M. R. Kessler, S. R. Sriram, E. N. Brown and S. Viswanathan, *Nature*, 2001, **409**, 794.
65. P. Martin, *Science*, 1997, **276**, 75.
66. K. S. Toohey, N. R. Sottos, J. A. Lewis, J. S. Moore and S. R. White, *Nat. Mater.*, 2007, **6**, 581.
67. B. Su, W. Guo and L. Jiang, *Small*, 2015, **11**, 1072.
68. S. Wang and L. Jiang, *Adv. Mater.*, 2007, **19**, 3423.
69. A. Lafuma and D. Quéré, *Nat. Mater.*, 2003, **2**, 457.
70. T. L. Sun, L. Feng, X. F. Gao and L. Jiang, *Acc. Chem. Res.*, 2005, **38**, 644.
71. X. M. Li, D. Reinhoudt and M. Crego-Calama, *Chem. Soc. Rev.*, 2007, **36**, 1350.
72. X. Zhang, F. Shi, J. Niu, Y. G. Jiang and Z. Q. Wang, *J. Mater. Chem.*, 2008, **18**, 621.
73. T. Darmanin, E. T. de Givenchy, S. Amigoni and F. Guittard, *Adv. Mater.*, 2013, **25**, 1378.
74. W. Barthlott and C. Neinhuis, *Planta*, 1997, **202**, 1.
75. L. Feng, S. H. Li, Y. S. Li, H. J. Li, L. J. Zhang, J. Zhai, L. Y. Song, B. Q. Liu, L. Jiang and D. B. Zhu, *Adv. Mater.*, 2002, **14**, 1857.
76. F. Xia and L. Jiang, *Adv. Mater.*, 2008, **20**, 2842.
77. T. Sun, G. Wang, L. Feng, B. Liu, Y. Ma, L. Jiang and D. Zhu, *Angew. Chem., Int. Ed.*, 2004, **43**, 357.
78. L. Chen, M. Liu, H. Bai, P. Chen, F. Xia, D. Han and L. Jiang, *J. Am. Chem. Soc.*, 2009, **131**, 10467.
79. H. S. Lim, J. T. Han, D. Kwak, M. H. Jin and K. Cho, *J. Am. Chem. Soc.*, 2006, **128**, 14458.
80. B. W. Xin and J. C. Hao, *Chem. Soc. Rev.*, 2010, **39**, 769.
81. S. Zhao, L. Zou, C. Y. Tang and D. Mulcahy, *J. Membr. Sci.*, 2012, **396**, 1.
82. Q. Ge, M. Ling and T.-S. Chung, *J. Membr. Sci.*, 2013, **442**, 225.
83. T.-S. Chung, S. Zhang, K. Y. Wang, J. Su and M. M. Ling, *Desalination*, 2012, **287**, 78.

84. M. M. Ling and T.-S. Chung, *J. Membr. Sci.*, 2011, **372**, 201.
85. Y. Kim, S. Han and S. Hong, *Water Sci. Technol.*, 2011, **64**, 469.
86. Y. Cai, W. Shen, R. Wang, W. B. Krantz, A. G. Fane and X. Hu, *Chem. Commun.*, 2013, **49**, 8377.
87. D. Li, X. Zhang, J. Yao, G. P. Simon and H. Wang, *Chem. Commun.*, 2011, **47**, 1710.
88. Y. Zeng, L. Qiu, K. Wang, J. Yao, D. Li, G. P. Simon, R. Wang and H. Wang, *RSC Adv.*, 2013, **3**, 887.
89. D. Li, X. Zhang, J. Yao, Y. Zeng, G. P. Simon and H. Wang, *Soft Matter*, 2011, **7**, 10048.
90. Q. Zhao, N. Chen, D. Zhao and X. Lu, *ACS Appl. Mater. Interfaces*, 2013, **5**, 11453.
91. Z. Chu, Y. Feng and S. Seeger, *Angew. Chem., Int. Ed.*, 2015, **54**, 2328.
92. Z. Xue, Y. Cao, N. Liu, L. Feng and L. Jiang, *J. Mater. Chem. A*, 2014, **2**, 2445.
93. B. Wang, W. Liang, Z. Guo and W. Liu, *Chem. Soc. Rev.*, 2015, **44**, 336.
94. Y. Zhu, D. Wang, L. Jiang and J. Jin, *NPG Asia Mater.*, 2014, **6**, e101.
95. L. Feng, Z. Zhang, Z. Mai, Y. Ma, B. Liu, L. Jiang and D. Zhu, *Angew. Chem.*, 2004, **116**, 2046.
96. M. Liu, S. Wang, Z. Wei, Y. Song and L. Jiang, *Adv. Mater.*, 2009, **21**, 665.
97. Z. Xue, S. Wang, L. Lin, L. Chen, M. Liu, L. Feng and L. Jiang, *Adv. Mater.*, 2011, **23**, 4270.
98. L. B. Zhang, Y. J. Zhong, D. Cha and P. Wang, *Sci. Rep.*, 2013, **3**, 2326.
99. Q. Wen, J. C. Di, L. Jiang, J. H. Yu and R. R. Xu, *Chem. Sci.*, 2013, **4**, 591.
100. F. Zhang, W. B. Zhang, Z. Shi, D. Wang, J. Jin and L. Jiang, *Adv. Mater.*, 2013, **25**, 4192.
101. L. B. Zhang, Z. H. Zhang and P. Wang, *NPG Asia Mater.*, 2012, **4**, e8.
102. B. Wang and Z. G. Guo, *Chem. Commun.*, 2013, **49**, 9416.
103. D. L. Tian, X. F. Zhang, Y. Tian, Y. Wu, X. Wang, J. Zhai and L. Jiang, *J. Mater. Chem.*, 2012, **22**, 19652.
104. B. L. Xue, L. C. Gao, Y. P. Hou, Z. W. Liu and L. Jiang, *Adv. Mater.*, 2013, **25**, 273.
105. Z. Xu, Y. Zhao, H. Wang, X. Wang and T. Lin, *Angew. Chem.*, 2015, **127**, 4610.
106. M. Guix, J. Orozco, M. García, W. Gao, S. Sattayasamitsathit, A. Merkoçi, A. Escarpa and J. Wang, *ACS Nano*, 2012, **6**, 4445.

CHAPTER 2

Smart Materials as Forward Osmosis Draw Solutes

SHUCHENG CHEN[a] AND XIANMAO LU*[a]

[a]Department of Chemical & Biomolecular Engineering, National University of Singapore, Singapore
*E-mail: chelxm@nus.edu.sg

2.1 Introduction

Sustainable water and energy supply have become two of the greatest global challenges of our time due to the fast-growing population and the depletion of fossil fuels.[1,2] Industrialization, water contamination, and climate change further plagues the problems of water scarcity. It is reported that over one-third of the current world population live in water-stressed areas, among which 1.2 billion people lack access to safe drinking water and 2.6 billion have inadequate sanitation.[3–5] The number of people that may be affected by water issues is projected to rise to over 3.5 billion by 2025. Though improved water conservation, maintenance of infrastructure, and novel design of catchment and distribution systems can alleviate these stresses, they can only improve the use of available fresh water rather than increase it. Desalination and water reuse are two of the few methods to increase water supply beyond the available amount from the hydrogeological cycle.[6] Therefore, it is of great importance to explore technologies to extract fresh water from seawater or wastewater efficiently.

RSC Smart Materials No. 20
Smart Materials for Advanced Environmental Applications
Edited by Peng Wang

Published by the Royal Society of Chemistry, www.rsc.org

In response to the growing water demand, substantial research efforts have been devoted to augment the water supply.[7] Because the production of fresh water is an energy-intensive process, the development of low-energy separation technologies is of great necessity to reduce the overall production cost.[8,9] Among the recently developed methods, the membrane separation process turns to be a promising one. Particularly in the field of seawater desalination, reverse osmosis (RO) as an exceedingly robust and effective process has been the most widely used technology. In a typical RO process, an external hydraulic pressure is applied to seawater to drive fresh water through a semipermeable membrane while retaining salt. However, the process requires a rather high hydraulic pressure to outweigh the inhibiting osmotic pressure originated by seawater. Though the energy consumption of current state-of-the-art RO processes has been brought down to about twice the minimum theoretical energy of de-mixing water from seawater (1.06 kWh m^{-3} for 50% water recovery from 3.5% seawater), it is still considerably high for sustainable water production.[2,6] Moreover, limitations such as membrane fouling and concentration polarization are difficult to overcome for RO processes.

Forward osmosis (FO) has emerged as a promising membrane technology for seawater desalination and wastewater treatment at low-energy cost.[10–13] It has also been investigated for the use of food processing,[14–16] power generation,[17–20] and fertilization.[21,22] FO is a naturally occurring osmosis-driven process. It consists of two key components: (i) a semipermeable membrane that allows the permeation of water while rejecting salts and other unwanted elements; and (ii) a draw solution that is able to generate sufficiently high osmotic pressure to draw water from the feed solution across the membrane (Figure 2.1). Compared with RO, FO holds great promise of low energy consumption if a suitable draw solute with an economical and efficient regeneration method is available.[2,6,23] In addition, FO possesses advantages such as low and reversible membrane fouling, robust rejection of various contaminants, high water flux and recovery, and reduced discharge of concentrated brine.[24–28] The ability to maintain the properties of the feed solution also

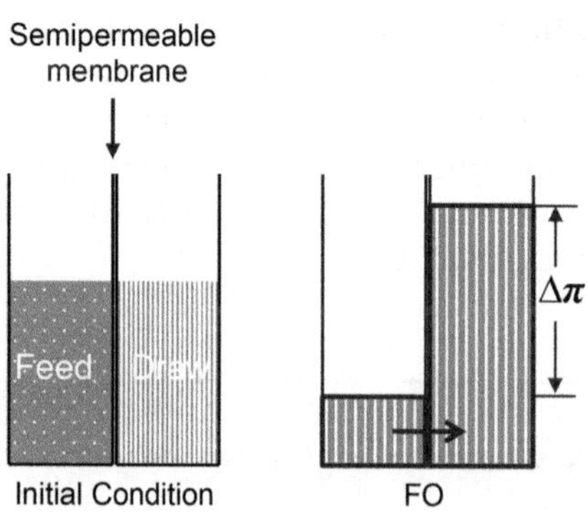

Figure 2.1 Schematic illustration of FO.

makes FO a good candidate for applications in the field of life sciences, such as protein enrichment and controlled drug release.[29–34]

In an FO process, solvent molecules move readily from a low concentration solution through a semipermeable membrane to a high concentration solution until their overall chemical potential reach equilibrium. The high concentration solution and low concentration solution are referred to as "draw solution" and "feed solution", respectively.[8,9,35] The difference in osmotic pressure ($\Delta\pi$) between the draw solution and feed solution acts as the driving force for the separation of feed solvent and solutes. When an asymmetric membrane is used, two modes of FO processes are defined depending on the membrane orientation: pressure retarded osmosis (PRO) where the draw solution faces the dense selective layer (active layer); and FO mode where the draw solution flows against the porous support layer. In osmotically driven membrane processes, concentration polarization is a readily encountered problem that is caused by the diluting effect of the draw solution and the concentrating effect of the feed solution.[36–38] The presence of both internal concentration polarization (ICP) and external concentration polarization (ECP), which occurs within the porous layer of the membrane and at the surface of the dense selective layer respectively, can substantially lower the water flux across the membrane compared with the theoretical value (Figure 2.2).[39–41] ECP occurs at the outer surface of the selective membrane and its effect on permeate flux can be mitigated by altering hydrodynamic conditions such as increasing the flow rate or turbulence.[42] In contrast to ECP, ICP occurs within the porous layer and it cannot be easily eliminated. It has been found that the flux decline in FO is primarily induced by ICP, which could deteriorate the water flux by over 80%.[43,44] Modeling results have revealed that both the structure of the membrane support layer and the diffusion coefficient of the

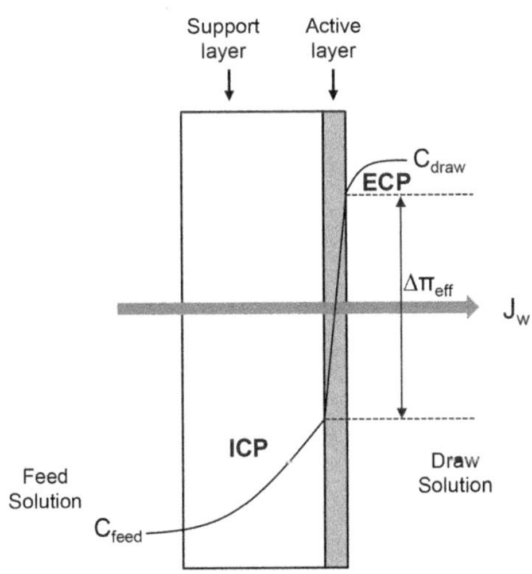

Figure 2.2 Schematic illustration of internal concentration polarization and external concentration polarization.

draw solution play predominant roles on ICP. The solute resistivity (K) is a parameter to measure the diffusion of solutes through the support layer:

$$K = t\tau/\varepsilon D = S/D$$

where t, τ, ε, and S are the membrane thickness, tortuosity, porosity, and structural parameter, respectively, and D is the diffusion coefficient of the solute. Therefore, ICP may be reduced by using a membrane with a low S value and/or a draw solution with a high diffusion coefficient.

The water flux J_w, with a unit of L m^{-2} h^{-1} (LMH), can be determined by the following equation:

$$J_w = A\sigma\Delta\pi$$

where A is the water permeability coefficient of the membrane and σ is the reflection coefficient.[8] $\sigma\Delta\pi$ refers to the effective osmotic pressure difference across the membrane after taking the concentration polarization into account.

An ideal FO membrane should have the following features: high water permeability, high rejection of solutes, significantly reduced internal concentration polarization (ICP), and high chemical stability and mechanical strength.[9] In addition to a suitable membrane with desired characteristics for the FO process, a satisfactory draw solution is crucial for its successful operation. An ideal draw agent has to meet the following basic criteria. First, it should be able to generate a high enough osmotic pressure to extract water from the feed solution. In the case of dilute salt solutions, the osmotic pressure can be estimated using the Van't Hoff equation:

$$\pi = iMRT$$

where π is the osmotic pressure in bars, i is the number of ions per dissociated solute molecule, M is the concentration of the solute in moles L^{-1}, and R and T are the gas constant (0.08314 L bar mol^{-1} K^{-1}) and absolute temperature in kelvin, respectively.[45] In the literature, iM is also referred as osmolality, the concentration of osmotically active particles in a solution with a unit of mOsm kg^{-1}. For typical seawater with a salinity of 3.5%, the osmolality is about 1200 mOsm kg^{-1} and the corresponding osmotic pressure is around 29 atm at room temperature. For the case of concentrated salt solutions, they deviate from the ideal solution and a factor φ, called the osmotic pressure coefficient, needs to be added into the equation to account for the non-ideal behavior. This equation indicates that suitable draw solutes should possess a high water solubility (high M), and a high degree of dissociation (high i). In addition, they should also cause minimal ICP by having a low diffusion coefficient, *i.e.* draw solutes with small molecular weight and low viscosity are preferred. Second, after the FO process, the draw solutes need to be easily recovered and regenerated from the diluted draw solution with a low energy input and operation cost. A FO process is usually accompanied with another step to produce clean water as well as recover the draw solutes (Figure 2.3). The draw solutes should be able to be completely removed to produce drinkable water, and they can retain their original performance after being regenerated for long-term operation. Third, the reverse flux of the draw solute has to be minimized. For some small draw solutes, the semipermeable membranes cannot prevent them from diffusing into the feed solution completely. The back-diffusion of these compounds will cause the loss of the draw solutes, and in turn decreased driving force. Subsequently, draw solutes have to be replenished to cover the loss. Moreover, the penetrated draw solutes may contaminate the feed solution, especially for proteins and pharmaceuticals which may lose their qualities upon exposure to these compounds.[9,29–31] On top of these criteria, draw solutes should also be non-toxic, have good chemical stability, and be compatible with the FO membrane.

Over the past few decades, a variety of compounds have been investigated as FO draw solutes. Earliest studies explored the use of a mixture of H_2O and volatile solutes, such as SO_2, or liquids, such as aliphatic alcohols, in FO application and these solutes were recovered by heating or distillation.[46,47] These processes were energy intensive and were not studied in detail. Moreover, the presence of SO_2 causes toxicity and acidity. Sugars, *i.e.* glucose or fructose, have also been used as draw agents to prepare a nutritious drink by extracting water from seawater, brackish water, or polluted water.[48–50] In this approach, the draw solutes are directly consumed and the recovery of the draw solution is avoided. Similarly, fertilizer has been investigated as a draw solute for the desalination of saline water for irrigation, and at the same time the diluted draw solution can be directly used for fertigation.[21,22] The study of new draw solutes for FO application has drawn rather great interest in recent years. Notably, Elimelech's group employed ammonium bicarbonate to draw

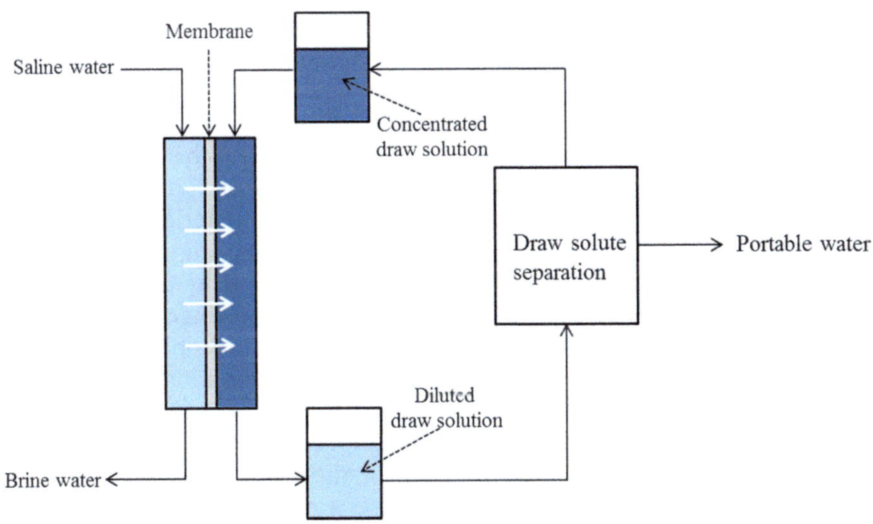

Figure 2.3 Schematic illustration of FO process with draw solute regeneration.

water from saline water.[10,11,23] The highly soluble ammonium bicarbonate was able to provide a high water flux as well as a high feed water recovery. After the FO process, moderate heating at about 60 °C was applied to induce the decomposition of ammonium bicarbonate into ammonia and carbon dioxide gases so that they can be recovered and further used as draw solutes. However, ammonia could not be completely removed and the presence of residual ammonium plagued the use of the product water as safe drinking water.

Many other chemicals, such as polyelectrolytes, 2-methylimidazole-based compounds, hexavalent phosphazene salts, hydroacid complexes, and sodium lignin sulfonate, have also been prepared and investigated as draw solutes in recent years.[51–56] These draw solutes were mainly regenerated with applied hydraulic pressure (*e.g.* RO, nanofiltration, and ultrafiltration) or heat input (*e.g.* membrane distillation). But these processes generally involve the consumption of a high amount of energy. Hence, draw agents with facile recovery methods have to be developed. Smart materials, which can respond to stimuli such as temperature, magnetic field, and light, offer opportunities as FO draw agents to lower the energy consumption of the process at a great potential; and recent years have witnessed the growing interests in developing such materials. So far, hydrophilic magnetic nanoparticles,[30,57–62] stimuli-responsive magnetic nanoparticles,[63,64] smart polymers and solvents,[65–67] and smart hydrogels,[68–75] have been investigated as draw agents in FO application. In the following sections, these smart draw agents are discussed.

2.2 Hydrophilic Magnetic Nanoparticles

The selective and efficient separation of nanoparticles from solution with an external magnetic field has provided magnetic separation apparent advantages over other separation techniques, such as centrifugation and filtration.[76–78] Functionalized magnetic nanoparticles can be easily prepared by coating multifunctional molecules or polymers onto a magnetic core. The ease of tailoring the functionalities of the magnetic nanomaterials have made them popular and intensively studied in wide research areas and applications, such as catalysis,[79] magnetic resonance imaging,[80] and biomedical applications.[81,82] The development of magnetic nanomaterials as FO draw solutes has attracted increasing interest in recent years. High osmotic pressure and easy separation can be concurrently achieved by functionalizing magnetic nanoparticles with a highly hydrophilic polymer. Several studies have been carried out on using functionalized hydrophilic magnetic nanoparticles for protein enrichment and seawater desalination in an FO process.[30,57,60–64,83] A schematic illustration of an FO process using magnetic nanoparticles as draw solutes is shown in Figure 2.4. Water is drawn from the saline water to draw solution through a semipermeable membrane. Then the diluted draw solution is passed through a magnetic field. During this process, the magnetic nanoparticles are recovered while clean product water is collected as the effluent.[59]

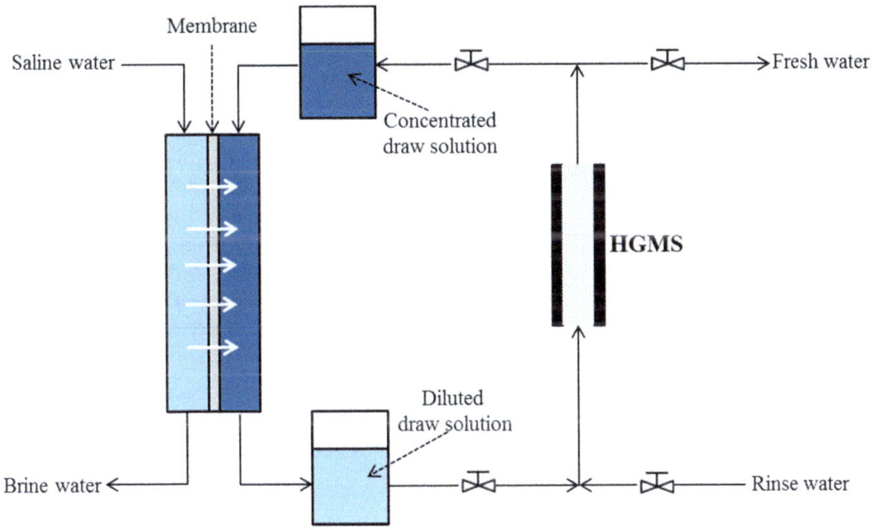

Figure 2.4 Schematic illustration of the FO process using magnetic nanoparticles as draw solute.

Hydrophilic magnetic nanoparticles as FO draw solutes have been prepared by several methods: one-pot polyol reaction,[30,57,60–63] multi-step reactions in water,[83] and post-coating of pre-synthesized polymer onto hydrophobic magnetic nanoparticles through ligand exchange.[64] Chung's group has prepared a series of hydrophilic magnetic nanoparticles functionalized with different surface groups using a one-pot synthesis through the reaction of an iron precursor – iron(III) acetylacetonate and the functional polymers in triethylene glycol (TEG) at the refluxing temperature (Figure 2.5).

Current research results have revealed that the surfactants, nanoparticle size, concentration, and the dissociation extent all have an important impact on the FO performance.[57,60–63] It is found a higher hydrophilic surface capping group can result in a much higher water flux. For example, among the four surfactants: 2-pyrrolidone (2-Pyrol), triethylene glycol (TREG), polyacrylic acid (PAA), and poly(*N*-isopropylacrylamide) (PNIPAM), the water flux exactly followed the order of their hydrophilicity: PAA > TREG > 2-Pyrol > PNIPAM (Figure 2.6). MNP-PAA produced a flux over 10 LMH in PRO mode when using DI water as the feed solution, while MNP-PNIPAM only gave a flux less than 2 LMH.[60,62] The water flux also increased with increasing draw solution concentration, and this was attributed to the increased effective osmotic pressure. However, the water flux did not increase proportionally with the draw solution concentration due to concentration polarization. Moreover, a higher solute concentration could result in greater viscosity.[57,84] The size of the nanoparticles can also affect the water flux. Nanoparticles with a smaller size have a larger surface area and a larger number of hydrophilic groups can anchor on the iron atoms on nanoparticle surface *via* chelation; hence, a larger osmotic pressure can be produced.[57,62,63] In the case of MNP-PAA,

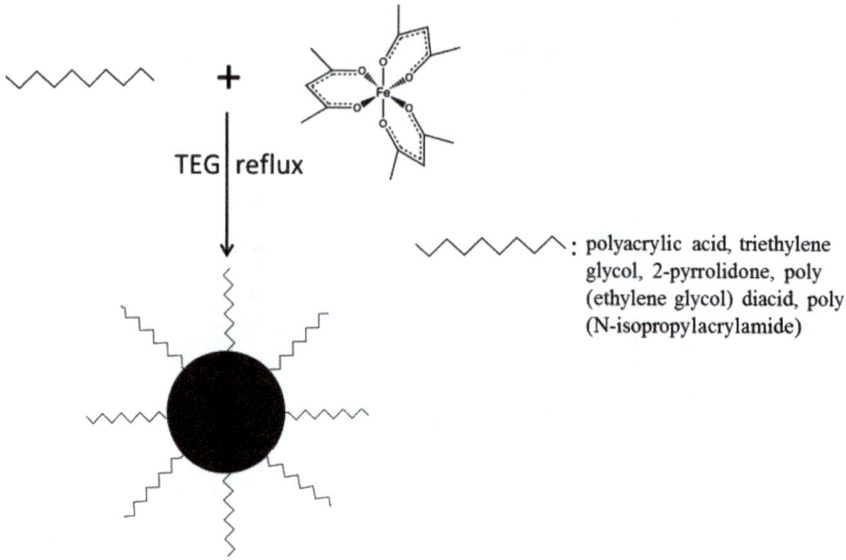

Figure 2.5 Synthesis of hydrophilic magnetic nanoparticles.

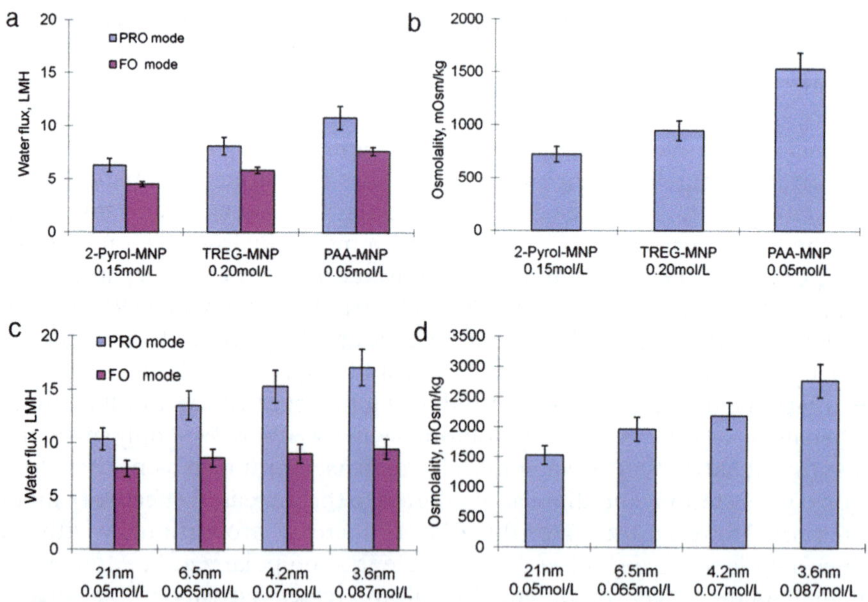

Figure 2.6 (a) Highest water flux and (b) osmolality of MNPs with different surfactants; effect of PAA-MNP size on (c) water flux and (d) osmolality. Reprinted (adapted) with permission from ref. 62. Copyright (2010) American Chemical Society.

nanoparticles with a diameter of 4 nm generate an osmotic pressure twice that of the 20 nm-sized nanoparticles (Figure 2.6).[62] However, the magnetic strength of the nanoparticles decreases as the size reduces and capturing of the nanoparticles with a high gradient magnetic separator (HGMS) is more difficult. Thus, a compromise has to be reached between high water flux and satisfactory recovery in terms of the nanoparticle size.

On top of that, the surface dissociation extent is found to be another factor that influences the FO performance. For example, PAA had limited dissociation in aqueous solutions at its inherent pH range, and the –COOH groups were the sole contributor that created osmotic pressure. Through the addition of NaOH, the PAA on nanoparticle surfaces could be further dissociated to generate a much higher osmotic pressure.[61] Besides, increasing the loading of hydrophilic groups onto the magnetic core could also increase the FO flux, but it would sacrifice the magnetic property of these nanoparticles.[62,64]

To recover the magnetic nanoparticles from the diluted draw solution, an external magnetic field needs to be applied and the generated magnetic forces on particles have to outweigh the opposing forces caused by Brownian motion, viscous drag, and sedimentation. In the case where the particles are too small and the field gradient is not high, the particles will not be captured.[78] Thus, to separate relatively small nanoparticles (less than a critical size of 20 nm), a high-gradient magnetic separator is required to create large applied fields.[59] The magnetic separator consists of an electromagnet in conjunction with an iron circuit with packed ferromagnetic stainless steel wool inside the column. During the recovery process, the magnetic field is first turned on to collect the nanoparticles by attracting and trapping them on the surfaces of the stainless steel wool. Ideally, clean water will pass through the column and be collected. Afterwards, the magnetic field is turned off and DI water is flushed through the matrix to remove the magnetic nanoparticles from the wool and these nanoparticles are then recycled as the draw solute for the next round of the FO process.

The use of HGMS to recover magnetic nanoparticles in the FO process exhibits several challenges: retaining the nanoparticles on the stainless steel wool after the flushing of water; the failure to capture relatively small nanoparticles; and the aggregation of nanoparticles after magnetic separation. Kim's group has found that the choice and replacement of stainless steel wool played an important role on the complete recovery of nanoparticles.[59] After the magnetic field was turned off, the stainless steel wool could not be completely de-magnetized due to its ferromagnetic feature and this residue magnetization prevented the complete removal of the nanoparticles *via* the flushing of water. These residue nanoparticles on the wool might negatively impact the performance of the recovery of nanoparticles in the following cycles. Thus, replacement of wool was required after each recycling process and this puts a limit on the practical application. On the other hand, the ferromagnetic property of the stainless wool was necessary to create a non-uniform field gradient to achieve the 100% absorption of nanoparticles. Therefore, high-performance magnetic separators that can achieve both complete retention and reuse of the nanoparticles have to be designed for future practical applications.

As mentioned earlier, as the critical size of the nanoparticles to be captured decreases, the required magnetic gradient increases substantially. When the prepared magnetic nanoparticles have a wide size distribution, it is likely that only large nanoparticles can be captured while the smaller outliers remain in the effluent water. This incomplete capture of small nanoparticles not only affects the quality of the water product, but also results in the loss of FO performance in the following cycles.[64,78] To address this problem, an additional treatment step such as ultrafiltration is required to recover the small nanoparticles as well as to provide a clean water product. Another solution is to pre-treat the nanoparticles before using them for FO application. The smaller nanoparticles can be separated through dialysis and the remaining mono-dispersed large nanoparticles can be further used as FO draw solutes. For some magnetic nanoparticles that are designated small in order to generate a high osmotic pressure and enhanced FO performance, they fall off the range that HGSM can capture and magnetic separation is no longer feasible to recover them, which is the case for PAA-MNP with an average diameter of 4 nm.[62] In this regard, different separation techniques have to be employed to recover them. Chung's group explored several methods, which included ultrafiltration, a novel dual-stage FO system, and an integrated electric field and nanofiltration system. Though these recovery methods demonstrated effective recovery of magnetic nanoparticles, they failed to employ the unique magnetic responsive properties of these nanoparticles. In that case, some other non-magnetic nanoparticles, such as carbon dots,[85] can also serve the role as a draw solute and provide better FO performance. It should be noted that current magnetic nanoparticle-based draw solutes all use Fe_3O_4 as the magnetic core. Other magnetic materials, such as CoFe and CoFe–Fe_3O_4, have higher magnetization and the use of them as the core may result in better magnetic separation.[86–88]

On top of the two challenges mentioned above, the most prominent problem encountered is the aggregation of the nanoparticles under the high strength of the HGMS. This will lead to the subsequent deterioration of the FO performance and steady water flux cannot be consistently achieved in multiple testing cycles. The earliest research on the aggregation behavior of nanoparticles by Ling *et al.* implied the gradual increase of nanoparticles after magnetic separation. For example, the sizes of 2-Pyrol-MNP, TREG-MNP, and PAA-MNP all increased after being recycled (from initial 28, 24, and 21 nm to 68.1, 58.8, and 50.8 nm, respectively).[62] Later study by Ge *et al.* on using poly(ethylene glycol) diacid-functionalized nanoparticles also revealed the gradual increase of nanoparticle sizes and the resultant loss of 21% water flux at the end of nine runs of recycle (Figure 2.7).[57] It was found that ultrasonication could break the agglomerated TREG-MNP nanoparticles without changing the surface chemistry and partially restored the lost FO performance caused by nanoparticle aggregation. However, sufficient ultrasonication time has to be provided to effectively reduce the nanoparticle size and this requires a lot of energy input.

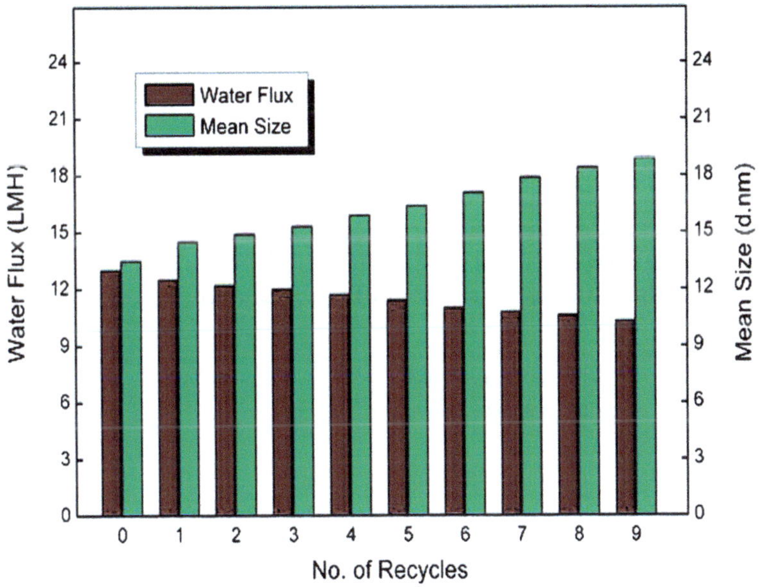

Figure 2.7 Water flux and particle size change after each FO run for poly(ethylene glycol) diacid-functionalized nanoparticles. Reprinted (adapted) with permission from ref. 57. Copyright (2010) American Chemical Society.

Moreover, there was a significant drop observed on the saturation magnetization of the nanoparticles, and it might be attributed to the conversion of the Fe_3O_4 core to Fe_2O_3 during the ultrasonication process.

2.3 Stimuli-Responsive Magnetic Nanoparticles

2.3.1 Introduction

As discussed, hydrophilic magnetic nanoparticles have the limitations of magnetic separation and aggregation. Relatively small nanoparticles can generate a high osmotic pressure, but they are difficult to capture by an external magnetic field. As the cycle of FO process proceeds, gradual aggregation of nanoparticles has also been observed. To solve these problems, stimuli-responsive hybrid magnetic nanoparticles can be used. With the assistance of another stimulus, reversible aggregation and redispersion of the magnetic nanoparticles can be consistently achieved. Moreover, the controlled aggregation of nanoparticles after the instillation of a second stimulus is able to facilitate the magnetic separation process and a low-gradient magnetic field will be sufficient to recover these nanoparticles. Thus far, different dual-responsive nanoparticles, hybrid nanoparticles, and heterodimers have been prepared and these magnetic nanomaterials can respond to either heat,[63,64] light,[89–93] or CO_2.[94,95] In particular, thermo-responsive magnetic nanoparticles have been attempted as FO draw solutes with promising results.

2.3.2 Thermo-Responsive Magnetic Nanoparticles

Chung's group first prepared thermo-responsive magnetic nanoparticles by reacting poly(*N*-isopropylacrylamide) (PNIPAM) with $Fe(acac)_3$ in triethylene glycol (TEG).[63] By partially functionalizing the magnetic nanoparticles with a thermo-responsive polymer (PNIPAM), the aggregation and redispersion of the nanoparticles could be efficiently achieved by tuning the temperature above and below their low critical solution temperature (LCST), respectively. TEG was also anchored on the surface of the nanoparticles to provide enhanced hydrophilicity. As shown in Figure 2.8a, the nanoparticles had a

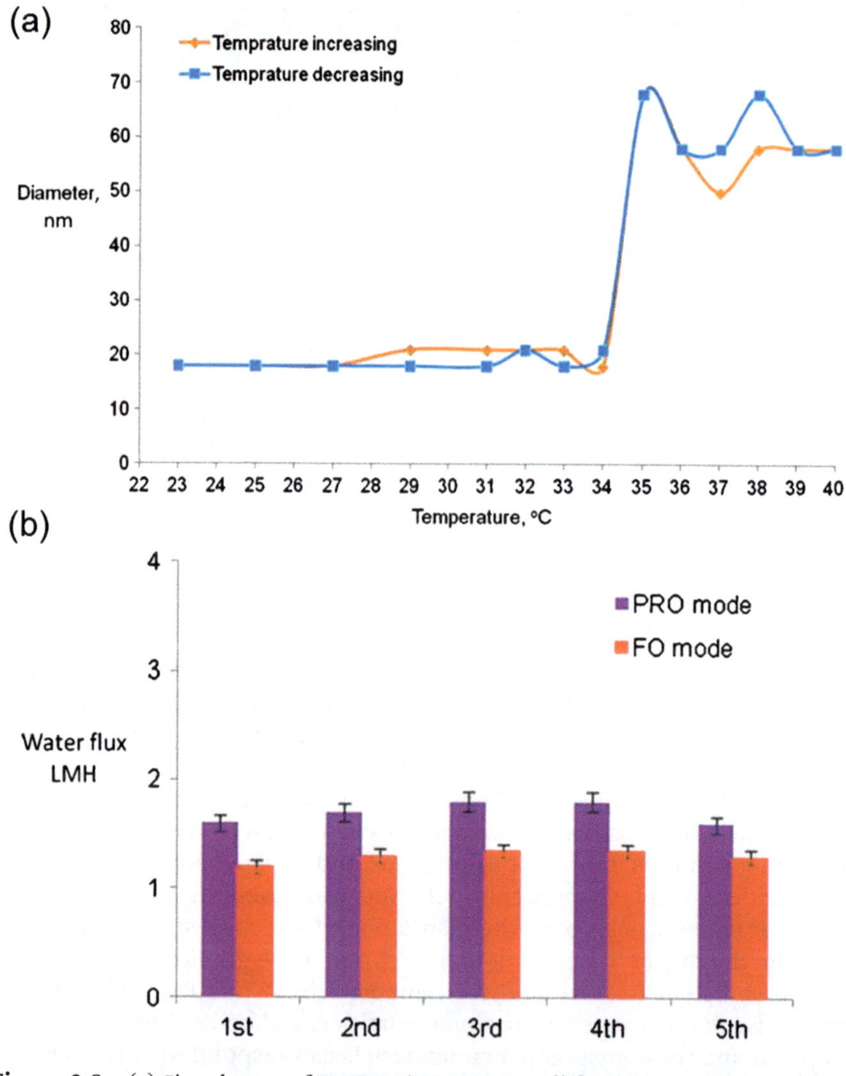

Figure 2.8 (a) Size change of PNIPAM/TEG-MNP at different temperatures. (b) FO performance of recycled PNIPAM/TEG-MNPs. Reproduced from ref. 63 with permission from the Royal Society of Chemistry.

size less than 20 nm at a temperature lower than 34 °C. When the temperature rose above this temperature, the nanoparticles formed larger aggregates at a size of about 60 nm. Besides, stable FO performance without deterioration after 5 cycles was obtained (Figure 2.8b). The steady water fluxes indicated the inclusion of the second stimuli (heat) effectively decreased the irreversible aggregation of nanoparticles caused by high strength magnetic fields. Though such magnetic nanoparticles demonstrated advantages of efficient regeneration of the draw solute and the recovery of water *via* heat-facilitated magnetic separation, the water flux obtained was low and the hydrophilic properties had to be greatly enhanced to make them feasible for seawater desalination.

To further improve the FO performance of thermo-responsive nanoparticles, our group attempted to anchor more hydrophilic polymers onto the nanoparticle surface.[64] Poly(sodium styrene-4-sulfonate)-*co*-poly(*N*-isopropylacrylamide) (PSSS-*co*-NIPAM) was pre-synthesized and grafted onto oleic acid-capped magnetic nanoparticles through ligand exchange, as illustrated in Figure 2.9. The copolymer consisted of 15 wt% PSSS, which could dissociate in water and provided a high osmotic pressure. In this work, nanoparticles with sizes of 5 nm and 9 nm were synthesized, *i.e.* MNP9-15SN and MNP5-15SN (Figure 2.9). After ligand exchange, the sizes of the nanoparticles slightly increased. A water flux as high as 15 LMH in PRO mode using DI water as the feed solution could be achieved with MNP5-15SN. MNP9-15SN produced a lower water flux at about 10 LMH due to the smaller available surface area.

2.3.3 Other Stimuli-Responsive Magnetic Nanoparticles as Potential FO Draw Solutes

2.3.3.1 Light-Responsive Magnetic Nanoparticles

For thermo-responsive magnetic nanoparticles, heat is applied as an external stimulus. However, since the nanoparticles are dissolved in aqueous solutions, additional thermal input is required to heat up the bulk solvent and this may involve a high energy cost. In this aspect, solar energy is advantageous as a stimulus to induce the reversible aggregation behavior of nanoparticles due to its renewable and low-cost feature. Therefore, as a preliminary work, our group specifically designed photo-responsive heterodimers that consisted of magnetic nanoparticles and plasmonic Ag nanoparticles functionalized with PNIPAM.[89] These nanoparticles were prepared by reacting CH_3COOAg and PNIPAM in the presence of pre-synthesized Fe_3O_4 nanoparticles as seeds. Under solar illumination, the Ag nanoparticles absorbed light at around a resonance wavelength of 420 nm due to plasmonic effect and induced localized heating based on the interband absorption mechanism. This heat would in turn increase the temperature of surrounding thermo-responsive polymers above the LCST and cause collapse of the PNIPAM and clustering of the dimers. Once the solar source was removed, the surrounding

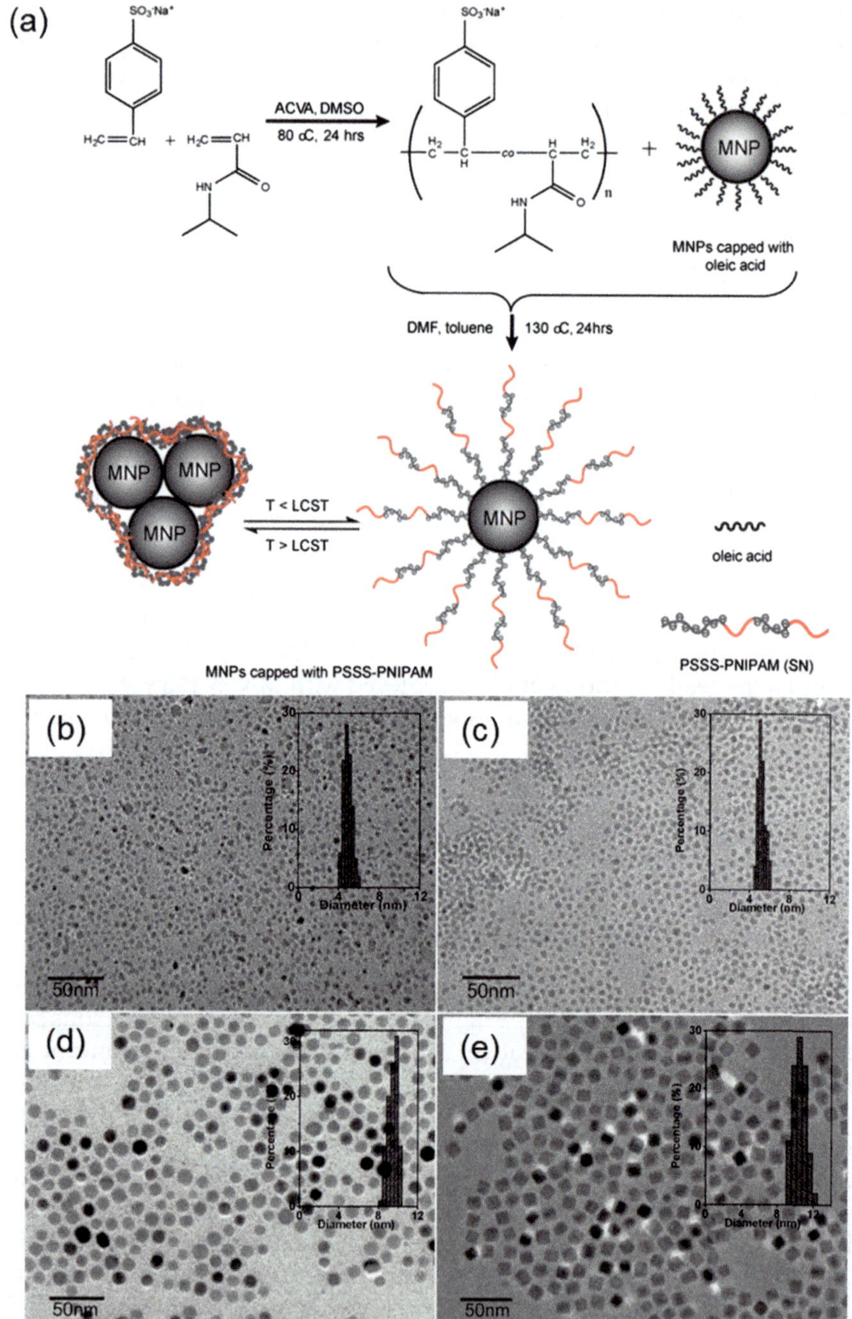

Figure 2.9 (a) Schematic illustration of the synthesis of MNPs functionalized with the thermo-responsive copolymer PSSS-PNIPAM; (b–e) TEM images and size distribution of 5 nm-MNPs, 5 nm-MNPs functionalized with PSSS-PNIPAM, 9 nm-MNPs, 9 nm-MNPs functionalized with PSSS-PNIPAM. Reprinted with permission from ref. 64. Copyright (2013) American Chemical Society.

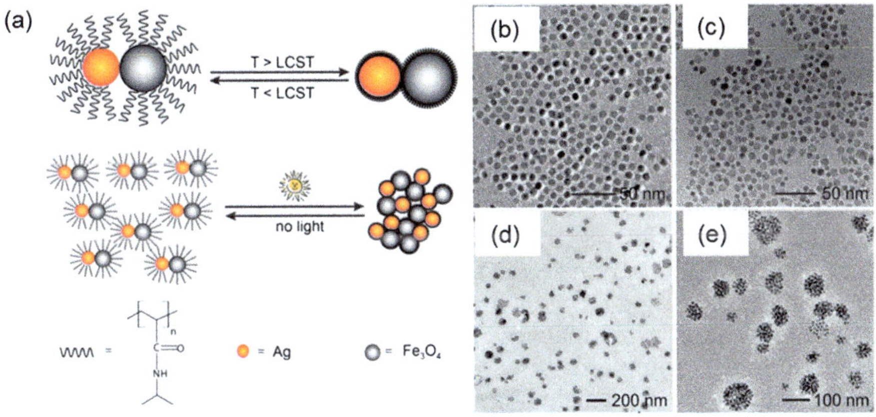

Figure 2.10 (a) Reversible clustering of PNIPAM–capped Ag–Fe_3O_4 heterodimers induced by solar illumination; TEM images of (b) Fe_3O_4 nanoparticles, (c) dispersed Ag–Fe_3O_4 heterodimers before solar illumination, (d and e) clustered heterodimers after solar illumination at different magnification. Reproduced from ref. 89 with permission from the Royal Society of Chemistry.

temperature would go down and the PNIPAM would expand and make the dimers re-disperse well in water (Figure 2.10a).

As seen in Figure 2.10b, without light, the dimers had an average size around 10 nm. Under solar illumination, the Ag–Fe_3O_4 heterodimers formed aggregates at around 100 nm and this would facilitate the magnetic separation process. These heterodimers demonstrated excellent reversible aggregation–redispersion behavior under repeated solar illumination. They hold great promise as a potential FO draw solute candidate. However, current heterodimers are functionalized with PNIPAM only and the expected hydrophilicity is poor. For the further work, the heterodimers have to be modified to incorporate more hydrophilic molecules. In the way, sufficiently high osmotic pressure can be generated for the application of seawater desalination. Moreover, magnetic nanoparticles functionalized with ligands containing light-responsive azobenzene groups have also been reported.[90–93] These nanoparticles exhibited good reversible aggregation behavior induced by light and they can also be potentially used as draw solutes for FO application.

2.3.3.2 CO_2-Responsive Magnetic Nanoparticles

As a safe, cheap, and abundant gas, CO_2 has advantages over other stimuli in terms of its benign reaction with responsive groups such as amidines and amines and its easy removal by mild heating or purging inert gas. Considering this, our group has successfully synthesized CO_2-responsive magnetic nanoparticles in a simple one-pot polyol synthesis.[95] In this work, 1,8-diaminooctane was used as a surfactant. While one primary amine group bound to the metal surface, the other free amine group facing the solution acted as the stimulus responder to provide CO_2-switchable hydrophilicity. Before purging CO_2, the primary amine was un-ionized and the hydrophobic interaction among the alkyl chains of the surfactants tended to make the nanoparticles aggregate in aqueous solutions. After purging CO_2, the primary amine could react with CO_2 to form alkylammonium carbonate and this ionization could give the nanoparticles good hydrophilicity and water solubility (Figure 2.11a). In this way, reversible aggregation/dispersion of the nanoparticles can be simply achieved by purging N_2/CO_2 respectively. Notably, magnetic nanoparticles after CO_2 purging formed a very stable solution and they could not be removed by an external magnetic field. After purging N_2, the nanoparticles could form aggregates as large as 1000 nm and they could be easily removed by an external magnetic field (Figure 2.11b). This would greatly facilitate the separation process. Future work involves improving the hydrophilicity of the nanoparticles by functionalizing them with polymers bearing more CO_2-responsive functional groups and increasing the polymer loading.

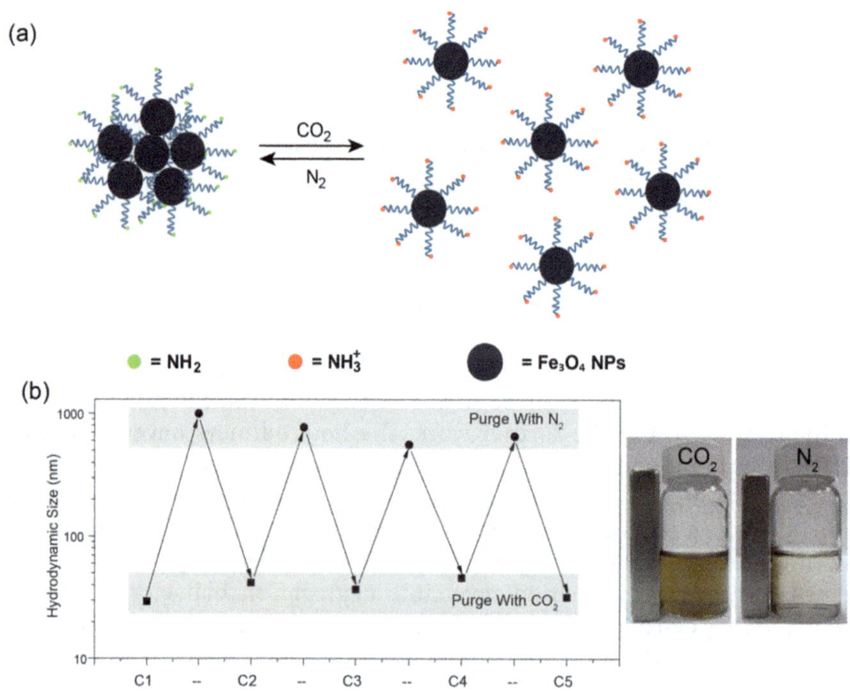

Figure 2.11 (a) Schematic illustration: the aggregation and dispersion of the 1,8-diaminooctane functionalized MNPs can be reversibly achieved by purging CO_2 and N_2 respectively; (b) the hydrodynamic sizes of 1,8-diaminooctane functionalized MNPs after different aggregation–dispersion cycles and images of the solutions under magnet. S. Chen, *et al.*, One Pot Synthesis of CO_2 Responsive Magnetic Nanoparticles with Switchable Hydrophilicity, *Chem.–Eur. J.*, 2014, **20** (43). Copyright © [2014].

2.4 Smart Polyelectrolytes and Solvents

2.4.1 Introduction

Polyelectrolytes have proved to be effective FO draw solutes since they not only can generate high osmotic pressures due to their structurally expanded configuration, they also possesses advantages over conventional ionic salts in terms of much lower reverse leakages and relative ease for regeneration *via* an ultrafiltration process. Ge *et al.* first explored the use of a series of polyacrylic acid sodium salt (PAA-Na) polyelectrolytes as draw solutes in FO process.[51,54] When 0.48 g mL^{-1} PAA-Na with a molecular weight of 1200 g mol^{-1} was used, more than 15 LMH water flux could be achieved with DI water as the draw solution. However, the viscosity of the polyelectrolytes increased with the concentration of draw solution and at the same time the extent of dissociation deceased. This would inhibit the osmotic pressure from increasing proportionally with the increased draw solution concentration. Meanwhile, the water production and the rejection of the polyelectrolytes decreased with the increase in the feed concentration. Nevertheless, a rejection over 98.5% could be achieved at a PAA-Na concentration of 0.12 g mL^{-1}. To further reduce the energy consumption in the polyelectrolyte-driven FO process and to make the water recovery process more efficient, stimuli-responsive polyelectrolytes, *i.e.*, thermo-sensitive polyelectrolytes and CO_2 switchable dual responsive polymers, have been researched as draw agents in FO application.[65,66,96] On top of that, switchable polarity solvents have also been researched as draw agents in FO applications.[67]

2.4.2 Thermo-Responsive Polyelectrolytes

Ou and coworkers fabricated thermo-sensitive polyelectrolytes through the copolymerization of NIPAM with different amounts of sodium acrylate.[66] 14.28 wt% PSA-*co*-NIPAM solutions were used as draw solutions. After FO, water was recovered from the diluted polyelectrolyte solution through hot ultrafiltration. As shown in Figure 2.12, the dilute polyelectrolyte solution was pumped into a stirred cell. The cell was immersed into a water bath where the dilute polyelectrolyte solution was heated above the polyelectrolytes' LCST so that they could go through phase transition and hydrophobic aggregation. An ultrafiltration membrane that could allow the passage of water while effectively retaining the polyelectrolyte aggregates was put at the bottom of the cell. An air compressor was connected to the stirred cell to provide a constant pressure at 2 bar. Notably, the heat-induced aggregation of polyelectrolytes effectively reduced the required pressure in the UF separation process and addressed the issues of reduced water production and polyelectrolytes rejection with increasing feed concentration. Based on the viscosity and dissociation of polyelectrolytes, different water fluxes could be induced by tuning the SA/NIPAM ratio in the polyelectrolyte. Among the synthesized PSA-*co*-NIPAM with different SA molar ratios, it was found the

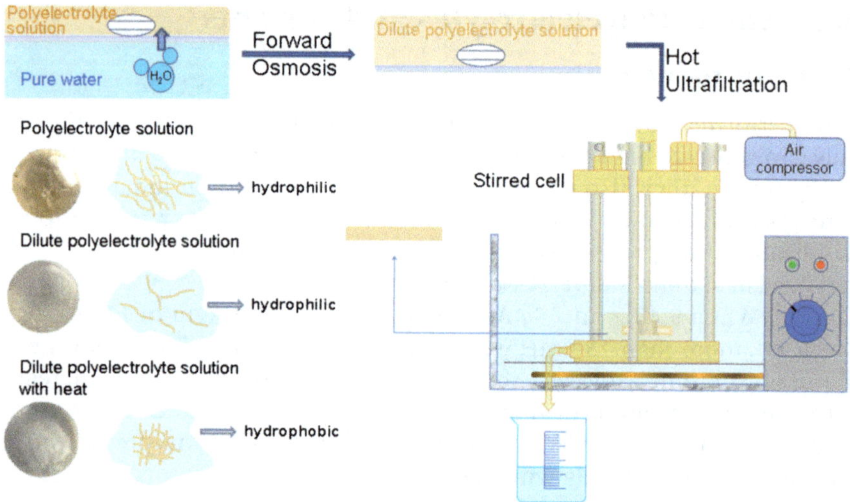

Figure 2.12 Polyelectrolyte solution as draw agent in FO process and the recovery of the water by hot ultrafiltration. Reprinted from R. Ou, *et al.*, Thermo-sensitive polyelectrolytes as draw solutions in forward osmosis process, *Desalination*, 2013, **318**, 48–55, Copyright (2013) with permission from Elsevier.

polymer containing 4% SA segment could give both satisfying FO flux and good water recovery. The resultant average water flux was 0.347 LMH when using pure water as the feed solution and its water recovery fraction was 65.2% at a separation temperature of 45 °C. This low flux value was attributed to the relatively low proportion of ionic groups in the polyelectrolytes. To achieve better FO flux performance, the chemical characteristics and structures of the polymers have to be further explored and refined.

In a later work by our group, poly(sodium styrene-4-sulfonate-*co*-*n*-isopropylacrylamide) (PSSS-PNIPAM) was employed as a draw solute.[96] PSSS was a strong polyelectrolyte and was able to produce a high osmotic pressure. After the FO process, the draw solute was then re-collected with membrane distillation (MD) at a temperature above the LCST of PNIPAM. The agglomeration of PNIPAM resulted in decreased osmotic pressure of the solution and consequently higher water vapor pressure. The combined FO-MD process is shown in Figure 2.13. PSSS-PNIPAM with different weight percentages of SSS (5%, 10%, 15%, 20%) – denoted as 5SN, 10SN, 15SN, and 20SN – were prepared. As expected, both the LCSTs and osmolalities increased with the SSS content in the copolymer since polyelectrolyte PSSS pre-dominantly contributed to the hydrophilicity of the copolymer and at the same time it could partially hinder the thermo-sensitivity of PNIPAM (Figure 2.13b). It is worth noting that the LCST for 20SN was not observed within the tested temperature range and it was not suitable as the draw solute. 15SN solution with a concentration of 33.3 wt% was run for FO tests and it produced a water flux as high as 4 LMH with simulated seawater (0.6 M NaCl solution) as the feed solution.

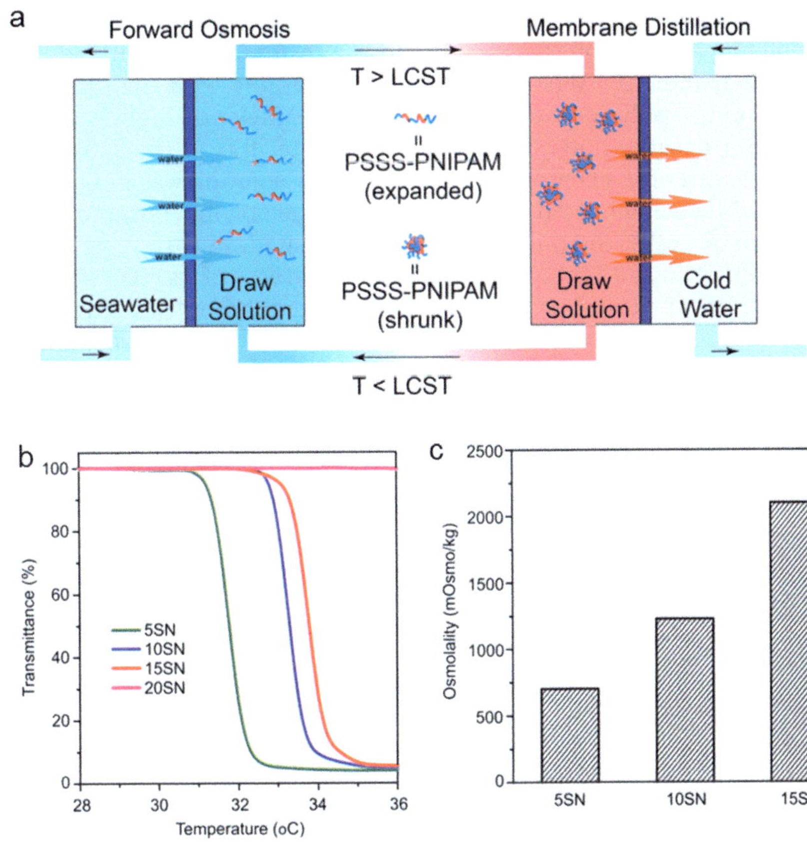

Figure 2.13 (a) Schematic illustration of the combined FO-MD process with PSSS-PNIPAM as the draw solute; (b) transmittance at 500 nm of the copolymers with different weight percentages of SSS (5%, 10%, and 15% for 5SN, 10SN, and 15SN), respectively; (c) osmolalities of PSSS-PNIPAM copolymers with different weight percentages of SSS at a concentration of 33.3 wt%. Reprinted from D. Zhao, *et al.*, Thermo-responsive copolymer-based draw solution for seawater desalination in a combined process of forward osmosis and membrane distillation, *Desalination*, 2014, **348**, 26–32, Copyright (2014) with permission from Elsevier.

2.4.3 CO_2 Switchable Dual Responsive Polymers

Hu's group has achieved high FO water flux with the development of a CO_2 and thermally dual responsive polymer, poly[2-(dimethylamino)ethyl methacrylate] (PDMAEMA).[65] The PDMAEMA were synthesized *via* atom transfer radical polymerization and three samples with number average molecular weights of 4000, 9000, and 13 000 g mol^{-1} were prepared. The protonation and deprotonation of this polymer can be reversibly switched by purging with CO_2 or an inert gas. The overall process of using this polymer as draw solute for water recovery is shown in Figure 2.14. Initially, the unprotonated polymer was dissolved in water and the resultant solution had a relatively low osmolality. When protonated *via* purging with CO_2, the draw solute became a polyelectrolyte and possessed sufficiently high osmolality for seawater desalination and induced high water flux in the subsequent FO process. After the FO process, the dilute draw solution was purged with inert gas while being heated above the LCST of the polymer (around 43 °C). During this process, the polymer was deprotonated, in turn re-gained its thermal sensitivity and precipitated, enabling the ease of its recovery. Then the precipitate and the supernatant were separated through a syringe filter. The remaining polymer in the permeate was further removed *via* an ultrafiltration membrane. Since the concentration of draw solute in the permeate was very low, a low pressure at 1.5 bar was enough to recover >95% water in the UF efficiently. Prominently, the osmolality of the polymer can be significantly enhanced after protonation *via* CO_2 purging. For example, after protonation, the osmolality of the draw solution with a concentration of 0.4 g g^{-1} increased by two times from 0.6 Osm kg^{-1}. This protonated draw solution generated a water flux over 8 LMH with DI water as the feed solution.

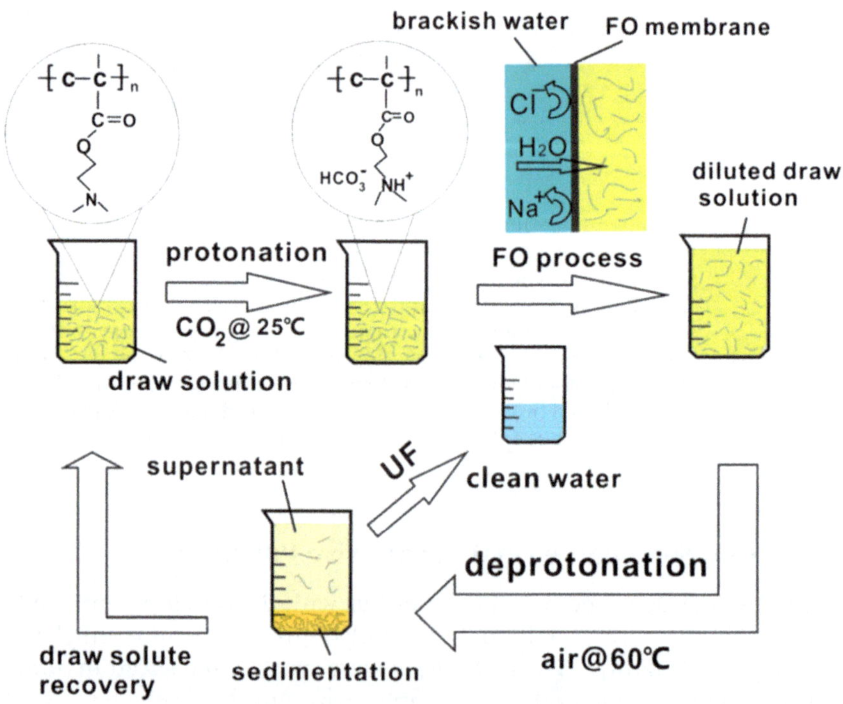

Figure 2.14 Schematic illustration of the dual responsive draw solute for FO desalination. Reproduced from ref. 65 with permission from the Royal Society of Chemistry.

2.4.4 Switchable Polarity Solvents

Recently, a series of reversible non-polar-to-polar solvents have been developed, and they have been tested to extract soybean oil from soybean flakes and to extract bio-oil from algae.[97–100] These "smart" solvents not only can be used to efficiently extract organic materials and as solvents for CO_2 detection, they can also be potentially applied in the "green" production of high-value chemicals such as pharmaceuticals. Typical switchable-polarity solvents involve the introduction of a trigger, CO_2, to transform the solvent from a low polarity form to a high polarity form and this process can be reversibly achieved by removing CO_2 through purging inert gas or mild heating. Both several amidines and tertiary amines have been proven as good candidates to prepare switchable-hydrophilicity solvents. These polarity-switchable solvents were attempted as novel draw solutes for FO by Mark and *et al.*[67] In this work, the solvent was prepared by exposing a steady stream of carbon dioxide to a mixture of deionized water and *N*,*N*-dimethylcyclohexylamine [N(Me)$_2$Cy]. This solvent can reversibly change from its non-polar water immiscible form to its polar water miscible form through the following mechanism:

$$\underset{\text{Water immiscible}}{NR_3 + CO_2 + H_2O} \rightleftharpoons \underset{\text{Water miscible}}{HNR_3^{+} + HCO_3^{-}}$$

In the water miscible form, the solvents formed highly concentrated ionic solutions that could generate sufficiently high osmotic pressure for FO process. Analytical results have indicated that a fully concentrated [HN(Me)$_2$Cy HCO$_3$] solution (59 wt%) was able to generate an osmotic pressure ten times that of seawater (3.5 wt%), which implies that the removal of 90% of the water from seawater is achievable. After the FO process, N_2 purging and a low grade heat of 60 °C were introduced into the dilute draw solution to induce phase separation between the purified water and the non-polar solvent, as shown in Figure 2.15. The draw solvent can be mechanically separated from the two phases and re-purged with CO_2 to be used in the next FO cycle. The separated water had a purity over 98% due to the presence of a trace amount of N(Me)$_2$Cy, which was limited by its water solubility. The trace amounts of draw solute can be further removed from the separated water through an RO process. The RO processes for conventional salt draw solute such as NaCl, KCl, and $MgCl_2$ have problems of concentration polarization and fouling when the feed solution volume is reduced. Subsequently, they can only produce a limited amount of water. The RO process associated with these polarity-switchable draw solutes avoids these problems, since once the solute reached saturation during the reduction of the volume of the RO feed solution, phase separation would occur. Moreover, residue protonated N(Me)$_2$Cy generated a much lower osmotic pressure than conventional draw solutes and this would minimize the required energy and pressure. However, degradation of the HTI cartridge membrane was observed when using the [HN(Me)$_2$Cy HCO$_3$] solvent as draw solute since their relatively high pH of 8.8

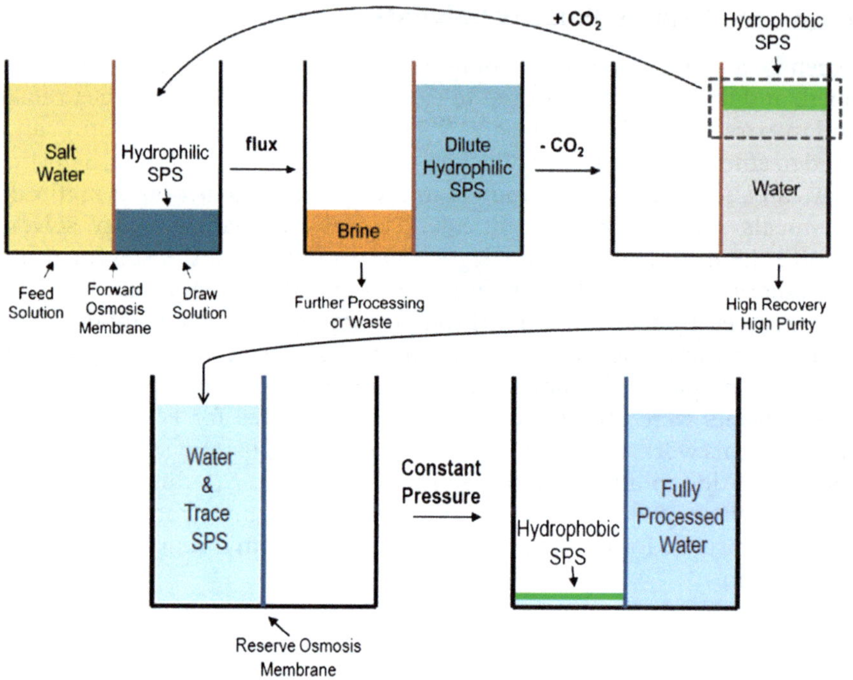

Figure 2.15 Schematic illustration of FO process with switchable polarity solvent. Reprinted from M. L. Stone, *et al.*, Switchable polarity solvents as draw solutes for forward osmosis, *Desalination*, 2013, **312**, 124–129, Copyright (2013) with permission from Elsevier.

fell off the normal pH operation range of the membrane. As a result, polyamide thin-film composite (TFC) membranes may be a better choice when using "smart" draw solutes that operate at acidic or basic conditions since these membranes can withstand severe operating conditions.

2.5 Smart Hydrogels

2.5.1 Introduction

Hydrogels are three-dimensional hydrophilic polymer networks that are cross-linked by either chemical or physical bonds (Figure 2.16). For chemical hydrogels, the covalent bonding of polymer chains prevents the dissolution of them in water. On the other hand, physical hydrogels are formed through the non-covalent interactions, such as hydrophobic and electrostatic interaction or hydrogen bridges between the synthetic or natural building blocks.[101] Hydrogels swell in aqueous media and can hold great amounts of water within their porous structure, with the water content ranging from 10% to several thousand times their own weight.[102] The water entrapping ability of hydrogels is mainly attributed to the presence of hydrophilic groups (amido,

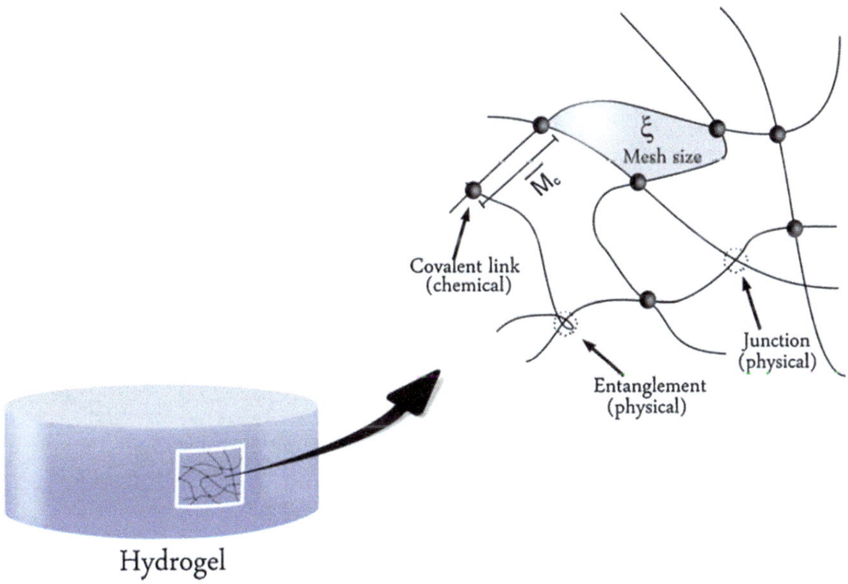

Figure 2.16 A typical cross-linked hydrogel structure. Reprinted from D. Buenger, F. Topuz and J. Groll, Hydrogels in sensing applications, *Prog. Polym. Sci.*, 2012, **37**, 1678–1719, Copyright (2012) with permission from Elsevier.

amino, carboxyl, hydroxyl groups, *etc.*) in polymer chains. The hydrophilicity of these polymer chains generates an osmotic pressure within the network that results in the swelling of the matrix in the presence of water. The swelling process consists of three steps: diffusion of water molecules through the matrix; relaxation of polymer chains *via* hydration; and expansion of polymer network upon relaxation.[103] Many physicochemical conditions or structural factors have been found to affect the swelling, with the most prominent ones being the nature of the polymer and the degree of cross-linking.[104–107] For example, more intensively cross-linked hydrogels have a smaller mesh size and swell less compared with lower cross-linked ones. Mesh size refers to the available diffusional space for the transfer of particles or molecules through the matrix of a hydrogel and a smaller mesh size corresponds to a decreased diffusion rate and thus a lesser degree of swelling.[108]

When the polymer network is functionalized with specific groups, it can undergo reversible changes in swelling of the network when exposed to some external stimuli. Typical stimuli include temperature, light, electric, and chemical or biochemical molecules.[109–112] These stimuli can induce changes in the characteristics of the hydrogel, such as the degree of cross-linking, composition of ionic species or chemical nature of the polymer backbone, and subsequently lead to altered swelling behavior. Stimuli-sensitive hydrogels have attracted considerable attention in wide applications in the development of biomaterials, drug delivery systems, and sensors.[101,109,111–114]

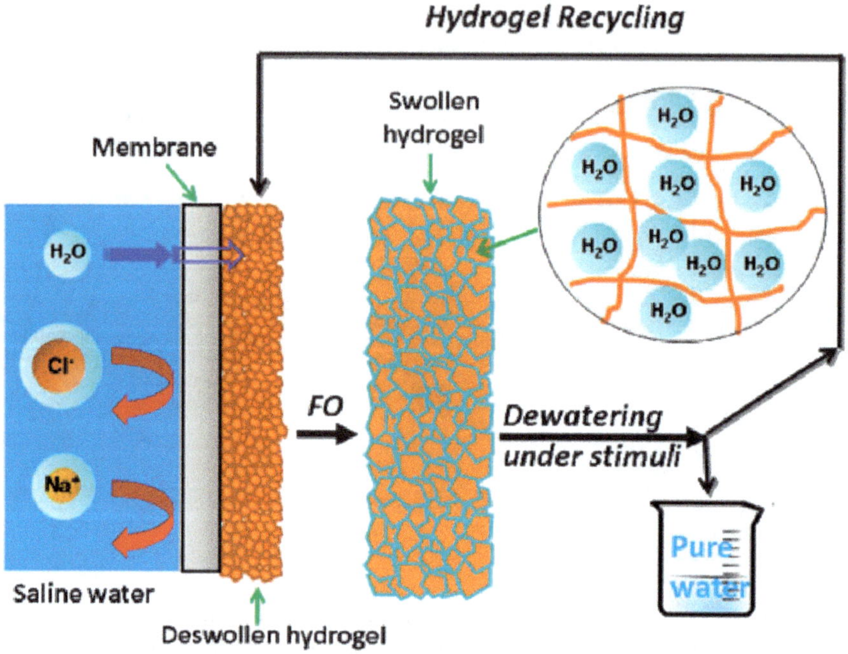

Figure 2.17 Polymer hydrogel for FO. Reproduced from ref. 75 with permission from the Royal Society of Chemistry.

The hydrogel's capacity of reversible volume change in response to external stimuli has made it a promising candidate as a draw agent in FO desalination. Wang's group pioneered in demonstrating the feasibility of using smart hydrogels as a draw agent in FO applications. Since then, multiple efforts have been devoted to design smart hydrogels for FO processes.[68–75] In a typical process, the polymer hydrogel and NaCl solution are placed on the active side and permeate side of a semipermeable FO membrane, respectively. During the FO process, water is drawn to the hydrogel through the membrane by the chemical potential difference created by the de-swollen hydrogel. After the hydrogel absorbs enough water, fresh water is collected *via* dewatering of the swollen polymer hydrogel under various stimuli (Figure 2.17).

2.5.2 Synthetic Methods and FO Performance

In general, three types of hydrogels have been prepared, including (1) pure hydrogels: poly(sodium acrylate) (PSA), polyacrylamide (PAM), poly(*N*-isopropylacrylamide) (PNIPAM), or poly(sodium acrylate)-*co*-poly(*N*-isopropyl-acrylamide) (PSA-*co*-NIPAM);[69,72,73,75] (2) composite hydrogels: PSA, PNIPAM, PSA-*co*-NIPAM and carbon particles composite, PSA, PSA-*co*-NIPAM and reduced grapheme oxide (rGO) composite, PSA-*co*-NIPAM and magnetic nanoparticle (γ-Fe_2O_3) composite;[70,71,74] (3) semi-interpenetrating network (semi-IPN): PNIPAM with PSA or polyvinyl alcohol (PVA).[68] These hydrogels were all prepared by free-radical polymerization of monomers in aqueous solution. The typical step to fabricate PSA-*co*-NIPAM and carbon particle composite hydrogel is shown in Figure 2.18. In the case of semi-IPN, the hydrogel was prepared by polymerizing NIPAM in the presence of linear PSA and PVA respectively, similar to the preparation of composite hydrogels.

It is found the swelling pressure is mainly affected by the chemical characteristics (degree of solubility, ionic strength), particle size, cross-linking density, and porosity of the hydrogels.[71–75,84] Hydrogels with a higher charge density induce higher swelling pressures and give better FO performance. For example, when PSA, PSA-*co*-NIPAM, and PNIPAM hydrogels were studied as the draw agents, the average fluxes in the first hour were 0.96, 0.55, and 0.30 LMH, respectively, indicating a clear decreasing trend with respect to the charge density. This can be explained by the fact that PSA and PNIPAM were charged and non-ionic hydrogels respectively; for the case of PSA, the covalently integrated anionic groups (–COO–) were electrostatically balanced by mobile counterions (Na^+) within the hydrogel phase, providing a higher driving force to draw water through the membrane; on the contrary, the absence of the charges in the neutral PNIPAM hydrogel was not able to produce such high osmotic pressure.

It has been reported that the incorporation of inorganic particles into polymer hydrogels can provide unique properties such as improved water absorbency, absorption rate, and responsiveness to external stimuli.[70] Wang's group first prepared a composite hydrogel incorporated with light-absorbing black carbon particles in order to facilitate the dewatering process. It turned out that such a composite hydrogel also exhibited superior swelling ratios

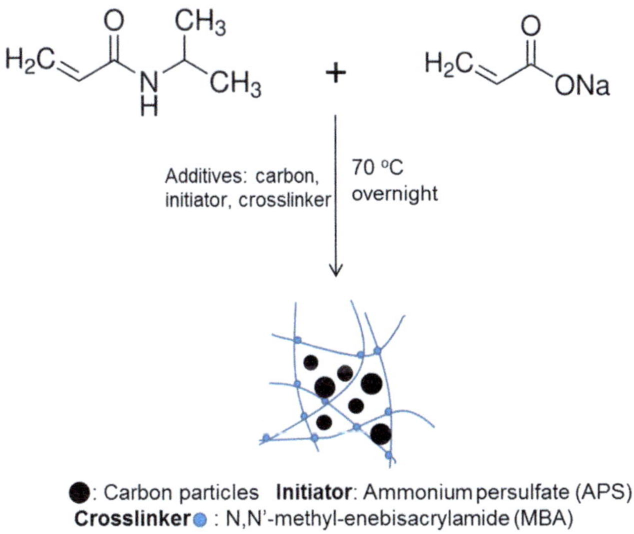

Figure 2.18 Synthesis of composite hydrogels consisting of PSA-*co*-NIPAM, and carbon particles.[70]

and FO performance than corresponding pure hydrogels. For example, the incorporation of the carbon into the PNIPAM increased the swelling ratio by 93.3%. In a later work by Zeng and *et al.*, polymer-reduced graphene oxide (rGO)-composite hydrogel was prepared and significantly enhanced water flux was achieved.[74] For example, the water flux produced by the composite PSA hydrogels with 1.2 wt% rGO (3.1 LMH) was 4.1 times that of the pure PSA (0.75 LMH) and the water flux of PSA-NIPAM with 1.2 wt% rGO (1.7 LMH) was 3.3 times that of the pure PSA-*co*-NIPAM (0.52 LMH), when 2000 ppm of NaCl was used as the feed solution. The improved swelling pressure of the composite hydrogels in the presence of a low amount of rGO can be ascribed to the following reasons: first, the inclusion of micro-sized graphene into the polymer network brings down the overall cross-linking density of the hydrogel while keeping the local cross-linking density between the graphene sheets. As a result, the polymer network can swell more easily by "freely" sliding on the graphene surface. Second, the existence of carboxyl groups and hydroxyl groups within the rGO sheets (roughly 9 wt%) can add additional water adsorption capacity to the composite hydrogel. Last, the negative charges borne by both rGO and the polymer network can lead to the expansion of the free volume between them through electrostatic repulsion, facilitating the transport of water molecules through the hydrogel particles and making room for more water molecules.

2.5.3 Dewatering Method and Performance

In addition to water flux, efficient dewatering and recovery of hydrogels are also important for an ideal hydrogel draw agent. Recently, PNIPAM has been chosen as the stimuli-responsive component within the hydrogel. At a temperature below its LCST, PNIPAM hydrogel absorbs water and swells. When the surrounding temperature reaches above its LCST, PNIPAM hydrogel undergoes a dramatic shrinkage, causing the quick deswelling and releasing of water.[75] Initially, hydraulic pressure at 30 bar and ambient temperature was applied for the recovery of water for PSA, PAM, PNIPAM, and PSA-*co*-NIPAM hydrogels. But the water recovery rate was very low (less than 5%) with the samples having 50% water content, which could be attributed to the strong molecular interaction between water and polymer chains. On the other hand, when hydraulic pressure was applied at an environmental temperature of 50 °C, the water recovery rate improved significantly for PNIPAM and PSA-*co*-NIPAM. The water recovery rate of PNIPAM and PSA-*co*-NIPAM swollen gel with 66.7% water content increased up to 70% and 17% respectively. Conversely, there was only slight increase for PAM and PSA hydrogel due to the absence of NIPAM units to provide thermo-responsivity.

In addition to direct heating, magnetic nanoparticles (γ-Fe_2O_3, <50 nm) have been employed to enable magnetic heating of the polymer network.[71] Unlike other conventional heating, magnetic heating can produce a notably lower temperature gradient and warm the whole volume of hydrogels uniformly;

and they can be precisely controlled to heat the target where required to avoid unnecessary energy wastage by heating all components of the membrane module. Superior dewatering performance of magnetic heating has been observed. For example, for PSA-*co*-NIPAM nanocomposite hydrogels (swelling ratio of 6 ± 0.8) with 16 wt% MNPs, magnetic field (400.5 A, 148 Oe and 372 kHz) induced heating could increase the hydrogel temperature to 65 °C, leading to around 70% water recovery rate. On the contrary, convection heating at 65 °C could only result in 15% water recovery rate.

2.6 Conclusions and Future Perspectives

Various smart materials, including hydrophilic magnetic nanoparticles, stimuli-responsive magnetic nanoparticles, polyelectrolytes and solvents, and hydrogels have been investigated as draw agents for FO application. These materials are advantageous over conventional draw solutes since they can be easily and efficiently regenerated with the assistance of external stimuli and they have great potential to reduce the operation and energy cost involved in an FO process.

Magnetic nanoparticles can be easily recovered through an external magnetic field with low energy consumption and they can effectively eliminate the problem of leakage or the back diffusion of draw solutes that has been observed for other conventional draw solutes, such as salts or low molecular weight organics. However, deterioration of FO performance due to nanoparticle aggregation during magnetic separation has to be addressed and a large-scale high performance magnetic separator has to be specifically designed for the recovery of these nanoparticles. In this aspect, stimuli-responsive magnetic nanoparticles are better candidates since the aggregation of the nanoparticles can be reversibly controlled with a second stimulus and the aggregation-facilitated magnetic separation can significantly reduce the required magnetic field strength and in turn the energy cost. However, the hydrophilicity of the dual-responsive nanoparticles has to be improved.

Compared with magnetic nanoparticles, polyelectrolytes and switchable-polarity solvents both have observable back diffusion. Especially for switchable-polarity solvents, the back diffusion is much worse due to the much smaller molecular size. Nevertheless, this problem can be potentially relieved by improved design of FO hollow membranes. Future improvements should be focused on the design of more suitable materials with high hydrophilicity, and optimizing the recovery process to minimize energy cost. A series of hydrogels have also been developed as FO draw agents but with poor FO flux. Intensive research has to be conducted to modify the structures and chemical characteristics of the hydrogels to significantly improve their FO performances. Nonetheless, the conceptual design of continuous water production using hydrogels imply that these hydrogels are promising for FO processes.

References

1. M. A. Shannon, P. W. Bohn, M. Elimelech, J. G. Georgiadis, B. J. Marinas and A. M. Mayes, *Nature*, 2008, **452**, 301–310.
2. M. Elimelech and W. A. Phillip, *Science*, 2011, **333**, 712–717.
3. M. A. Montgomery and M. Elimelech, *Environ. Sci. Technol.*, 2007, **41**, 17–24.
4. R. F. Service, *Science*, 2006, **313**, 1088.
5. M. Elimelech, *Aqua*, 2006, **55**, 3–10.
6. R. L. McGinnis and M. Elimelech, *Environ. Sci. Technol.*, 2008, **42**, 8625–8629.
7. L. Chekli, S. Phuntsho, H. K. Shon, S. Vigneswaran, J. Kandasamy and A. Chanan, *Desalin. Water Treat.*, 2012, **43**, 167–184.
8. S. Zhao, L. Zou, C. Y. Tang and D. Mulcahy, *J. Membr. Sci.*, 2012, **396**, 1–21.
9. Q. Ge, M. Ling and T.-S. Chung, *J. Membr. Sci.*, 2013, **442**, 225–237.
10. J. R. McCutcheon, R. L. McGinnis and M. Elimelech, *Desalination*, 2005, **174**, 1–11.
11. J. R. McCutcheon, R. L. McGinnis and M. Elimelech, *J. Membr. Sci.*, 2006, **278**, 114–123.
12. E. R. Cornelissen, D. Harmsen, K. F. de Korte, C. J. Ruiken, J.-J. Qin, H. Oo and L. P. Wessels, *J. Membr. Sci.*, 2008, **319**, 158–168.
13. R. W. Holloway, A. E. Childress, K. E. Dennett and T. Y. Cath, *Water Res.*, 2007, **41**, 4005–4014.
14. K. B. Petrotos, P. C. Quantick and H. Petropakis, *J. Membr. Sci.*, 1999, **160**, 171–177.
15. K. B. Petrotos and H. N. Lazarides, *J. Food Eng.*, 2001, **49**, 201–206.
16. K. B. Petrotos, P. Quantick and H. Petropakis, *J. Membr. Sci.*, 1998, **150**, 99–110.
17. A. Achilli, T. Y. Cath and A. E. Childress, *J. Membr. Sci.*, 2009, **343**, 42–52.
18. A. Seppälä and M. J. Lampinen, *J. Membr. Sci.*, 1999, **161**, 115–138.
19. E. M. Garcia-Castello, J. R. McCutcheon and M. Elimelech, *J. Membr. Sci.*, 2009, **338**, 61–66.
20. K. L. Lee, R. W. Baker and H. K. Lonsdale, *J. Membr. Sci.*, 1981, **8**, 141–171.
21. S. Phuntsho, H. K. Shon, S. Hong, S. Lee and S. Vigneswaran, *J. Membr. Sci.*, 2011, **375**, 172–181.
22. S. Phuntsho, H. Shon, S. Hong, S. Lee, S. Vigneswaran and J. Kandasamy, *Rev. Environ. Sci. Biotechnol.*, 2012, **11**, 147–168.
23. R. L. McGinnis and M. Elimelech, *Desalination*, 2007, **207**, 370–382.
24. J. L. Cartinella, T. Y. Cath, M. T. Flynn, G. C. Miller, K. W. Hunter and A. E. Childress, *Environ. Sci. Technol.*, 2006, **40**, 7381–7386.
25. T. Y. Cath, N. T. Hancock, C. D. Lundin, C. Hoppe-Jones and J. E. Drewes, *J. Membr. Sci.*, 2010, **362**, 417–426.
26. B. Mi and M. Elimelech, *Environ. Sci. Technol.*, 2010, **44**, 2022–2028.
27. B. Mi and M. Elimelech, *J. Membr. Sci.*, 2010, **348**, 337–345.

28. A. Achilli, T. Y. Cath, E. A. Marchand and A. E. Childress, *Desalination*, 2009, **239**, 10–21.
29. K. Y. Wang, M. M. Teoh, A. Nugroho and T.-S. Chung, *Chem. Eng. Sci.*, 2011, **66**, 2421–2430.
30. M. M. Ling and T.-S. Chung, *J. Membr. Sci.*, 2011, **372**, 201–209.
31. Q. Yang, K. Y. Wang and T.-S. Chung, *Sep. Purif. Technol.*, 2009, **69**, 269–274.
32. A. G. Thombre, J. R. Cardinal, A. R. DeNoto, S. M. Herbig and K. L. Smith, *J. Controlled Release*, 1999, **57**, 55–64.
33. G. Santus and R. W. Baker, *J. Controlled Release*, 1995, **35**, 1–21.
34. C.-Y. Wang, H.-O. Ho, L.-H. Lin, Y.-K. Lin and M.-T. Sheu, *Int. J. Pharm.*, 2005, **297**, 89–97.
35. T.-S. Chung, S. Zhang, K. Y. Wang, J. Su and M. M. Ling, *Desalination*, 2012, **287**, 78–81.
36. Z. Liu, H. Bai, J. Lee and D. D. Sun, *Energy Environ. Sci.*, 2011, **4**, 2582–2585.
37. B. Mi and M. Elimelech, *J. Membr. Sci.*, 2008, **320**, 292–302.
38. C. Y. Tang, Q. She, W. C. L. Lay, R. Wang and A. G. Fane, *J. Membr. Sci.*, 2010, **354**, 123–133.
39. W. A. Phillip, J. S. Yong and M. Elimelech, *Environ. Sci. Technol.*, 2010, **44**, 5170–5176.
40. G. T. Gray, J. R. McCutcheon and M. Elimelech, *Desalination*, 2006, **197**, 1–8.
41. J. R. McCutcheon and M. Elimelech, *J. Membr. Sci.*, 2006, **284**, 237–247.
42. T. Y. Cath, A. E. Childress and M. Elimelech, *J. Membr. Sci.*, 2006, **281**, 70–87.
43. G. D. Mehta and S. Loeb, *J. Membr. Sci.*, 1978, **4**, 261–265.
44. G. D. Mehta and S. Loeb, *J. Membr. Sci.*, 1978, **4**, 335–349.
45. A. Yokozeki, *Appl. Energy*, 2006, **83**, 15–41.
46. G. W. Batchelder, *US Pat.*, US3171799 A, 1965.
47. D. N. Glew, *US Pat.*, US3216930 A, 1965.
48. J. Yaeli, *US Pat.*, US5098575 A, 1992.
49. K. Stache, *US Pat.*, US4879030 A, 1989.
50. J. O. Kessler and C. D. Moody, *Desalination*, 1976, **18**, 297–306.
51. Q. Ge, J. Su, G. L. Amy and T.-S. Chung, *Water Res.*, 2012, **46**, 1318–1326.
52. S. K. Yen, N. F. Mehnas Haja, M. Su, K. Y. Wang and T.-S. Chung, *J. Membr. Sci.*, 2010, **364**, 242–252.
53. M. L. Stone, A. D. Wilson, M. K. Harrup and F. F. Stewart, *Desalination*, 2013, **312**, 130–136.
54. Q. Ge, P. Wang, C. Wan and T.-S. Chung, *Environ. Sci. Technol.*, 2012, **46**, 6236–6243.
55. Q. Ge and T.-S. Chung, *Chem. Commun.*, 2013, **49**, 8471–8473.
56. J. Duan, E. Litwiller, S.-H. Choi and I. Pinnau, *J. Membr. Sci.*, 2014, **453**, 463–470.
57. Q. Ge, J. Su, T.-S. Chung and G. Amy, *Ind. Eng. Chem. Res.*, 2010, **50**, 382–388.

58. R. Y. Hong, B. Feng, L. L. Chen, G. H. Liu, H. Z. Li, Y. Zheng and D. G. Wei, *Biochem. Eng. J.*, 2008, **42**, 290–300.
59. Y. Kim, S. Han and S. Hong, *Water Sci. Technol.*, 2011, **64**, 469–476.
60. M. M. Ling and T.-S. Chung, *Desalination*, 2011, **278**, 194–202.
61. M. M. Ling and T.-S. Chung, *Ind. Eng. Chem. Res.*, 2012, **51**, 15463–15471.
62. M. M. Ling, K. Y. Wang and T.-S. Chung, *Ind. Eng. Chem. Res.*, 2010, **49**, 5869–5876.
63. M. M. Ling, T.-S. Chung and X. Lu, *Chem. Commun.*, 2011, **47**, 10788–10790.
64. Q. Zhao, N. Chen, D. Zhao and X. Lu, *ACS Appl. Mater. Interfaces*, 2013, **5**, 11453–11461.
65. Y. Cai, W. Shen, R. Wang, W. B. Krantz, A. G. Fane and X. Hu, *Chem. Commun.*, 2013, **49**, 8377–8379.
66. R. Ou, Y. Wang, H. Wang and T. Xu, *Desalination*, 2013, **318**, 48–55.
67. M. L. Stone, C. Rae, F. F. Stewart and A. D. Wilson, *Desalination*, 2013, **312**, 124–129.
68. Y. Cai, W. Shen, S. L. Loo, W. B. Krantz, R. Wang, A. G. Fane and X. Hu, *Water Res.*, 2013, **47**, 3773–3781.
69. D. Li, X. Zhang, G. P. Simon and H. Wang, *Water Res.*, 2013, **47**, 209–215.
70. D. Li, X. Zhang, J. Yao, Y. Zeng, G. P. Simon and H. Wang, *Soft Matter*, 2011, **7**, 10048–10056.
71. A. Razmjou, M. R. Barati, G. P. Simon, K. Suzuki and H. Wang, *Environ. Sci. Technol.*, 2013, **47**, 6297–6305.
72. A. Razmjou, Q. Liu, G. P. Simon and H. Wang, *Environ. Sci. Technol.*, 2013, **47**, 13160–13166.
73. A. Razmjou, G. P. Simon and H. Wang, *Chem. Eng. J.*, 2013, **215–216**, 913–920.
74. Y. Zeng, L. Qiu, K. Wang, J. Yao, D. Li, G. P. Simon, R. Wang and H. Wang, *RSC Adv.*, 2013, **3**, 887–894.
75. D. Li, X. Zhang, J. Yao, G. P. Simon and H. Wang, *Chem. Commun.*, 2011, **47**, 1710–1712.
76. J. Hubbuch, D. Matthiesen, T. Hobley and O. T. Thomas, *Bioseparation*, 2001, **10**, 99–112.
77. S.-C. Chang, T. Anderson, S. Bahrman, C. Gruden, A. Khijniak and P. Adriaens, *J. Ind. Microbiol. Biotechnol.*, 2005, **32**, 629–638.
78. C. T. Yavuz, J. T. Mayo, W. W. Yu, A. Prakash, J. C. Falkner, S. Yean, L. Cong, H. J. Shipley, A. Kan, M. Tomson, D. Natelson and V. L. Colvin, *Science*, 2006, **314**, 964–967.
79. W. Wang, Y. Xu, D. I. C. Wang and Z. Li, *J. Am. Chem. Soc.*, 2009, **131**, 12892–12893.
80. O. Veiseh, J. W. Gunn and M. Zhang, *Adv. Drug Delivery Rev.*, 2010, **62**, 284–304.
81. Z. Huang, N. Pei, Y. Wang, X. Xie, A. Sun, L. Shen, S. Zhang, X. Liu, Y. Zou, J. Qian and J. Ge, *Biomaterials*, 2010, **31**, 2130–2140.
82. J. Gao, H. Gu and B. Xu, *Acc. Chem. Res.*, 2009, **42**, 1097–1107.

83. H. Bai, Z. Liu and D. D. Sun, *Sep. Purif. Technol.*, 2011, **81**, 392–399.
84. D. Li and H. Wang, *J. Mater. Chem. A*, 2013, **1**, 14049–14060.
85. C. X. Guo, D. Zhao, Q. Zhao, P. Wang and X. Lu, *Chem. Commun.*, 2014, **50**, 7318–7321.
86. Z. Li, Y. Ma and L. Qi, *Mater. Res. Bull.*, 2013, **48**, 3157–3163.
87. C. Wang, S. Peng, L.-M. Lacroix and S. Sun, *Nano Res.*, 2009, **2**, 380–385.
88. G. S. Chaubey, C. Barcena, N. Poudyal, C. Rong, J. Gao, S. Sun and J. P. Liu, *J. Am. Chem. Soc.*, 2007, **129**, 7214–7215.
89. H. Han, J. Y. Lee and X. Lu, *Chem. Commun.*, 2013, **49**, 6122–6124.
90. J. H. Schenkel, A. Samanta and B. J. Ravoo, *Adv. Mater.*, 2014, **26**, 1076–1080.
91. S. Das, P. Ranjan, P. S. Maiti, G. Singh, G. Leitus and R. Klajn, *Adv. Mater.*, 2013, **25**, 422–426.
92. M. Taguchi, K. Yamada, K. Suzuki, O. Sato and Y. Einaga, *Chem. Mater.*, 2005, **17**, 4554–4559.
93. R. Mikami, M. Taguchi, K. Yamada, K. Suzuki, O. Sato and Y. Einaga, *Angew. Chem., Int. Ed.*, 2004, **43**, 6135–6139.
94. J. Guo, N. Wang, J. Wu, Q. Ye, C. Zhang, X.-H. Xing and J. Yuan, *J. Mater. Chem. B*, 2014, **2**, 437–442.
95. S. Chen, C. X. Guo, Q. Zhao and X. Lu, *Chem.–Eur. J.*, 2014, **20**, 14057–14062.
96. D. Zhao, P. Wang, Q. Zhao, N. Chen and X. Lu, *Desalination*, 2014, **348**, 26–32.
97. P. G. Jessop, S. M. Mercer and D. J. Heldebrant, *Energy Environ. Sci.*, 2012, **5**, 7240–7253.
98. P. G. Jessop, L. Kozycz, Z. G. Rahami, D. Schoenmakers, A. R. Boyd, D. Wechsler and A. M. Holland, *Green Chem.*, 2011, **13**, 619–623.
99. P. G. Jessop, L. Phan, A. Carrier, S. Robinson, C. J. Durr and J. R. Harjani, *Green Chem.*, 2010, **12**, 809–814.
100. P. G. Jessop, D. J. Heldebrant, X. Li, C. A. Eckert and C. L. Liotta, *Nature*, 2005, **436**, 1102.
101. D. Buenger, F. Topuz and J. Groll, *Prog. Polym. Sci.*, 2012, **37**, 1678–1719.
102. Y. Samchenko, Z. Ulberg and O. Korotych, *Adv. Colloid Interface Sci.*, 2011, **168**, 247–262.
103. F. Ganji, S. Vasheghani-Farahani and E. Vasheghani-Farahani, *Iran Polym. J.*, 2010, **19**, 375–398.
104. I. Donati, Y. A. Mørch, B. L. Strand, G. Skjåk-Bræk and S. Paoletti, *J. Phys. Chem. B*, 2009, **113**, 12916–12922.
105. J. M. G. Swann, W. Bras, P. D. Topham, J. R. Howse and A. J. Ryan, *Langmuir*, 2010, **26**, 10191–10197.
106. M. Gao, K. Gawel and B. T. Stokke, *Soft Matter*, 2011, **7**, 1741–1746.
107. A. Pourjavadi and H. Salimi, *Ind. Eng. Chem. Res.*, 2008, **47**, 9206–9213.
108. G. D. Nicodemus and S. J. Bryant, *Tissue Eng., Part B*, 2008, **14**, 149–165.
109. I. Tomatsu, K. Peng and A. Kros, *Adv. Drug Delivery Rev.*, 2011, **63**, 1257–1266.

110. L. A. Lyon, Z. Meng, N. Singh, C. D. Sorrell and A. St. John, *Chem. Soc. Rev.*, 2009, **38**, 865–874.
111. R. Kulkarni and S. Biswanath, *J. Appl. Biomater. Biomech.*, 2006, **5**, 125–139.
112. T. Miyata, T. Uragami and K. Nakamae, *Adv. Drug Delivery Rev.*, 2002, **54**, 79–98.
113. L. Zha, B. Banik and F. Alexis, *Soft Matter*, 2011, **7**, 5908–5916.
114. K. Gawel, D. Barriet, M. Sletmoen and B. T. Stokke, *Sensors*, 2010, **10**, 4381–4409.

CHAPTER 3

Superwetting Nanomaterials for Advanced Oil/Water Separation: From Absorbing Nanomaterials to Separation Membranes

SHOUJIAN GAO[a] AND JIAN JIN*[a]

[a]Nano-Bionics Division and *i*-Lab, Suzhou Institute of Nano-Tech and Nano-Bionics, Chinese Academy of Sciences, Suzhou, 215123, China
*E-mail: jjin2009@sinano.ac.cn

3.1 Introduction

Oily water is widely generated during global industrialization processes and seriously harms the environment and people's health in a wide range of ways.[1,2] Accidents involving oil spills, from the Lakeview Gusher oil spill in 1910 to the Deepwater Horizon oil spill in 2010, release millions of tonnes of crude oil into the environment, especially the marine areas, every year and produce large quantities of oily water.[3,4] Besides the petroleum industry, oily water is also widely generated by diverse industrial technologies, such as the food, textile, leather, steel and metal finishing industries *etc.*[1] In a broader sense, oily water can be produced by the releasing of organic reagents into

RSC Smart Materials No. 20
Smart Materials for Advanced Environmental Applications
Edited by Peng Wang

Published by the Royal Society of Chemistry, www.rsc.org

water as well.[5] The direct discharge of oily water kills large numbers of creatures, seriously damages people's physical health, and even further aggravates the shortage crisis of drinkable freshwater.[6–8] The unsuitable treatment of oil pollution, such as *in situ* burning, will give rise to a fire hazard and finally bring about dire air pollution.[4] Separating oil from water is widely recognized as a right and sustainable strategy and has become a new worldwide subject.[9] However, this subject is always tough and challenging as oily water is prone to be present in multiple forms under differing conditions: free layered oil/water mixtures and oil-in-water or water-in-oil emulsions by formation, surfactant-free emulsions and surfactant-stabilized emulsions by components, micrometre-scale emulsions and nanometre-scale emulsions by oil droplet size.[10]

Traditional oil/water separation techniques, such as oil skimmers, centrifuges, settling tanks and flotation technologies, are time-consuming, energy-intensive, complex in operation and usually involved in secondary pollution.[11–16] To efficiently separate the increasing amount of oily water, new advanced techniques must be developed. As oil/water separation is an interfacial challenge, people believe separating oil and water using superwetting (including superhydrophilic–underwater superoleophobic, superhydrophobic–superoleophilic, superhydrophobic–superoleophobic, *etc.*) materials is an effective and facile way.[17–19] Over the past decade, nanomaterials and nanotechnologies have rapidly transformed from an academic pursuit to commercial reality.[20] Superwetting nanomaterials which are constructed to combine surface chemical composition and nanostructured surface roughness have been extensively developed and provide a brand new strategy for effective, energy-efficient and process-simple oil/water separation – while still having a series of advanced functions, such as high permeability, catalytic reactivity, self-cleaning ability and fouling resistance.

Since the superoleophilic–superhydrophobic mesh-based film was first prepared by Jiang *et al.* in 2004, a series of advanced works have been carried out by designing and fabricating superwetting nanomaterials including superwetting absorbing nanomaterials and superwetting separation membranes to achieve oil/water separation.[21–26] As superwetting nanomaterials exhibit excellent water-favouring and oil-repellent or oil-favouring and water-repellent properties in water-rich or oil-rich conditions, they can be used to effectively remove oil or water from oil/water mixtures and emulsions by selective absorption or permeation. To be specific, the superhydrophilic–underwater superoleophobic nanomaterials commonly show a very high water affinity but a very low underwater oil adhesion force, which enables water go through but effectively repels oil under water, simultaneously cut down the oil fouling issue to a minimum. On the contrary, the superoleophilic–superhydrophobic nanomaterials exhibit a very high oil affinity and very high water-repellent ability, which enables oil go through or absorbs oil but rejects water, consequently achieving the effective separation of oil and water. Besides, absorbing materials and filtration membranes are widely regarded as better technologies than traditional techniques from an operational point of view.[20,27] Hence, the superwetting absorbing nanomaterials and separation membranes show big advantages in operation and separation efficiency compared with traditional oil/water separation techniques when treating oily water.

In this chapter, we present a brief review of recent advances for the separation of both free and emulsified oil/water mixtures by superwetting nanomaterials, including superwetting absorbing nanomaterials and superwetting separation membranes. Firstly, a theoretical principle to construct nanomaterials with superwetting surfaces is introduced in Section 3.2. This is followed by the recent advanced works in the separation of free oil/water mixtures using superwetting sponge-, foam- and textile-based absorbing nanomaterials, introduced in Section 3.3. Particularly, the superwetting separation membranes including mesh-based films, polymer-dominated filtration membranes and nanomaterial-based ultrathin films for advanced separation of oil/water emulsions are discussed in Section 3.4. Finally, a summary about this brief review and a perspective for oil/water separation in future are described in Section 3.5.

3.2 How to Construct Nanomaterials with Superwetting Surfaces

3.2.1 Theoretical Basis of Wettability of Solid Materials

The wetting behaviour of a liquid on a solid surface is a function of the interfacial free energies among the solid/liquid (γ_{SL}), solid/vapour (γ_{SV}) and liquid/vapour (γ_{LV}) interfaces. This can be described by the liquid contact angle (CA) and the classic model by Thomas Young when a liquid droplet is presented onto an ideal flat solid surface (Figure 3.1a and eqn (3.1)).[28] That is to say, the wettability of an ideal flat solid surface is determined by the surface chemical composition. Commonly, when the solid surface shows a liquid CA less than 90°, this solid material is defined as lyophilic, when the solid surface shows a liquid CA more than 90°, this solid material is defined as lyophobic. For lyophilic materials, $\cos\theta_{\text{CA-ideal}} > 0$, $\gamma_{SV} > \gamma_{SL}$, these materials commonly have a high surface energy. For lyophobic materials, $\cos\theta_{\text{CA-ideal}} < 0$, $\gamma_{SV} < \gamma_{SL}$, these materials commonly have a low surface energy.

$$\cos\theta_{\text{CA-ideal}} = (\gamma_{SV} - \gamma_{SL})/\gamma_{LV} \tag{3.1}$$

However, most of the real solid surfaces are not truly flat, but relatively rough. Therefore, the wettability of a solid surface is determined by both the surface chemical composition and the surface roughness factor. For a solid material with a rough surface, the classic Wenzel model and Cassie model are proposed to explain and evaluate the wettability. In the Wenzel model, the liquid completely pins the grooves of the rough surface in

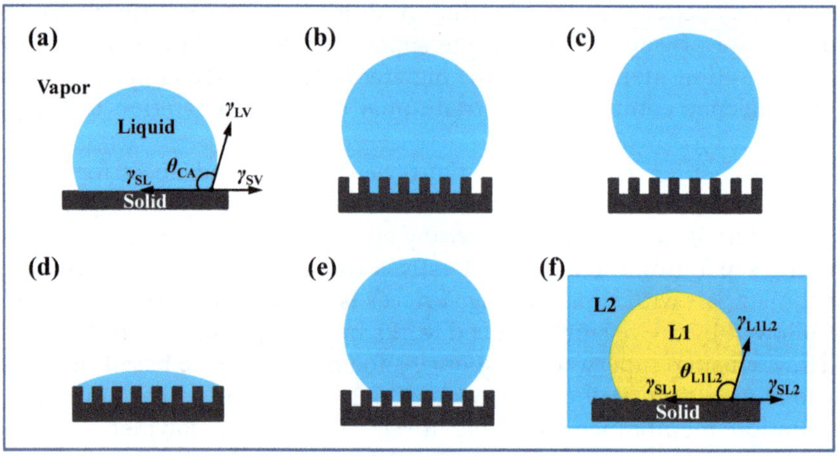

Figure 3.1 Schematic showing the wetting mechanism of a liquid droplet on a solid surface in vapour or under another liquid. (a) A liquid droplet on an ideal flat lyophobic solid surface in vapour. A liquid droplet on a rough lyophobic solid surface in a wet-contact state (b) or a non-wet-contact state (c) in vapour. (d) A liquid droplet on a rough lyophilic solid surface in vapour. (e) A liquid droplet on a rough lyophobic solid surface in a transition state in vapour. (f) A droplet of liquid (L1) on a solid surface under another liquid (L2).

a wet-contact state where they contact as illustrated in Figure 3.1b and described by eqn (3.2).[29]

$$\cos\theta_{CA} = r\cos\theta_{CA\text{-ideal}} = r(\gamma_{SV} - \gamma_{SL})/\gamma_{LV} \tag{3.2}$$

where θ_{CA} is the apparent liquid CA in the Wenzel mode, $\theta_{CA\text{-ideal}}$ is the liquid CA for an ideal flat solid surface, and r is the surface roughness factor. As r is always larger than 1, the surface roughness factor will provide an enhanced surface wettability and even generate a superlyophobic state with the liquid CA more than 150° or a superlyophilic state with the liquid CA less than 5°. In brief, for a lyophilic solid material, $\theta_{CA} < \theta_{CA\text{-ideal}}$. For a lyophobic solid material, $\theta_{CA} > \theta_{CA\text{-ideal}}$.

Another classic wetting model is the Cassie model, as illustrated in Figure 3.1c. Vapour is assumed to be trapped in the grooves of the rough surface underneath the liquid, giving a composite surface. In this composite state, the vapour parts of the surface can be considered perfectly non-wetting, the liquid is thus assumed to contact the surface through the top of the asperities in a non-wet-contact mode. In the Cassie model, the wettability of a solid material can be correlated to the chemical heterogeneity of the rough surface and described by eqn (3.3).[30]

$$\cos\theta_{CA} = f_S(r_S\cos\theta_{CA\text{-ideal}} + 1) - 1 = f_S[r_S(\gamma_{SV} - \gamma_{SL})/\gamma_{LV} + 1] - 1 \tag{3.3}$$

where θ_{CA} is the apparent liquid CA in the Cassie mode, $\theta_{CA\text{-ideal}}$ is the liquid CA for an ideal flat solid surface, f_S is the area fraction of the solid on the surface, and r_S is the roughness factor of the solid parts of the surface with which the liquid contact. For a lyophobic solid material, the grooves in rough surface trap vapour pockets and reduce the value of f_S. As a result, θ_{CA} increases greatly ($\theta_{CA} > \theta_{CA\text{-ideal}}$) and the surface wettability is enhanced even to be superlyophobic in the vapour. And for a lyophilic solid material, the liquid will permeate inside the structural grooves (Figure 3.1d), leading to an enhanced lyophilic state ($\theta_{CA} > \theta_{CA\text{-ideal}}$).

The Cassie state reflects the lowest energy state in the open-vapour regime, also known as the metastable composite states, while the Wenzel state represents the absolute minimum energy in a wetted lyophobic state.[31,32] According to the difference between the Wenzel model and the Cassie model or eqn (3.2) and eqn (3.3), for a solid surface with same roughness, it should exhibit two distinct apparent liquid CAs. Generally, the liquid droplet in the Wenzel state always shows a stronger contact angle hysteresis than in the Cassie state.[33] It has also been reported that the contact mode in the solid/liquid/vapour system will change and generate a transition model between the Cassie state and the Wenzel state when the liquid droplet is under pressure, vibration, impact or electrical field (Figure 3.1e).[34–38] A threshold liquid CAθ_T ($\theta_T > 90°$) between the Cassie state and the Wenzel state can be calculated by eqn (3.4).[33,39]

$$\cos\theta_T = (f_S - 1)/(r - f_S) \tag{3.4}$$

When $\theta_{CA\text{-ideal}} < \theta_T$, the as-trapped vapour pockets in the surface grooves are unstable, and the solid/vapour contact state will change to the Wenzel state. When $\theta_{CA\text{-ideal}} > \theta_T$, the as-trapped vapour pockets are stable in the surface grooves, and the apparent CA of a liquid droplet will agree with the Cassie mode.[40]

Besides the solid/liquid/vapour system, in the solid/liquid/liquid system, the liquid (named L1) wetting behaviour on the solid surface under another immiscible liquid (named L2) can be illustrated in Figure 3.1f and described by eqn (3.5).[32,41]

$$\cos\theta_{L1L2} = (\gamma_{SL2} - \gamma_{SL1})/\gamma_{L1L2} = (\gamma_{L1A}\cos\theta_{L1} - \gamma_{L2A}\cos\theta_{L2})/\gamma_{L1L2} \tag{3.5}$$

where θ_{L1L2} is the apparent L1 CA on the solid surface under L2. θ_{L1}, θ_{L2} are the apparent liquid 1 CA and liquid 2 CA on the solid surface in air, respectively. γ_{SL2}, γ_{SL1}, γ_{L1L2}, γ_{L1A}, γ_{L2A} and γ_{L1L2} are the interfacial free energies of the solid/L2, solid/L1, L1/L2, L1/air, L2/air and L1/L2 interfaces, respectively. As for the solid/oil/water system, the oil CA on a solid surface under water can be described by eqn (3.6).[42]

$$\cos\theta_{OW} = (\gamma_{SW} - \gamma_{SO})/\gamma_{OW} = (\gamma_{OA}\cos\theta_O - \gamma_{WA}\cos\theta_W)/\gamma_{OW} \tag{3.6}$$

where θ_{OW} is the apparent oil CA on the solid surface under water. θ_O, θ_W are the apparent oil CA and water CA on the solid surface in air, respectively.

γ_{SW}, γ_{SO}, γ_{OW}, γ_{OA}, γ_{WA} and γ_{OW} are the interfacial free energies of the solid/water, solid/oil, oil/water, oil/air, water/air and oil/water interfaces, respectively.

In summary, a superwetting (superlyophilic or superlyophobic) material can be constructed in combination of surface chemical composition (lyophilic with a high surface energy or lyophobic with a low surface energy) and surface roughness.

3.2.2 Theoretical Principle to Construct Superwetting Nanomaterials

According to the latest definition by Volger *et al.*, a new watershed CA to divide the hydrophilicity and hydrophobicity of solid materials should be 65° rather than 90° by considering the actual chemical and structural state of water droplets.[32,43,44] Hence, if a solid material shows a water CA less than 65° on an ideal flat surface, we can construct a superhydrophilic state of this material by generating enough surface roughness. If a solid material shows a water CA more than 65° on an ideal flat surface, we can construct a superhydrophobic state of this material by magnifying its surface roughness to the extreme. Commonly, materials with high surface energy tend to behave hydrophilically and are chosen as the effective components for fabricating superhydrophilic materials. On the contrary, low surface energy materials tend to behave hydrophobically and are commonly chosen to fabricate superhydrophobic materials by increasing surface roughness. Similarly, superoleophilic materials can be constructed by increasing the surface roughness of oleophilic materials with slightly high surface energy. Superoleophobic materials can be constructed from materials with extremely low surface energy. According to the surface energy of various surface functional groups reported by Zisman and co-workers, the surface energy decreases in the order $-CH_2 > -CH_3 > -CF_2 > -CF_2H > -CF_3$.[45] Hence, fluorinated and perfluorinated materials, including perfluorinated phosphates, perfluorinated silanes, fluorinated monomers, polymers and copolymers, and other fluorinated precursors, are the logical materials to be used to construct superoleophobic materials.[46,47] To fabricate highly rough surfaces, the micro- and nanostructures are extremely important and indispensable.

It is worth noting that most of the hydrophilic materials simultaneously show an oleophilic property in air because γ_{SO} is usually lower than γ_{SW} (usually $\gamma_{SO} < \gamma_{SW}$, if $\gamma_{SA} > \gamma_{SW}$, then $\gamma_{SA} > \gamma_{SO}$), but behave oleophobically underwater. Some hydrophobic materials, such as most of the fluorine-free hydrophobic materials like polydimethylsiloxane (PDMS) and some of the fluorine-containing hydrophobic materials like polyvinylidene fluoride (PVDF) and polytetrafluoroethylene (PTFE), can simultaneously show a oleophilic property if γ_{SA} is between γ_{SW} and γ_{SO} ($\gamma_{SW} > \gamma_{SA} > \gamma_{SO}$). The oleophobic materials which are composed of materials with extremely low surface energy usually show a hydrophobicity at the same time (usually $\gamma_{SO} < \gamma_{SW}$, if $\gamma_{SA} < \gamma_{SO}$, then $\gamma_{SA} < \gamma_{SW}$). Hence, it is possible to construct superwetting materials with both superhydrophilicity and superoleophilicity or superhydrophilicity and underwater superoleophobicity or superhydrophobicity and superoleophilicity or superhydrophobicity and superoleophobicity by choosing the correct materials and generating enough surface roughness.

3.2.3 Natural and Artificial Examples of Superwetting Nanomaterials

A classic and natural example of a superhydrophobic material is the lotus leaf (Figure 3.2a), which is composed of micro- and nanometre-scale hierarchical structures and hydrophobic cuticular wax (Figure 3.2b and c).[48] People believe that the main reason for its superhydrophobicity is the combination of the hydrophobic surface chemical composition with low surface energy

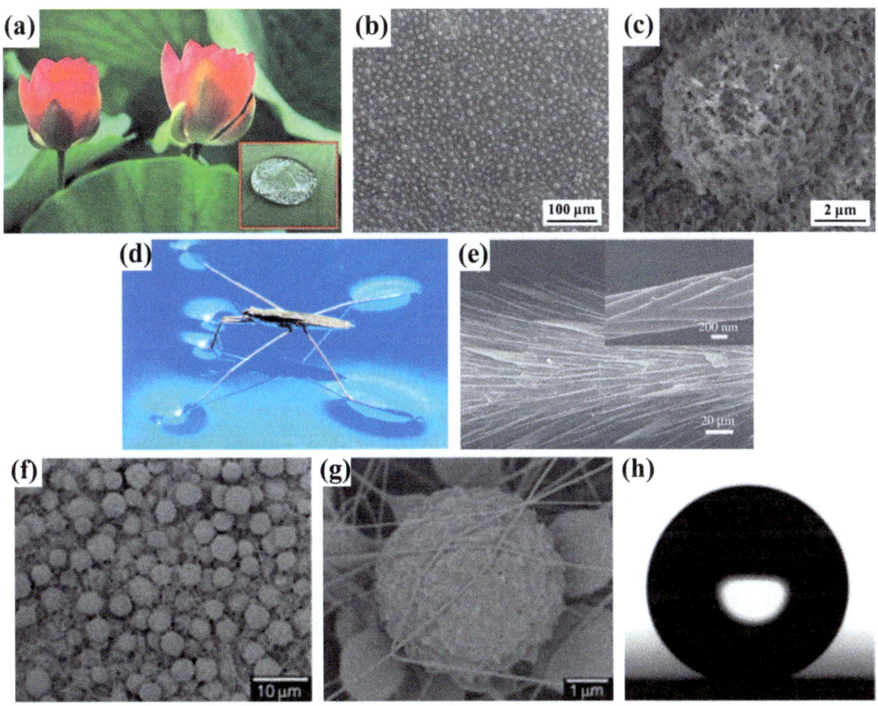

Figure 3.2 (a)–(c) Digital photograph and SEM images of the superhydrophobic lotus leaf.[48] Reprinted with permission from ref. 48. Copyright 2008, AIP Publishing LLC. (d) Digital photograph of a water strider which stands on the water surface.[49] (e) SEM images of the water strider's legs. Reprinted by permission from Macmillan Publishers Ltd: *Nature* (ref. 49), copyright 2004. (f) and (g) SEM images of the porous microspheres and nanofibres of the superhydrophobic PS film.[50] (h) A water droplet on the superhydrophobic PS film. L. Jiang, *et. al.*, A lotus-leaf-like superhydrophobic surface: a porous microsphere/nanofibre composite film prepared by electrohydrodynamics, *Angew. Chem., Int. Ed.*, 2004, **43**, 33. Copyright © 2004 John Wiley & Sons, Inc.

and the micrometre- and nanometre-scale surface roughness. Another classic and natural example of a superhydrophobic material is the water strider's non-wetting legs which can enable the water strider to stand on the water surface effortlessly (Figure 3.2d).[49] The superhydrophobicity of the water strider's legs has been traced to the micrometre-scale needle-shaped setae with numerous nanometre-scale grooves and the secreted hydrophobic wax on the elaborate structures (Figure 3.2e). This is the result of the combination of the hydrophobic surface chemical composition and the micrometre- and nanometre-scale surface roughness, too. Inspired by these natural superhydrophobic materials, and combining the theories of wettability, people have invented a series of artificial superhydrophobic nanomaterials by generating highly rough surfaces of low surface energy materials or integrating low surface energy coatings on as-fabricated rough surfaces. Jiang *et al.* developed a lotus leaf-like superhydrophobic polystyrene (PS) film with a novel composite structure composed of porous microspheres and nanofibres (Figure 3.2f) by a electrohydrodynamics technique.[50] Like the hydrophobic cuticular wax on the lotus leaf, the PS is a low surface energy and hydrophobic material as well. Sizes of the microspheres and nanofibres are 3–7 μm and 60–140 nm, respectively. There are also many protuberances and cavities with dimensions of tens to hundreds of nanometres on the surfaces of the microspheres (Figure 3.2g). The porous microspheres play the leading role in the superhydrophobicity of the film by increasing surface roughness, and the nanofibres interweave to form a 3D network and act as a skeleton to reinforce the composite film. This PS composite film exhibits an excellent water-repellent property with a water CA of 160.4 ± 1.28° (Figure 3.2h).

Seeger and co-workers presented a simple method of growing polysiloxane nanofilaments with diameter ranging from 20 to 50 nm (Figure 3.3a) on various substrates to fabricate superhydrophobic surfaces.[51] The combination of the hydrophobicity of the silicone and the nanostructured surface roughness formed by the nanofilament contributes to the superhydrophobicity. What is worth mentioning is that the two main requirements for superhydrophobic property of a material, namely a low surface energy composition and the nanometre-scale roughness are obtained in a single fabrication step. A frosted glass coated with these silicone nanofilaments shows a water CA of 166 ± 3° and a water sliding angle (SA) lower than 3°. Other coated substrates like cotton fabric, wood, silicone rubber and aluminium, *etc.* all show a superhydrophobicity (Figure 3.3b). Most importantly, no fluorine-containing reagents are needed in the fabrication of this superhydrophobic coating. However, the silicone nanofilament coating still behaves superoleophilically. When the silicone nanofilaments are further modified with a lower surface material 1*H*,1*H*,2*H*,2*H*-perfluorodecyltrichlorosilane (PFDTS, Figure 3.3c), the nanostructured coating exhibits a superoleophobic property for various non-polar liquids, such as mineral oil, toluene, hexadecane, decane and cyclohexane (Figure 3.3d and e).[52] Owing to the introduction of fluoroalkyl groups, the glass coated with these superoleophobic nanofilaments show oil CAs higher than 155° and oil SAs lower than 6° for all of the oils investigated.

Oil droplets can easily roll off from the slightly tilted (3°) glass. Even jets of toluene and decane can bounce off the nanofilament-coated glass without leaving a trace (Figure 3.3f). This superoleophobic nanofilament coating also possesses excellent transparency, chemical and environmental stability.

A series of typical superlyophobic nanomaterials containing fluorodecyl polyhedral oligomeric silsesquioxane (f-POSS) have been reported by Tuteja

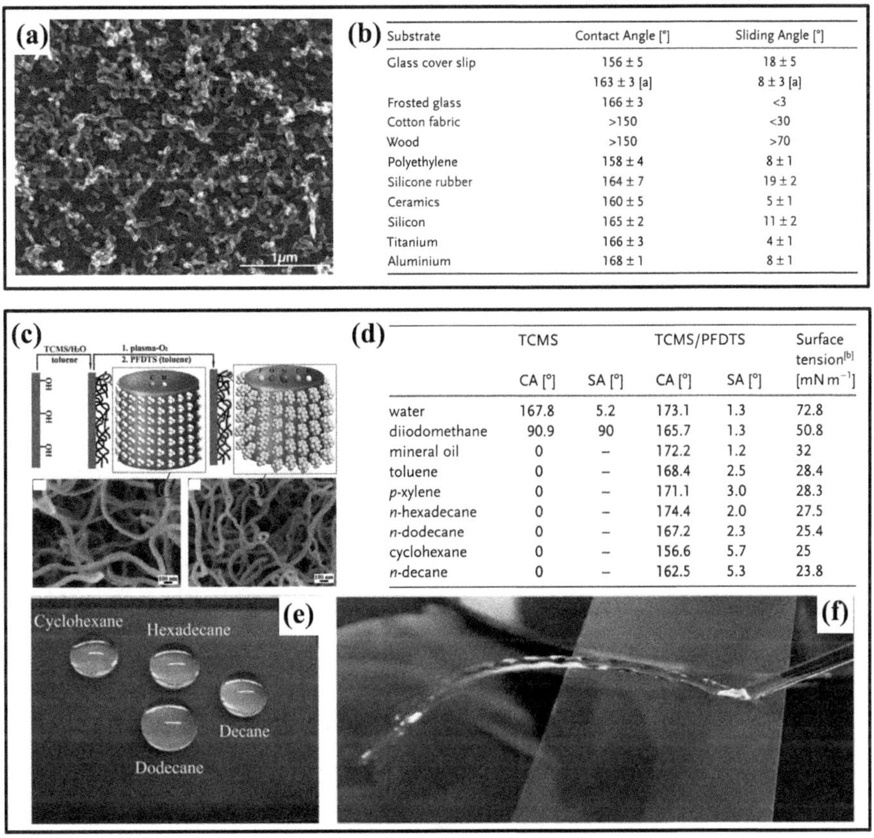

(b)

Substrate	Contact Angle [°]	Sliding Angle [°]
Glass cover slip	156 ± 5	18 ± 5
	163 ± 3 [a]	8 ± 3 [a]
Frosted glass	166 ± 3	<3
Cotton fabric	>150	<30
Wood	>150	>70
Polyethylene	158 ± 4	8 ± 1
Silicone rubber	164 ± 7	19 ± 2
Ceramics	160 ± 5	5 ± 1
Silicon	165 ± 2	11 ± 2
Titanium	166 ± 3	4 ± 1
Aluminium	168 ± 1	8 ± 1

(d)

	TCMS		TCMS/PFDTS		Surface tension[b]
	CA [°]	SA [°]	CA [°]	SA [°]	[mN m^{-1}]
water	167.8	5.2	173.1	1.3	72.8
diiodomethane	90.9	90	165.7	1.3	50.8
mineral oil	0	–	172.2	1.2	32
toluene	0	–	168.4	2.5	28.4
p-xylene	0	–	171.1	3.0	28.3
n-hexadecane	0	–	174.4	2.0	27.5
n-dodecane	0	–	167.2	2.3	25.4
cyclohexane	0	–	156.6	5.7	25
n-decane	0	–	162.5	5.3	23.8

Figure 3.3 (a) SEM image of the polysiloxane nanofilaments.[51] (b) Water CAs and SAs of a series of substrates coated by the polysiloxane nanofilaments. G. R. J. Artus, *et. al.*, Silicone nanofilaments and their application as superhydrophobic coatings, *Adv. Mater.*, 2006, **18**, 20. Copyright © 2006 John Wiley & Sons, Inc. (c) Schematic showing the growth of silicone nanofilaments using (trichloromethylsilane) TCMS and subsequent modification with PFDTS.[52] The SEM images below are the silicone nanofilaments before and after PFDTS modification. (d) CAs and SAs for water and various non-polar liquids on the TCMS and TCMS/PFDTS nanofilaments-coated glasses. Digital images of non-polar liquid droplets (e) and a jet of toluene (f) on the TCMS/PFDTS nanofilament-coated glass. W. Choi, *et. al.*, Superoleophobic coatings with ultralow sliding angles based on silicone nanofilaments, *Angew. Chem., Int. Ed.*, 2011, **50**, 29. Copyright © 2011 John Wiley & Sons, Inc.

and co-workers.[53–56] The f-POSS possesses an extremely low surface energy of about 10 mN m^{-1}, which plays an important role in the superoleophobicity of the blends. Among these materials, a superlyophobic coating of cross-linked poly(dimethylsiloxane) (PDMS) + 50 wt% f-POSS ($\gamma_{SA} \approx 11.5$ mN m^{-1}) was fabricated on a stainless steel mesh with hierarchically structured surface (Figure 3.4a–c).[54] The fabrication process was achieved by a simple technique based on electrospun coating. The reason for choosing PDMS in this work is its excellent chemical resistance against a range of different chemicals upon cross-linking. This superoleophobic coating exhibits very high apparent CAs, ultralow SAs and ultralow CA hysteresis for a range of polar and non-polar

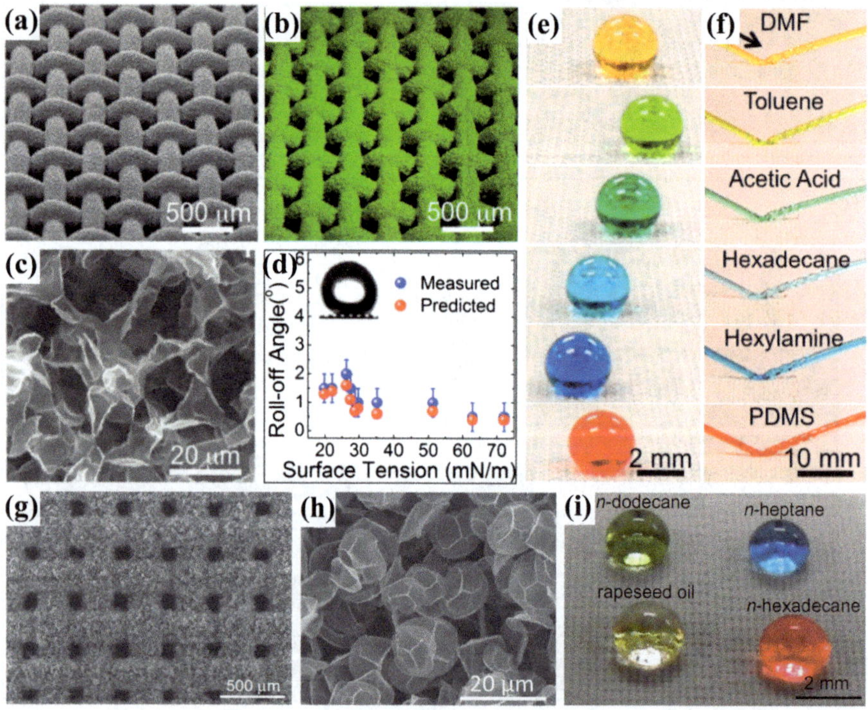

Figure 3.4 (a) SEM image of the hierarchically structured surface illustrating the electrospun coating of x-PDMS + 50 wt% f-POSS on a stainless steel mesh.[54] (b) Elemental mapping of fluorine on the hierarchically structured surface. (c) SEM image illustrating the re-entrant curvature of the electrospun texture. (d) Roll-off angles for various Newtonian liquids on the surface shown in (a). Droplets (e) and jets (f) of various low surface tension Newtonian liquids on the surface shown in (a). Reprinted with permission from ref. 54. Copyright 2013 American Chemical Society. (g) SEM image of a stainless steel mesh coated with electrospun microbeads of 50 wt% f-POSS + PMMA blend.[55] (h) SEM images of the electrospun microbeads obtained using AK225 with 5 vol% DMF. (i) Droplets of different low surface tension liquids on the surface shown in (g). A. K. Kota, *et. al.*, Hierarchically structured superoleophobic surfaces with ultralow contact angle hysteresis, *Adv. Mater.*, 2011, **24**, 43. Copyright © 2011 John Wiley & Sons, Inc.

low surface tension Newtonian liquids, including various acids, bases and solvents, due to the hierarchical texture along with re-entrant curvature and the ultralow surface energy of the coating (see Figure 3.4d and e). Due to the low CA hysteresis, even jets of different Newtonian liquids can easily bounce on the surface of the PDMS/f-POSS coating (Figure 3.4f). They also prepared a superoleophobic surface by employing an electrospun coating of 50 wt% f-POSS + poly(methylmethacrylate) (PMMA) on a textured substrate (Figure 3.4g and h).[55] The hierarchical texture and the low surface energy of this material render the surface superoleophobic with very high apparent CAs (Figure 3.4i) and ultralow CA hysteresis even with extremely low surface tension liquids such as *n*-heptane (apparent CA = 155°, CA hysteresis = 4°).

For most of the superhydrophilic materials, they simultaneously exhibit a superoleophilic property in air as is introduced in Section 3.2.2, but behave superoleophobically underwater, as is widely reported by researchers.[57–60] A typical natural example of a hydrophilic and underwater superoleophobic material is fish scales, with an underwater oil CA of 156.4 ± 3.0°.[61] The surface of fish scales is composed of a thin hydrophilic mucus layer with a hierarchical structure of oriented micropapillae and nanostructured roughness. Jiang *et al.* believes that the water surrounding the fish scales leads to wettability reversion. Inspired by this, a novel strategy to generate underwater superoleophobic materials without the assistance of fluoride compounds was extensively studied and the oil–water–air–solid coexisting four-phase system was discovered.[61–63] Jiang and co-workers prepared a series of underwater superoleophobic polyacrylamide (PAM) hydrogel films by a casting technique.[61] Different PAM hydrogel films with fish-scale structures, microstructures or micro/nanostructures all behave underwater superoleophobically with underwater oil CAs as high as 174.8 ± 2.3°. Superhydrophilic–underwater superoleophobic materials, especially separation membranes which possess opposite superwetting behaviours for water and oil in a water-rich environment, are thus widely explored in the field of oil/water separation.

Besides the superwetting nanomaterials introduced above, superwetting absorbing nanomaterials including sponge-, foam- and textile-based nanomaterials, and superwetting separation membranes with nanostructure morphologies including meshes and textile-based films, polymer-dominated membranes as well as the one-dimensional (1D) nanomaterial-based ultrathin films with nanometre-scale thickness are also widely prepared and studied by researchers to achieve the separation of oil/water mixtures and emulsions. These works are discussed in Section 3.3 and Section 3.4 in detail.

3.3 Superwetting Absorbing Nanomaterials for Separation of Free Oil/Water Mixtures

Superwetting absorbing nanomaterials, such as superhydrophobic–superoleophilic sponge-, foam- and textile-based nanomaterials, are ideal materials for separation of large amounts of free oil/water mixtures from oil spill and leakage by selective absorption for oil and repellent for water.[64–68] These nanomaterials are commonly constructed with hierarchical micro- and nanostructures and with huge specific surface areas.

3.3.1 Sponge- and Foam-Based Superwetting Absorbing Nanomaterials

Sponges and foams are very common and commercially available three-dimensional (3D) porous materials and have an excellent absorbing ability for various liquids, including water and organic solvents, due to its porous structures and huge specific surface area. However, raw sponges and foams usually absorb both water and oil. Hence they can't be used for separation of oil and water by selective absorption due to their initial wettability. After proper surface modification with low surface energy nanostructured materials, various sponge- and foam-based superhydrophobic–superoleophilic absorbing nanomaterials can be obtained and used to achieve the separation of free oil/water mixtures. Recently, a variety of surface modification methods to prepare superhydrophobic–superoleophilic sponges and foams, such as *in situ* growing,[69,70] dip coating,[65,66,71–76] chemical vapour deposition (CVD),[77] and block copolymer grafting,[78] have been adopted by researchers to fabricate the sponge- and foam-based superwetting absorbing materials.

Guo and co-workers reported some superhydrophobic sponges by the coating of thiol-ligand VIII and IB nanocrystals on commercial sponge surfaces *via* a simple two-step dip coating method.[66] The VIII and IB metals includes Fe, Co, Ni, Cu and Ag. The interaction between the nanocrystals and thiol plays a significant role in the formation of the special wetting surface. It is worth noting that these superhydrophobic nanocoatings show a good durability towards hot water, surfactant aqueous washing and ultrasonic treatment in non-polar solvents and could be used for oil/water separation for a long time. They further presented an *in situ* growing method to fabricate robust superhydrophobic–superoleophilic sponges, instead of the two-step dip coating method.[69] The authors believe the micrometre- and nanometre-scale hierarchical structures magnify the wettability of the modified sponge surfaces to the extreme. The as-prepared superwetting sponges from the *in situ* growing method realized selective absorption of oils from water with absorption capacity up to 25 g g^{-1} and could be used repeatedly many times.

Tai *et al.* reported a new strategy to fabricate a superhydrophobic–superoleophilic graphene-based sponge and used it as an effective absorbing material for oil/water separation.[75] The superwetting graphene-based sponge was fabricated using commercially available sponge and hydrophobic graphene nanosheets *via* a facile and inexpensive dip coating method. In detail, an as-cleaned sponge was dipped into an ethanol solution of graphene nanosheets and then dried in a vacuum oven at 100 °C for 2 hours. Finally, the graphene nanosheets were physically coated onto the sponge skeleton due to the strong van der Waals interactions between the sponge and the flexible graphene nanosheets. The micro- and nanostructure of graphene nanosheets and the microstructure of the sponge skeleton build the hierarchical and rough surface of the resulting graphene-based sponge, consequently switching the wettability of the raw sponge from hydrophilicity to superhydrophobicity while maintaining its superoleophilicity. The graphene loading greatly determines the wettability of the graphene-based sponge and a superhydrophobic sponge can be obtained when the graphene loading is more than 7.3%. This superwetting sponge is also very stable in a broad range of pH conditions. Benefiting from the superhydrophobic–superoleophilic property and huge specific surface area, the graphene-based sponge exhibits excellent selective absorption for a variety of oils and organic solvents with absorption capacity ranging from 54 to 165 times its weight while repelling water completely. On the other hand, the graphene-based sponges can absorb soybean oil with 1.12 ton m^{-3} and chloroform with 1.86 ton m^{-3}. The absorption capacity is higher than nanowire membranes and micro-porous polymers and is comparable to carbon nanotube sponges.

Deng and co-workers presented a kind of superhydrophobic–superoleophilic conjugated microporous polymers (CMP) with excellent adsorption performance for oils and organic solvents.[76] In their report, HCMP-1 was prepared *via* the Pd^{II}/Cu^{I}-catalysed homocoupling polymerization of 1,3,5-triethynylbenzene. The HCMP-1 has a porous 3D network structure which is shown by the computational simulation and TEM image (Figure 3.5a and b). BET surface area of the as-prepared HCMP-1 was measured to be 955 $m^2\ g^{-1}$.

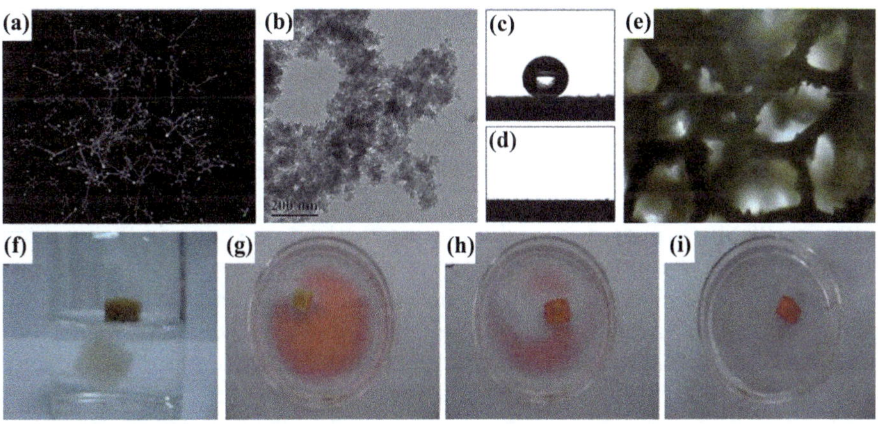

Figure 3.5 Simulation structure (a) (Materials Studio, polymer builder) and TEM image (b) of the HCMP-1, scale bar: 200 nm.[76] Optical images of a water droplet (c) and a diesel oil droplet (d) on the HCMP-1. (e) Optical microscopy image of the HCMP-1 treated sponge. (f) Digital photograph of the raw sponge which sinks into water and the HCMP-1 treated sponge which floats on water surface. (g)–(i) Snapshots showing the adsorption of a 12 cm^{-2} octane film (dyed red) by a piece of HCMP-1 treated sponge. Reproduced from ref. 76 with permission of The Royal Society of Chemistry.

The HCMP-1 shows a superhydrophobic property with a water CA of 167° and a water SA of 2.1° (Figure 3.5c). However, the HCMP-1 can quickly absorb oil with an oil CA of nearly 0° (Figure 3.5d). Benefiting from the superwetting property and open pore structures, the HCMP-1 can easily absorb oils and non-polar organic solvents without adsorption of water, which endows the HCMP-1 with the capacity for oil/water separation. The absorption capacity of the HCMP-1 for a variety of polar organic solvents and toxic organic solvents ranges from 7 to 15 times its weight. When the HCMP-1 was loaded on a hydrophilic sponge by a dip coating method, the surface wettability of the sponge transformed from hydrophilicity to superhydrophobicity (Figure 3.5e and f). The as-treated superwetting sponge exhibits an excellent ability for removing oils and organic solvents from water with an absorption capacity of 23 g g^{-1} for octane and 33 g g^{-1} for nitrobenzene (Figure 3.5g–i). Specifically, absorption capacity for the HCMP-1 loaded sponge reaches up to 0.6 t m^{-3} for octane and 0.9 t m^{-3} for nitrobenzene.

Pan and co-workers fabricated a kind of robust superhydrophobic–superoleophilic polyurethane (PU) sponge through a one-step solution immersion method involving the hydrolysis of reactive methyltrichlorosilane.[65] Water CA of the as-prepared superwetting sponge is 157°, while a drop of lubricating oil can completely spread into the pores of the sponge within 1 s. The superwetting sponge shows high selectivity and high absorption capacity of 15–25 times its own weight when employed as an oil-absorption material for collecting oils and non-polar solvents from the surface of water. Moreover, the superhydrophobic sponge shows a very high reusability of more than 300 cycles in oil/water separation without losing its superhydrophobicity and elasticity.

Yu's group presented an oil collection system based on the combination of porous hydrophobic and oleophilic materials (PHOMs) with pipes and a self-priming pump (Figure 3.6a) to realize the continuous and *in situ* oil/water separation and oil collection.[27] The PHOM used in their report is a commercially available polymer sponge modified with a hydrophobic coating of SiO_2 NPs and PDMS through a facial dip coating method. As shown in Figure 3.6b, the skeleton of the modified polymer sponge is covered with a thick layer of hydrophobic SiO_2 NPs/PDMS composite coating with micrometre- and nanometre-scale roughness. The hydrophobic polymer sponge (HPS) shows high resistance to water permeation (Figure 3.6c) and good absorption capacity for oil (Figure 3.6d). Based on the different behaviours of the HPS towards oil and water, the authors built up a bench-scale oil collection apparatus (Figure 3.6e). This oil collection apparatus can continually and quickly separate and collect the oil from the oil/water mixture. Such a self-controlled behaviour means that a small piece of HPS can collect a large area of floating oil, which provides a simple and efficient strategy to treat the oil spills by a ship pulling a floating collection system made of HPSs, pipes and pumps instead of carrying a large amount of oil-absorption materials.

Besides the superwetting and selective absorbing capacity discussed above, some other functions like magnetic responsivity and pH responsivity can also be integrated on the sponge- and foam-based absorbing nanomaterials. Calcagnile's group reported a magnetic and superwetting foam-based absorbing material to separate oils from water.[64] The composite absorbing material was fabricated using a commercially available PU foam and functionalized by oleic acid-capped superparamagnetic iron oxide nanoparticles (NPs) and submicrometre PTFE particles *via* a simple and easy scalable technique. This PU/PTFE/NPs foam behaves with superhydrophobic, superoleophilic and magnetic properties, which endows it the functions of

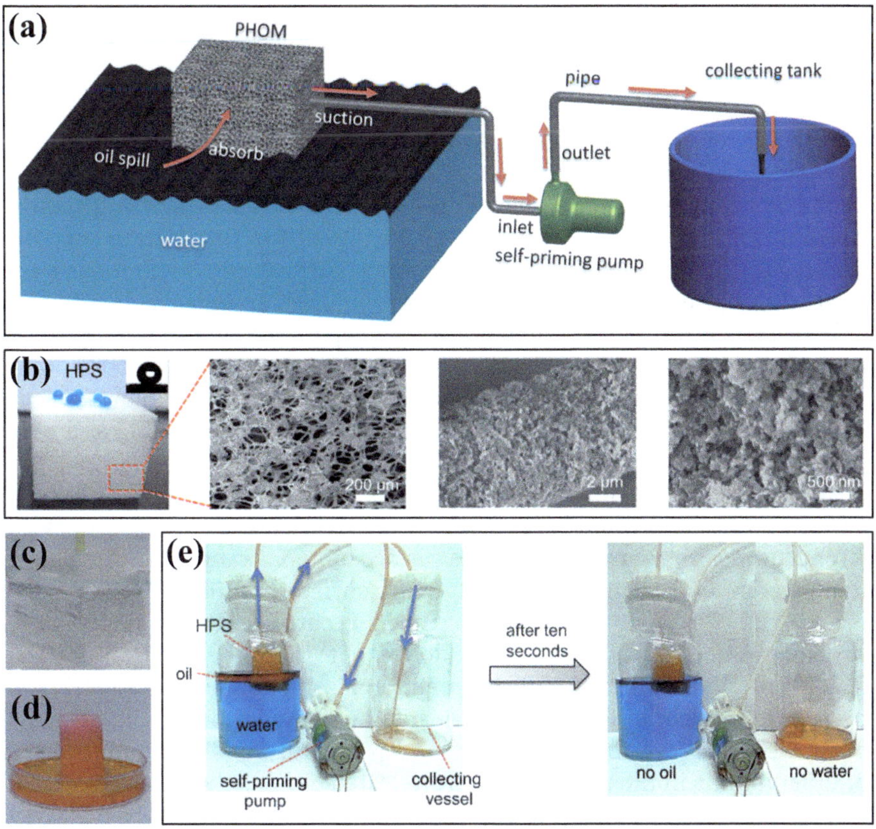

Figure 3.6 (a) Schematic illustration for a pumped PHOM device for oil spill remediation.[27] (b) Photograph and SEM images of the as-synthesized hydrophobic polymer sponge (HPS). Insert in (b) is an optical image of a water droplet on the HPS. (c) Photograph of a HPS immersed in water showing a silver mirror-like surface and indicating its high resistance to water. (d) Photograph of the HPS placed on the thin layer of *n*-hexane (dyed red) and indicating its high absorption capacity for oil. (e) Photographs of continuously absorbing *n*-hexane *in situ* from a water surface with an apparatus equipped with HPS. J. Ge, *et. al.*, Pumping through porous hydrophobic/oleophilic materials: an alternative technology for oil spill remediation, *Angew. Chem., Int. Ed.*, 2014, **53**, 14. Copyright © 2014 John Wiley & Sons, Inc.

water-repellence, oil-absorption and magnetic responsivity. The authors find that the combined functionalization of PTFE and iron oxide particles significantly increases the speed of oil absorption. Due to the magnetic responsivity of the PU/PTFE/NPs foam, people can drive this superwetting absorbing material to treat large oil spills by a magnetic field from a distance toward the polluted areas with minimal energy consumption.

Another superhydrophobic–superoleophilic, magnetic and durable sponge-based absorbing material was fabricated by CVD of tetraethoxysilane (TEOS) in the presence of the Fe_3O_4 NPs and dip coating in a fluoropolymer (FP) aqueous solution.[77] The CVD of TEOS forms a thin layer of SiO_2, which immobilizes the Fe_3O_4 NPs on the skeleton of the PU sponge. The in-turn binding of Fe_3O_4 NPs, SiO_2 layer and FP not only endows the sponge with superhydrophobicity (water CA = 157°), superoleophilicity (oil CA ≈ 0°) and magnetic properties, it also improves the elongation at break. The PU@Fe_3O_4@SiO_2@FP sponge thus has an excellent oil-absorption capacity, magnetic responsivity and high mechanical stability. This superwetting sponge-based absorbing material can be magnetically driven to separate both oils and organic solvents floating on the water surface and high-density oils under water. It is interesting that the functionalized sponge can be used not only as an absorbing material, but also as a separation membrane. With the help of a vacuum pump, the functionalized sponge can achieve a continuous separation of large amounts of oils, which makes it a very promising material for practical oil absorption and oil/water separation in a feasible and energy-efficient way.

Inspired by the bio-adhesion of marine mussels, Pan's group also reported a direct and robust immobilization method to anchor nanoparticles of different compositions and sizes on the surface of foams ranging from polymer to metals in an dopamine solution.[79] After surface modification, Fe_3O_4 NPs of ~450 nm were densely and homogenously coated on the outer and inner surface of the PU sponge. Adding *n*-dodecanethiol to the solution can lead to the superhydrophobicity of the functionalized sponge. The superhydrophobic–superoleophilic sponge fabricated by this method exhibited a high selectivity and recyclability for oil/water separation. The absorption capacity of different oils in the report range from 18 to 26 times the sponge's weight. After separation, this absorbing material can be easily collected by a magnet.

Wang's group developed a series of smart superwetting surfaces which can alter between superoleophilicity and superoleophobicity in aqueous media by a block copolymer-grafting strategy on non-woven textiles and PU sponges.[78] The block copolymer is comprised of pH-responsive poly(2-vinylpyridine) and hydrophobic–oleophilic polydimethylsiloxane (*i.e.* P2VP-*b*-PDMS). The P2VP black can switch its conformation and its wettability *via* protonation and deprotonation in response to the pH condition, which in turn provides controllable and switchable access of oil by the oleophilic PDMS block, realizing the smart surface with switchable oil wettability in aqueous media (Figure 3.7a). The P2VP-*b*-PDMS-functionalized porous material trends to behave superoleophilically and superhydrophobically in water

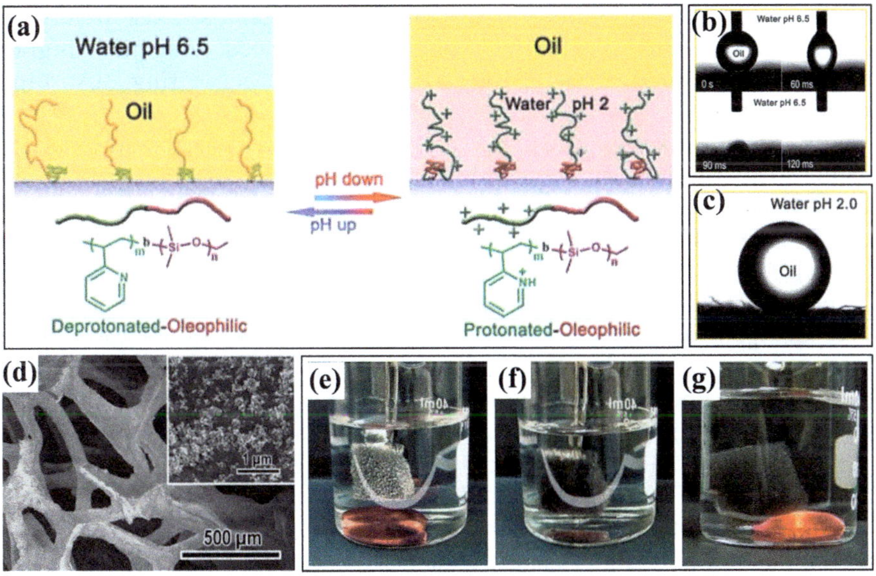

Figure 3.7 Schematic illustration (a) and optical images (b and c) showing the switchable oil wettability of the P2VP-*b*-PDMS-grafted materials in water with a pH of 6.5 (b) and 2.0 (c).[78] (d) SEM image of the P2VP-*b*-PDMS-functionalized PU sponge. Snapshots showing the absorption of 1,2-dichloroethane (dyed red) by a P2VP-*b*-PDMS-functionalized sponge in water with a pH of 6.5 (e and f) and the release of the captured 1,2-dichloroethane from the same sponge in acidic water with a pH of 2.0 after compression (g). Reprinted by permission from Macmillan Publishers Ltd: *NPG Asia Mater.* (ref. 78), copyright 2012.

with pH of 6.5 and the oil (1,2-dichloroethane) droplet can be immediately sucked into the functionalized porous material within 0.12 s (Figure 3.7b). In contrast, the P2VP-*b*-PDMS-functionalized porous material trends to behave superhydrophilically and superoleophobically in acidic water with an oil CA of 165.3° (Figure 3.7c). Based on the block copolymer grafting strategy, the authors fabricated the P2VP-*b*-PDMS functionalized PU sponges (Figure 3.7d) with the switchable superoleophilicity and superoleophobicity for highly controllable oil/water separation processes (Figure 3.7e–g). Specifically, when the functionalized PU sponges were used in water with a pH of 6.5, the sponges can quickly absorb the oil without any residue left behind. While in an acidic condition (pH 2.0), the oil was repelled by the functionalized sponges.

Except for the traditional sponge- and foam-based superwetting absorbing materials which are composed of commercial sponges and foams, a series of superwetting sponges, foams and aerogels which are composed of carbon-based nanomaterials like carbon nanotube (CNT)[80,81] have been explored and fabricated for highly efficient oil/water separation. These carbon-based absorbing nanomaterials have a big advantage with their ultralow density, ultrahigh porosity and huge specific surface area, with the result that their absorbing capacity is enhanced and much higher than the traditional sponge- and foam-based superwetting absorbing nanomaterials.

Wu and co-workers fabricated a superhydrophobic–superoleophilic CNT sponge with ultralow density, ultrahigh porosity, ultrahigh structural flexibility and robustness.[80] The CNT sponges were fabricated by a CVD method and consisted of self-assembled and interconnected CNT skeletons (Figure 3.8a and b). This as-fabricated CNT sponge is very flexible and robust and can be bent to a large degree or twisted without breaking apart. Moreover, the CNT sponge with an ultralow density of 5–10 mg cm^{-3} exhibits a water-repellent property with a water CA of 156°. Due to the ultrahigh porosity (>99%) and huge specific surface area of the CNT sponge, it is demonstrated to be an excellent superwetting absorbing nanomaterial for oil/water separation. A small densified CNT sponge can quickly absorb and remove a spreading oil film with an area up to 800 times that of the sponge (Figure 3.8c and d), suggesting its potential application for clean-up of large-scale oil spills. The absorption capacity of the CNT sponge reaches 80–180 times its own weight

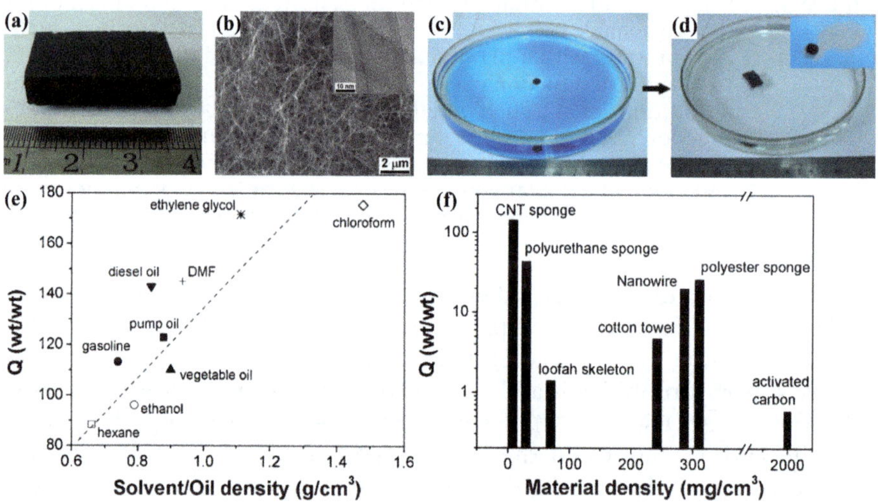

Figure 3.8 (a) Digital photograph of a monolithic CNT sponge with a size of 4 cm × 3 cm × 0.8 cm and a bulk density of 7.5 mg cm^{-3}.[80] (b) Cross-sectional SEM image of the CNT sponge showing a porous morphology and overlapped CNTs. The insert is a TEM image of large-cavity, thin-walled CNTs. (c) and (d) Digital photographs showing the large-area oil clean-up ability of the CNT sponge. The oil area is about 800 times that of the projected sponge. (e) Absorption capacity of the CNT sponge measured for a range of oils and organic solvents. (f) Summary of the absorption capacity for diesel oil measured from CNT sponges, natural products, polymeric sponges and activated carbon. X. Gui, *et. al.*, Carbon nanotube sponges, *Adv. Mater.*, 2010, **22**, 5. Copyright © 2010 John Wiley & Sons, Inc.

for a wide range of oils and organic solvents (Figure 3.8e) simultaneously with excellent selectivity and recyclability. For diesel oil, the absorption capacity of the CNT sponge is several times that of polymeric sponges, 35 times that of cotton and loofah, two orders of magnitude higher than that of activated carbon (Figure 3.8f).

A kind of carbon nanofibre (CNF) aerogel was also reported by Yu's group as an absorbing material for oil/water separation with high adsorption capacity and excellent recyclability.[81] The as-prepared CNF aerogel has a self-assembled, interconnected, 3D network structure with an ultralow density of *ca.* 10 mg cm^{-3} and an ultrahigh porosity of >99%. This CNF aerogel with high hydrophobicity and superoleophilicity can absorb a variety of oils and organic solvents from the water surface with adsorption capacity approaching 139 times its own weight. In addition, the CNF aerogel can maintain its oil-absorption ability over a wide temperature range, from liquid nitrogen temperature up to *ca.* 400 °C, providing a simple strategy for oil clean-up under harsh temperatures.

3.3.2 Textile-Based Superwetting Absorbing Nanomaterials

Apart from the sponge- and foam-based materials, textile is another good candidate to be modified with superhydrophobic–superoleophilic coatings and designed as an absorbing material to achieve oil/water separation. However, the textiles and fabrics are more often functionalized to be superwetting separation membranes for the separation of free oil/water mixtures.

Segeer's group reported a simple and one-step method to prepare the superhydrophobic–superoleophilic polyester textile *via* CVD of trichloromethylsilane for oil/water separation.[68] After surface modification, a dense layer of silicone nanofilaments are randomly grown on the microfibres of the polyester textile. The initial hydrophobic–oleophilic property of silicone and micrometre- and nanometre-scale hierarchical structures of the silicone-coated textile surface contribute to the superhydrophobicity and superoleophilicity of the resulting textile. But the silicone-coated textile isn't good for direct oil absorption due to its thin and two-dimensional structure, which seriously limits its oil absorption capacity. Based on the good flexibility and mechanical stability of the modified textile, the authors designed a functionalized absorbing bag containing an outer superhydrophobic–superoleophilic textile layer and inner absorbent material (a sponge was chosen) to selectively absorb oil and consequently achieve the oil/water separation. The functionalized absorbing bag can hold a large amount of water without any leakage owing to its excellent water-repellent property and can be fabricated in any size depending on practical conditions. The authors also investigated the practical absorption capacity of the functionalized absorbing bag by immersing the bag in the mixture of crude oil and water. The bag could quickly and selectively absorb almost all of the crude oil within 10 min and no water was absorbed by the bag.

3.4 Superwetting Separation Membranes for Oil/Water Separation

In addition to the superwetting absorbing nanomaterials, mesh- and textile-based superwetting films can also be used for the separation of oil/water mixtures. However, for emulsified oil/water mixtures including oil-in-water emulsions and water-in-oil emulsions, surfactant-free emulsions and surfactant-stabilized emulsions, superwetting absorbing materials and mesh/textile-based superwetting films are incapable of achieving the separation.[62,64–79,82–84] Instead, superwetting filtration membranes are acknowledged as an advanced and better technology for the separation of various oil/water emulsions, especially for surfactant-stabilized emulsions, to meet discharge standards and with a relatively simple process.[85–88] In this section, we will introduce these superwetting separation membranes including mesh- and textile-based films, polymer-dominated membranes, and 1D nanomaterial-based ultrathin films as well giving details on their separation performance for oil/water emulsions.

3.4.1 Mesh- and Textile-Based Superwetting Films for Separation of Oil/Water-Free Mixtures and Emulsions

Metal meshes, such as stainless steel meshes and copper meshes, are very common and commercially available 2D pore-structured materials composed of weaved metal strands. These metal meshes usually possess a very high mechanical strength and are widely adopted by researchers as the substrates to fabricate superwetting filtration films for oil/water separation. During the recent decade, a series of mesh-based superwetting filtration films were intensely fabricated by modifying them with special surface composition and hierarchical roughness.[54,55,62,89–97] The superhydrophobic–superhydrophilic mesh-based films are usually modified with low surface energy materials and used for removing water from oil-rich mixtures. The superhydrophilic–underwater superoleophobic mesh-based films are usually modified with high surface energy materials and used for removing oil from water-rich mixtures. To endow the mesh-based films with superwettabilities, the micro- and nanoscale hierarchical roughness is indispensable.

The superhydrophobic–superoleophilic PTFE coating mesh film which is constructed by micro- and nanostructured rough surfaces of a low surface energy fluorine-containing material on a stainless steel mesh is the first reported mesh-based superwetting film for oil/water separation.[21] This coating mesh film was prepared by a spray-and-dry method. Ball-like PTFE with a diameter of 2–5 μm is uniformly coated on the skeleton of stainless steel mesh and no coating materials exists in the pores of the mesh. Some balls are glued to and embedded in each other, aggregating to form the blocks. Craters with a diameter of about 71 ± 8 nm are densely and evenly distributed on the surface of each ball, forming a structure that resembles that of a golf ball. The PTFE coating mesh film exhibits an excellent water-repellent property with a water CA of 156.2 ± 2.8° and a water SA of 4°. Moreover, the PTFE coating mesh film shows a superoleophilicity with a diesel oil CA of 0 ± 1.38°. A droplet of diesel oil can spread quickly on the film and permeate thoroughly within only 240 ms. This unique superwetting property provides the PTFE coating mesh film with the ability for the effective separation of a diesel oil/water mixture, which allows diesel oil to permeate through the film but the film to reject water.

Feng and co-workers reported a mesh-based film with high hydrophobicity and superoleophilicity for highly efficient separation of oil/water mixtures.[94] This superwetting mesh-based film was fabricated by combining mussel-inspired chemistry and Michael addition reaction. To be specific, a polydopamine (PDA) layer was firstly coated on the strands of the stainless steel mesh and then *N*-dodecyl mercaptan (NDM) was conjugated with PDA. The as-prepared PDA–NDM mesh film exhibits a water CA of 144°, but a diesel oil droplet can quickly spread and permeate the film with an oil CA of 0°. The opposite wettability for water and oil endows the mesh-based film with an excellent capacity to remove a series of oils, including hexane, gasoline, diesel, *etc.*, from water driven only by gravity. Separation efficiencies for these oils are all above 98.12% and remain very high after 30 times use. Furthermore, the relatively high intrusion pressure (2.2 kPa) guarantees the PDA–NDM mesh film can treat a large amount of oil/water mixtures.

Since the hydrophilic fish scales were reported to exhibit a superoleophobic property underwater in 2009, a strategy to fabricate superhydrophilic and underwater oil-repellent materials without the assistance of fluoride compounds has been followed.[57–61] Feng and Jiang reported a superhydrophilic–underwater superoleophobic hydrogel-coated mesh film made up of a nanostructured hydrogel coating and a stainless steel mesh.[62] This superwetting mesh-based film was fabricated by the photo-initiated *in situ* radical polymerization of polyacrylamide (PAM) with *N*,*N*′-methylene bisacrylamide as a chemical cross-linker. After the polymerization process, a dense layer of PAM hydrogel with random papillae structures of 80–500 nm is uniformly coated on the stainless steel strands (Figure 3.9a–c). Thickness of the PAM hydrogel layer is about 1.2 μm and pore size of the PAM hydrogel-coated mesh film is approximately 50 μm, which is very similar to that of raw stainless steel mesh (300 mesh size). The microporosity of the PAM-coated mesh film can be controlled by selection of different raw mesh sizes. The hydrophilicity of the PAM hydrogel and the micro- and nanoscale hierarchical roughness give rise to the superhydrophilicity and underwater superoleophobicity (underwater oil CA = 155°) of the resulting PAM-coated mesh film (Figure 3.9d). When the mesh film is immersed in water, the oil droplet of 1,2-dichloroethane is unstable and can easily roll off the film surface (Figure 3.9e). Meanwhile, the mesh film exhibits an ultralow oil adhesion force of 0.8 ± 0.3 μN. When an oil/water mixture was poured onto the PAM-coated mesh film, water quickly permeated through the film while the oil was retained above the film due to the superwetting property of the film (Figure 3.9f and g). The separation efficiencies for a variety of oils, including vegetable oil,

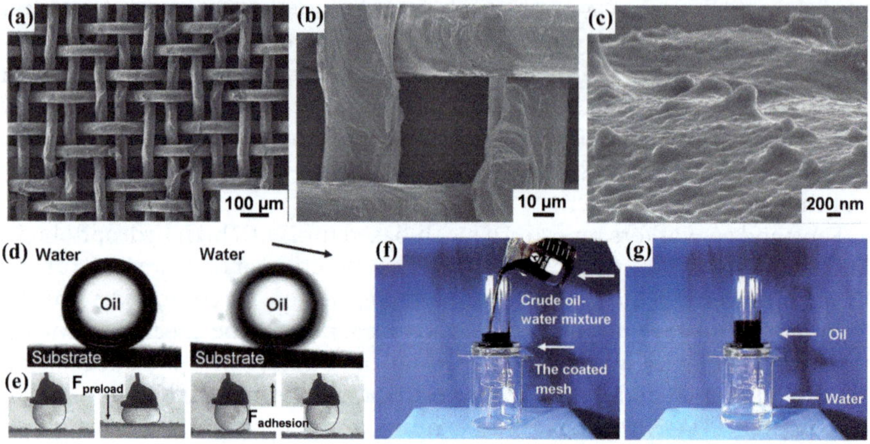

Figure 3.9 (a) and (b) SEM images of the PAM hydrogel-coated stainless steel mesh.[62] (c) Higher magnification image of the PAM hydrogel, in which nanostructured papillae can be clearly observed. (d) Photographs of an oil droplet (1,2-dichloroethane, 2 μL) on the hydrogel-coated mesh with an oil CA of 155.3° ± 1.8° and oil SA of 2.6° ± 0.5°. (e) Photographs of the dynamic underwater oil-adhesion measurements on the hydrogel-coated mesh. (f) and (g) Digital photographs showing the oil/water separation studies of the PAM hydrogel-coated mesh. Z. X. Xue, *et. al.*, A novel superhydrophilic and underwater superoleophobic hydrogel-coated mesh for oil/water separation, *Adv. Mater.*, 2011, **23**, 37. Copyright © 2011 John Wiley & Sons, Inc.

gasoline, diesel and even crude oil, are all above 99%, indicating its potential for practical separation of oil/water mixtures.

Besides the coating method, Jin's group reported a different strategy to fabricate an all-inorganic superwetting $Cu(OH)_2$ nanowire-haired film *via* an *in situ* growing method for separation of free oil/water mixtures and surfactant-free oil-in-water emulsions.[95] The all-inorganic mesh-based film is fabricated by the surface oxidation of copper and *in situ* growing of $Cu(OH)_2$ nanowires in an alkaline aqueous solution with $(NH_4)_2S_2O_8$. As shown in Figure 3.10a and b, $Cu(OH)_2$ nanowires with length of 10–15 μm and diameters of 200–500 nm are thickly and uniformly grown on the copper skeleton. The micro- and nano-structured hierarchical surface roughness enhances the hydrophilicity of $Cu(OH)_2$ to the extreme and generates the superhydrophilicity (water CA = 0°) and underwater ultralow adhesive superoleophobicity (underwater oil CA > 150°, underwater oil adhesion force < 1 μN for all the oils investigated, see Figure 3.10c) of the $Cu(OH)_2$ nanowire-haired film. The authors also expressed the underwater superoleophobicity of the mesh-based film according to the Cassie model. Pore size of the $Cu(OH)_2$ nanowire-haired mesh film can also be controlled by the selection of different raw meshes (mesh numbers 200, 300, 400 and 500). This superwetting mesh-based film can effectively separate a series of free oil/water mixtures and oil-in-water emulsions solely driven by gravity, with extremely high separation efficiency

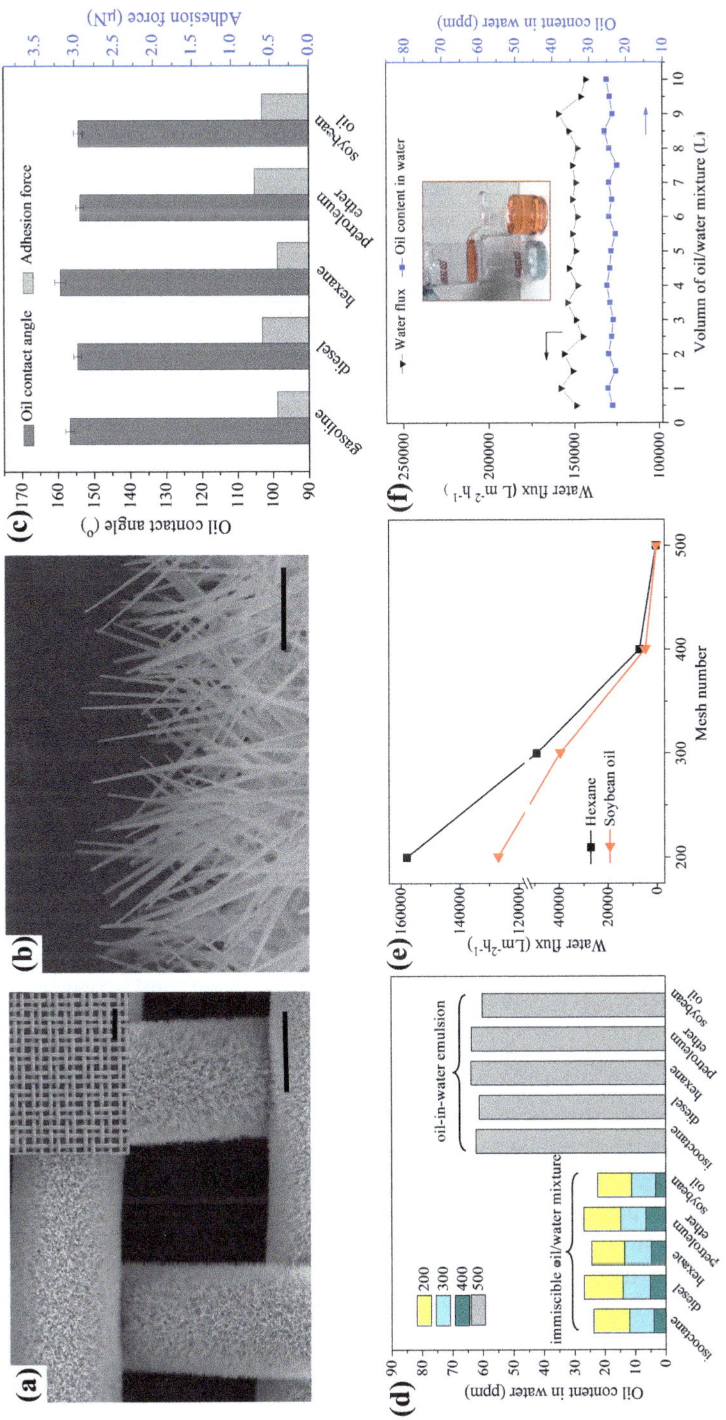

Figure 3.10 (a) SEM image of nanowire-haired membrane, scale bar: 50 μm. The inset is a low-magnification SEM image with scale bar of 200 μm.[95] (b) SEM image of the $Cu(OH)_2$ nanowires, scale bar: 5 μm. (c) Underwater CA and adhesion force of a series of oils on the nanowire-haired membrane. (d) and (e) Oil/water separation results of a nanowire-haired membrane. (f) Real-time monitoring of the variation of flux and oil content in the filtrate as a function of permeated volume of the oil/water mixture up to 10 L. F. Zhang, *et. al.*, Nanowire-haired inorganic membranes with superhydrophilicity and underwater ultralow adhesive superoleophobicity for high-efficiency oil/water separation, *Adv. Mater.*, 2013, **25**, 30. Copyright © 2013 John Wiley & Sons, Inc.

and separation fluxes (Figure 3.10d and e). Moreover, this film exhibits an excellent anti-oil fouling property for long-term use, which can continuously separate 10 L of an oil/water mixture without a decrease in flux (Figure 3.10f). The excellent separation performance, low cost, easily scaled-up fabrication process, solvent resistance and antifouling property of the mesh film indicate its promising application in treating large amounts of oil/water mixtures and emulsions from industry and oil spills.

Except for the frequently introduced superwetting and superantiwetting properties, some other functions like self-cleaning and photo-catalytic abilities are also extensively explored and integrated on the mesh-based superwetting films.[93,96] Wang and co-workers fabricated an all-inorganic underwater superoleophobic mesh film with a self-cleaning property by layer-by-layer assembly of sodium silicate and TiO_2 NPs on a stainless steel mesh.[93] The sodium silicate and TiO_2 NPs are densely and uniformly coated on the skeleton of the mesh film, which alters the wettability of the mesh film from hydrophobicity (water CA = 127.5°) to hydrophilicity (water CA = 21.9°). The silicate/TiO_2-coated mesh film shows an underwater superoleophobic property for a series of typical oils with underwater oil CA larger than 150° and can separate these oils from the oil/water mixtures effectively driven by gravity. The integration of the self-cleaning property into the silicate/TiO_2-coated mesh film endows the film with removal ability for contaminants by UV illumination, and enables the facile recovery of separation performance of an oil-fouled mesh. Feng's group reported a double-layer TiO_2-based mesh film with the multifunctions of oil/water separation and soluble pollutant degradation due to the photo-catalysis of TiO_2.[96] The upper layer is a superhydrophilic TiO_2-coated mesh film fabricated by a hydrothermal approach with micro- and nanostructures. The lower layer is a superhydrophobic–superoleophilic (water CA = 165°) TiO_2-coated mesh film modified with octadecyl phosphonic acid (ODP). During the separation process, the double-layer TiO_2-based mesh film was obliquely fixed to be easily contacted by the oil in the oil/water mixture. Hence, the oil can penetrate the mesh film and flow down while water is repelled above the film because of the superhydrophobic–superoleophilic property of the upper layer, consequently achieving the oil/water separation process. After ultraviolet (UV) light illumination, the ODP is degraded because of the photo-catalysis of TiO_2 and the upper layer becomes superhydrophilic. Hence, the repelled water permeates the mesh film and flow down. The soluble pollutants in water are simultaneously degraded during the UV illumination and the water purification process is completed. The superhydrophobic–superoleophilic lower layer can be re-obtained by the modification of ODP again.

Textiles and fabrics are another kind of common and commercially available micro-textured 2D porous materials with similar structures to metal meshes, and are widely chosen as the substrates to fabricate superwetting textile- and fabric-based separation films for oil/water separation. Wang's group grafted the as-mentioned pH-responsive P2VP-*b*-PDMS on a non-woven textile and used it for controllable separation of free oil/water mixtures.[78] The P2VP-*b*-PDMS-functionalized textile behaves superhydrophilically and superoleophobically in acidic water (pH = 2.0) with an underwater oil CA of 165.3°, but behaves superoleophilically and superhydrophobically in water with a pH of 6.5. Based on the pH-responsive property of the P2VP-*b*-PDMS-functionalized textile, it was used as a separation film to achieve a controllable separation of gasoline and water from their mixture. When the pH of the mixture was 6.5, the gasoline passed through the textile quickly, but water was kept above the textile because of its superoleophilicity and superhydrophobicity. However, when the textile was firstly wetted by acidic water with a pH of 2.0 and used under the same the pH condition, the opposite separation result was realized due to its superhydrophilicity and underwater superoleophobicity.

A breakthrough was achieved by Tuteja and co-workers who fabricated the novel blend of f-POSS and cross-linked poly(ethylene glycol) diacrylate (x-PEGDA).[97] Mesh- and fabric-based separation films modified with this f-POSS + x-PEGDA blend *via* a dip coating method exhibit a superhydrophilic and superoleophobic property both in air and under water (Figure 3.11a and b), which are barely reported. A f-POSS + x-PEGDA blend-coated mesh film (or fabric film) can achieve the gravity-driven separation of both surfactant-stabilized oil-in-water emulsions and surfactant-stabilized water-in-oil emulsions with very high separation efficiency above 99.9% in a single-unit operation (Figure 3.11c–f). The superoleophobicity under water and in air is very essential for the separation of oil-in-water and water-in-oil emulsions, respectively. In addition, the superwetting mesh-based film can continually separate several litres of oil/water mixtures (Figure 3.11g) or continually separate the mixtures for 100 hours without a decrease in flux (Figure 3.11h) by using a scaled-up apparatus.

However, these mesh- and textile-based superwetting films remain with limitations for oil/water separation. The major shortcoming is that they are only capable for free oil/water mixtures and oil/water emulsions with oil or water droplets larger than the pore size of these films usually at micrometre scale. For example, the f-POSS + x-PEGDA blend-coated mesh film (mesh 400) can only remove the oils or water with droplet size above 30 μm in the separation process.[97] For the oil/water emulsions with oil or water droplets at submicrometre and nanometre scale, these mesh- and fabric-based films don't work well.[62,82–84,94,96]

3.4.2 Polymer-Dominated Superwetting Filtration Membranes for Separation of Oil/Water Emulsions

For free oil/water mixtures, superwetting absorbing nanomaterials and mesh-/fabric-based superwetting films can be used to achieve separation by selective absorption or permeation. Unfortunately, real oily wastewater is not always well layered, but more frequently exists in the form of an emulsion. For emulsified oil/water mixtures, especially for surfactant-stabilized

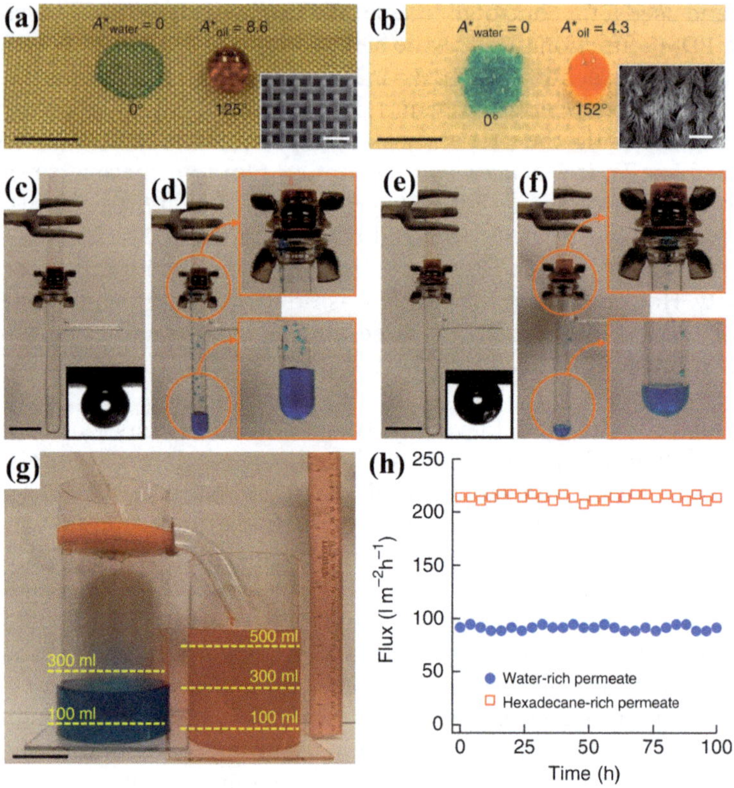

Figure 3.11 Droplets of water (dyed blue) and rapeseed oil (dyed red) on stainless steel mesh 100 (a) and polyester fabric (b).[97] Both surfaces have been dip coated with a 20 wt% f-POSS + x-PEGDA blend, scale bars, 5 mm. Insets are morphologies of the respective dip coated mesh and fabric surfaces, scale bars, 500 µm. Batch separation of hexadecane-in-water (c and d) and water-in-hexadecane (e and f) emulsions by a stainless steel mesh 400 modified with 20 wt% f-POSS + x-PEGDA blend. Inserts show a hexadecane droplet on the modified surface under water. (g) Photographs showing the continuous separation of water-in-oil emulsions by a scaled-up apparatus. (h) Measured fluxes as a function of time. Reprinted by permission from Macmillan Publishers Ltd: *Nat. Commun.* (ref. 97), copyright 2012.

emulsions with oil or water droplets at the submicrometre and nanometre scale, the above-mentioned separation technologies are not effective. Instead, filtration membranes are capable and widely used to separate various oil/water emulsions, especially for surfactant-stabilized emulsions.[20,98–103] However, traditional polymer-dominated filtration membranes in various industrial fields for treating oily wastewater still have drawbacks. The major problem is that most of these filtration membranes are not superwetting and thus suffer from oil-fouling and quick decline of permeation flux and separation efficiency due to pore plugging by oil droplets.[104–107] This weakness of traditional filtration membranes bring about high costs for treating large effluent volumes of oily water. Developing new advanced polymer-dominated filtration membranes with superwetting and antifouling properties is extremely critical for efficient separation of emulsified oil/water mixtures and other separation applications. To improve the hydrophilicity and antifouling performance of polymer-dominated membranes, a variety of methods such as blending, block copolymer grafting and surface coating with hydrophilic components have been adopted by researchers.[108–112] However, the hydrophilicity of the membranes obtained *via* these methods are usually nondurable because of the releasing of the water-favouring additives during long-term use. Recently, a series of advanced works on developing inherently superwetting polymeric membranes have been carried out by controlling the surface morphology with micro-/nano-scale hierarchical structures.[85–88]

Jin's group reported a novel salt-induced phase-inversion method to fabricate superhydrophilic and underwater superoleophobic poly(acrylic acid)-grafted polyvinylidene fluoride (PAA-*g*-PVDF) membranes for effective separation of oil-in-water emulsions.[88] Completely different from the general approach of adding salts as a pore-forming additive to change the pore structure of the membrane,[113,114] in their report, the salt (NaCl) added into the coagulation bath at a nearly saturated concentration functions as nucleater during the solvent exchange and induces the assembly of PAA-*g*-PVDF micelles around them (as shown in Figure 3.12a). The micelle assembly on the surface of the PAA-*g*-PVDF membrane gives rise to a micro-/nano-scale hierarchical structure (Figure 3.12b and c) and endows the membrane with superhydrophilic and underwater superoleophobic properties. This superwetting PAA-*g*-PVDF membrane can separate both surfactant-free and surfactant-stabilized oil-in-water emulsions with oil droplet size from micrometre to nanometre, either under a small applied pressure (0.1 bar) or solely driven by gravity, with very high permeation flux (up to 2300 L m^{-2} h^{-1}, see Figure 3.12d) and very high separation efficiency (>99.99 wt% after one-time separation, see Figure 3.12e). Due to the underwater superoleophobicity of this membrane, it also exhibits an excellent anti-oil fouling property and is easily recycled for long-term use. The outstanding separation performance of this membrane for oil-in-water emulsions and its industrially scalable fabrication process indicate its great potential for treating real oily water. They also reported a facile method to fabricate superhydrophobic and superoleophilic PVDF membrane *via* an inert solvent-induced phase-inversion process (Figure 3.13a).[85] The surface of the as-prepared PVDF membrane was composed of spherical micro-particles and rather rough (Figure 3.13b), which results in the superhydrophobicity (Figure 3.13c) and superoleophilicity (Figure 3.13d) of the membrane. This PVDF membrane can effectively separate both surfactant-free and surfactant-stabilized water-in-oil emulsions solely driven by gravity, with very high separation efficiency (oil purity after one-time separation >99.95%) and very high permeation flux up to 3400 L m^{-2} h^{-1} (Figure 3.13e and f).

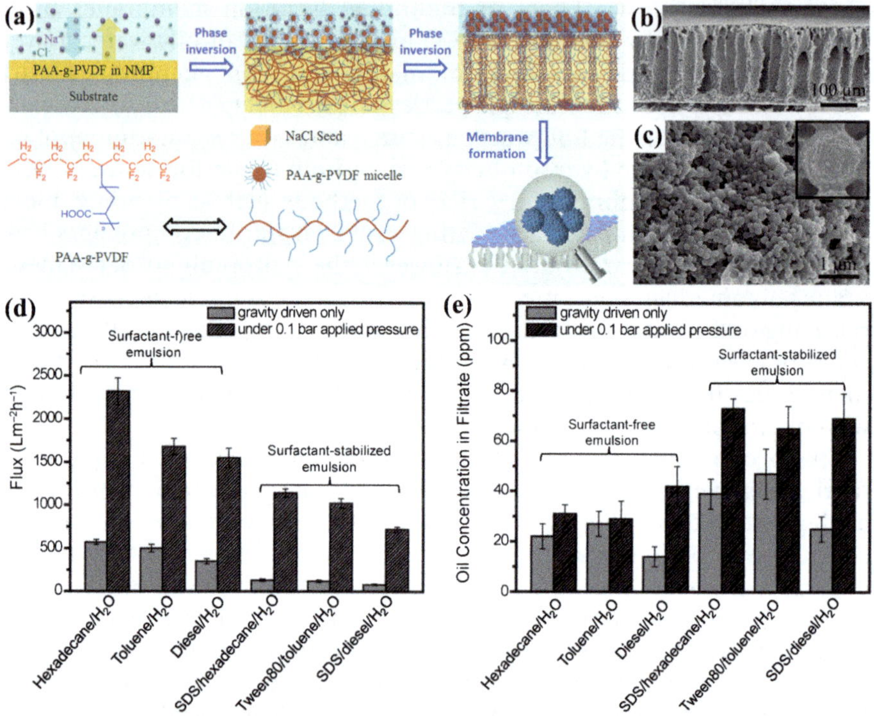

Figure 3.12 (a) Schematic showing the formation of the superhydrophilic–underwater superoleophobic PAA-*g*-PVDF membrane by a salt-induced phase-inversion process.[88] (b) and (c) are the cross-section and top-view SEM images of the membrane, respectively. Permeation flux (d) and oil concentration in the corresponding filtrate (e) for a series of surfactant-free and surfactant-stabilized oil-in-water emulsions permeating the membrane. W. B. Zhang, *et. al.*, Salt-induced fabrication of superhydrophilic and underwater superoleophobic PAA-*g*-PVDF membranes for effective separation of oil-in-water emulsions, *Angew. Chem., Int. Ed.*, 2014, **53**, 3. Copyright © 2014 John Wiley & Sons, Inc.

A novel superhydrophilic–superoleophilic PVDF membrane with multiscale surface structure was reported by Liu and co-workers.[87] This PVDF membrane was obtained by two steps including fabrication of the porous PVDF membrane on a non-woven fabric (NWF) *via* phase inversion and then peeling off the PVDF membrane from the NWF. Discrete pores of 50–100 μm in width are disorderly distributed on the membrane surface with numerous randomly oriented grooves of ~10 μm in width (Figure 3.14a and b). Petaloid structures with sizes of hundreds of nanometres form the tightly packed spherulites with size of several micrometres (Figure 3.14c). The authors believe the NWF is essential for the formation of membrane surface topography. This superhydrophilic and superoleophilic PVDF membrane exhibits a low adhesive superhydrophobicity under oil and an ultralow adhesive superoleophobicity under water (Figure 3.14d). Hence, the superwetting PVDF

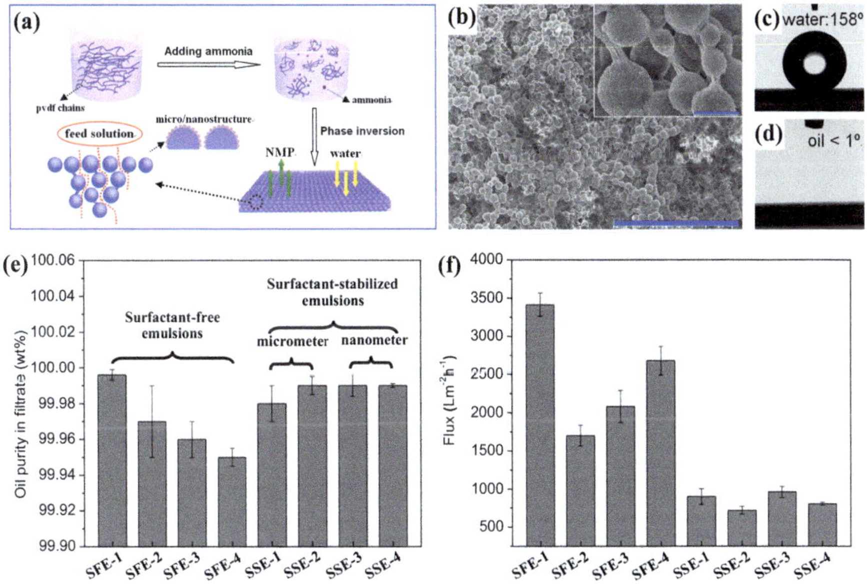

Figure 3.13 (a) Schematic illustration of the formation of a superhydrophobic–superoleophilic PVDF membrane *via* a modified phase-inversion process.[85] (b) SEM image of the membrane, scale bars: 50 μm. The insert is a higher magnification image of membrane, scale bars: 1 μm. (c) and (d) are optical images of a water droplet and an oil droplet on the membrane in air, respectively. Oil purity in filtrate (e) and permeation flux (f) for various water-in-oil emulsions separated by the PVDF membranes. W. B. Zhang, *et. al.*, Superhydrophobic and superoleophilic PVDF membranes for effective separation of water-in-oil emulsions with high flux, *Adv. Mater.*, 2013, **25**, 14. Copyright © 2013 John Wiley & Sons, Inc.

membrane can be used to separate both oil-in-water and water-in-oil emulsions (Figure 3.14e) due to the switchable transport property and it displays excellent permeation flux (Figure 3.14f), separation efficiency, recyclability and antifouling performance.

Wang and co-workers developed a dual-scaled perforated nitrocellulose (p-NC) membrane with superhydrophilicity and underwater superoleophobicity by a simple and reproducible perforating method using a desktop robot punch system equipped with a home-made needle owing a conical tip.[86] A series of p-NC membranes with different pore sizes were obtained *via* controlling needle punching depth. The dual-scaled hierarchical porosity improves the permeation flux of the NC membrane and ensures the rapid separation of water from a variety of oil/water mixtures including gasoline, diesel, hexane, petroleum ether and even high-viscosity crude oil/water mixture without external power and with high separation efficiency (>99%). Separation time and intrusion pressure of the superwetting p-NC membrane can be readily tuned by controlling the sizes of the perforated micro-pores. The superwetting p-NC membrane also possesses good environmental stability and recyclability.

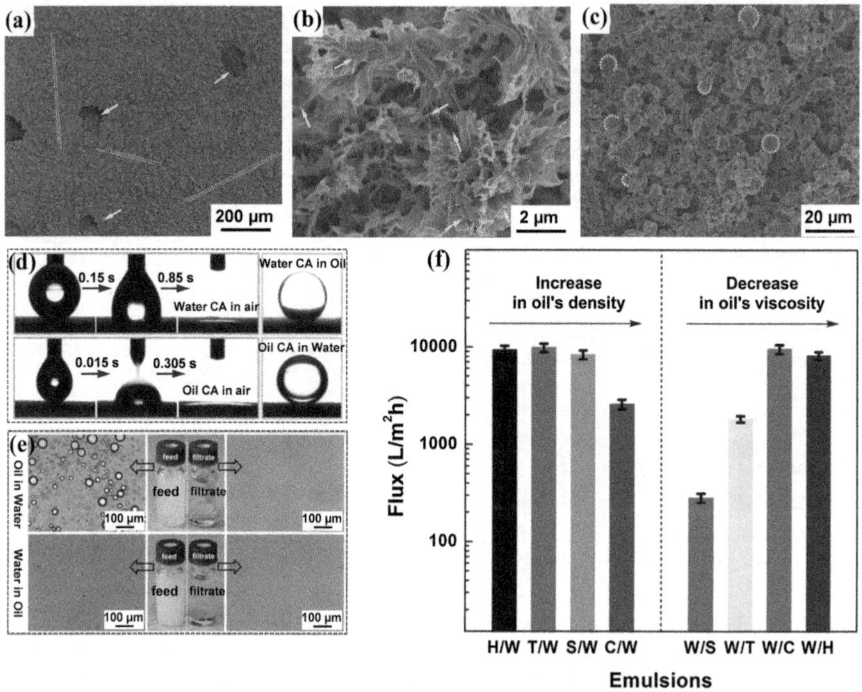

Figure 3.14 (a)–(c) SEM images for the membrane surface with scale bar of 200 μm, 20 μm and 2 μm, respectively.[87] (d) Wettability of the membrane toward water (upper) and oil (lower). (e) Separation results for oil-in-water and water-in-oil emulsions by the PVDF membrane. (f) Permeate flux for various oil-in-water and water-in-oil emulsions by the PVDF membrane. M. Tao, *et. al.*, An intelligent superwetting PVDF membrane showing switchable transport performance for oil/water separation, *Adv. Mater.*, 2014, **26**, 18. Copyright © 2014 John Wiley & Sons, Inc.

Besides the inherently superwetting membranes discussed above, some other advanced works on fabricating polymer-dominated superwetting membranes are also worth introducing.[115–118] A superhydrophilic and underwater superoleophilic nanosilica-decorated polypropylene (PP) microfiltration (MF) membrane with a mussel-inspired intermediate layer was reported by Xu's group for separation of oil-in-water emulsions.[115] The superwetting property of the membrane was obtained by regulating and controlling the surface chemistry and morphology, too. Herein, PDA is introduced as an adhesive component in the intermediate layer, polyethylenimine (PEI) is co-deposited onto the membrane surface for the polycondensation of silicic acid *via* a biomineralization process. After the surface modification, silica NPs are uniformly decorated on the membrane surface and inherent micropores, resulting in the superhydrophilicity (water CA = 0°) and the underwater oil-repelling property (underwater oil CA > 152° for a series of oils) of the nanosilica-decorated PP membrane. Because of the robust superwetting property, this membrane exhibits very high permeation fluxes up to 1400 L m^{-2} h^{-1} and very high oil rejections (>99%) for four kinds of surfactant-stabilized oil-in-water emulsions under a very low trans-membrane pressure of 0.04 MPa. The nanosilica-decorated membrane also possesses excellent flux recovery after simply rising with water. Lu and co-workers reported a mussel-inspired hybrid coating on PVDF MF membranes *via* the simultaneous polymerization of dopamine and hydrolysis of silane by one step to fabricate superhydrophilic and underwater superoleophilic PVDF MF membranes.[117] The superwetting coating endows the PVDF MF membranes with the ability for separation of oil-in-water emulsions with high water flux and excellent antifouling performance. Poly(3-(*N*-2-methacryloxyethyl-*N*,*N*-dimethyl) ammonatopropanesultone) (PMAPS), a novel zwitterionic polyelectrolyte, was fabricated by Jin's group and grafted on commercially available PVDF MF membranes for separation of oil-in-water emulsions.[118] Instead of the hydrophobicity of the original PVDF membrane, the PMAPS-*g*-PVDF membrane behaves superhydrophilically and underwater superoleophilically with ultralow underwater oil adhesion force. Because of the superwetting property, the PMAPS-*g*-PVDF membrane can thoroughly separate oils from oil-in-water emulsions with ultrahigh separation efficiency (oil content after one-time separation less than 10 ppm).

3.4.3 Nanomaterial-Based Ultrathin Superwetting Films for Separation of Oil/Water Emulsions

Besides the surface wettability, another major limitation of traditional polymer- and ceramic-dominated filtration membranes is their relatively thick separation layer, which results in their low permeation fluxes for various separation applications. Only if this problem was solved, will people be able to separate the global wastewater quickly and provide enough fresh water for the growing population. According to the classical Hagen–Poiseuille equation:

$$J = \varepsilon\pi r_p^{\,2}\Delta p/8\mu L \tag{3.7}$$

where the permeation flux J is described as a function of the surface porosity ε, the pore radius r_p, the pressure drop Δp, the liquid viscosity μ and the total distance L of the liquid running through.[119] It predicts that the permeation flux of a membrane is directly proportional to the square of the effective pore size, and inversely proportional to the membrane thickness. Therefore, an ideal filtration membrane should have a membrane thickness as thin as possible and doesn't sacrifice its effective pore size.[120,121] However, it is difficult to achieve this goal in the conventional system of polymeric- and ceramic-dominated membranes. Designing and developing a new type of ultrathin superwetting membrane or film with advanced structures to achieve ultrahigh permeation flux is highly desirable for oil/water separation and other separation applications.

During recent years, a series of novel ultrathin filtration films constructed by nanomaterials with film thickness even at nanometre scale have been reported for various separation applications.[119,122–134] Benefiting from the ultrathin film thickness, these nanomaterial-based films exhibit excellent separation performance, including both very high permeation flux and very high separation efficiency. Among these ultrathin films, superwetting films made up of 1D nanomaterials such as CNTs, nanofibres and nanowires have been successfully developed and used for oil/water separation.

Jin's group fabricated a series of hydrophobic and superoleophilic single-walled carbon nanotube (SWCNT) network films with tuneable thickness and pore size at nanometre scale (Figure 3.15a and b), which are assembled by SWCNTs *via* a simple vacuum-filtering method.[131] This hydrophobic–superoleophilic SWCNT network film can allow oil to permeate through the film, but reject water above the film (Figure 3.15c), eventually achieving the separation of a water-in-oil emulsion. It is worth noting that the thickness of the film can be tuned from 30 to 120 nm and the pore size of the film can be tuned from 200 to 20 nm by adjusting the dosage of SWCNT. The ultrathin SWCNT network films with thickness of tens of nanometres can effectively separate both micrometre- and nanometre-sized surfactant-free

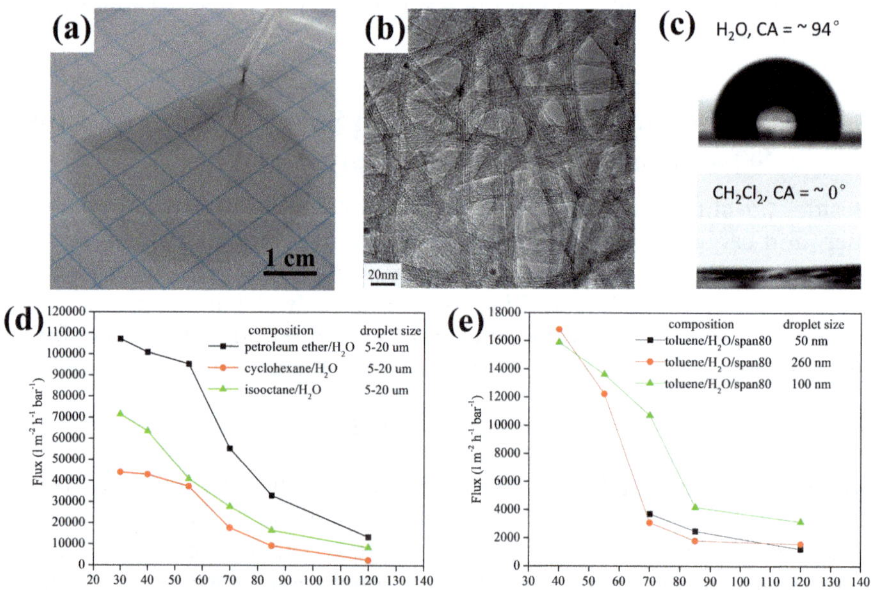

Figure 3.15 (a) and (b) Digital photograph and TEM image of a 70-nm-thick SWCNT film.[131] (c) Optical images of a water droplet and an oil (dichloromethane) droplet on the SWCNT film in air. Plots of fluxes for surfactant-free (d) and surfactant-stabilized (e) water-in-oil emulsions as a function of the thickness of SWCNT films. Z. Shi, *et. al.*, Ultrafast separation of emulsified oil/water mixtures by ultrathin free-standing single-walled carbon nanotube network films, *Adv. Mater.*, 2013, **25**, 17. Copyright © 2013 John Wiley & Sons, Inc.

and surfactant-stabilized water-in-oil emulsions with ultrahigh fluxes up to 100 000 L m^{-2} h^{-1} bar^{-1} (Figure 3.15d and e), which are 2–3 orders of magnitude higher than the fluxes of traditional filtration membranes with similar separation performance, and with very high separation efficiency (above 99.95 wt%) and excellent antifouling ability.

Because of the inherent hydrophobicity and superoleophilicity of SWCNT film, it is incapable of separating oil-in-water emulsions. Jin's group then designed the photo-induced superhydrophilic and underwater superoleophobic SWCNT/TiO_2 ultrathin film by coating TiO_2 NPs onto SWCNT network film *via* a sol–gel process and used it for an ultrafast separation of oil-in-water emulsions.[132] Anatase-type TiO_2 NPs are uniformly and robustly coated on every SWCNT. After UV irradiation, the SWCNT/TiO_2 nanocomposite film exhibits a superhydrophilic and underwater superoleophobic property ascribed to photo-responsive wettability of TiO_2 NPs. A SWCNT/TiO_2 nanocomposite film with thickness of ~60 nm and pore size of 20–60 nm can effectively separate both surfactant-free and surfactant-stabilized oil-in-water emulsions in an ultrafast manner with fluxes up to 30 000 L m^{-2} h^{-1} bar^{-1}, which is 2 orders of magnitude higher than commercial UF membranes with similar separation performance, and with very high separation efficiency (oil content after one-step separation is lower than 40 ppm, meeting the standards for wastewater discharge). Moreover, this SWCNT/TiO_2 nanocomposite film exhibits excellent self-cleaning performance duo to the photo-catalytic property of anatase-type TiO_2 NPs. Jin and co-workers also designed and fabricated an inherently superhydrophilic–underwater superoleophobic polymer-decorated SWCNT ultrathin film with modification of PDA and PEI.[133] The SWCNT/PDA/PEI nanocomposite film with thickness of ~160 nm and pore size of ~10 nm can effectively separate the nano-sized oils and oil-in-water nanoemulsions with ultrahigh permeation fluxes up to 6000 L m^{-2} h^{-1} bar^{-1}, which is about 10 times the fluxes of traditional ultrafiltration membranes with a similar rejection property. Meanwhile, the SWCNT/PDA/PEI nanocomposite film exhibits excellent pH stability and antifouling property. These works point a direction for designing and fabricating new types of ultrathin superwetting films by using 1D nanomaterials for emulsion separation.

Besides the SWCNT-based separation films, Wang and co-workers fabricated a MnO_2 nanowire membrane *via* a vacuum-filtering method.[135] This MnO_2 nanowire membrane behaves hydrophilically because of the hydrogen bond between water molecules and oxygen atoms exposed on the surfaces of MnO_2 nanowires. The hydrophilic MnO_2 nanowire membrane with pore size of tens of nanometres can be used to achieve the separation of both free and emulsified oil/water mixtures. These SWCNT- and nanowire-based ultrathin superwetting films have a promising potential for treating real emulsified oily wastewater in large quantities, as an alternative to traditional polymer- and ceramic-dominated membranes.

3.5 Summary and Perspective

Oil spills and other oil-related industrial or daily activities generate tremendous quantities of oily water every year. The direct discharge of oily water seriously harms the environment and people's health and even further aggravates the shortage crisis of drinkable fresh water. As oil/water separation is an interfacial problem, superwetting nanomaterials offer an encouraging opportunity for people to develop a brand new, energy-efficient and process-simple strategy to treat oily water due to their selective behaviours for oil and water. Theoretically, superwetting nanomaterials can be constructed by a combination of surface chemical composition and micro/nanostructured surface roughness. Guided by this theoretical principle, a series of remarkable projects on designing and fabricating superwetting nanomaterials, including absorbing nanomaterials and separation membranes as well as nanomaterial-based ultrathin films, have been carried out for the separation of free oil/water mixtures and oil/water emulsions. These superwetting nanomaterials exhibit excellent separation abilities and hold a promising potential for treating real oily water from industries and daily life in large quantities.

Although these superwetting materials introduced in this chapter behave excellently and have an encouraging potential for practical oily water treatment, there are still a lot of difficulties and challenges hindering the industrialization and large-scale application of these materials.

- Firstly, these superwetting materials are commonly composed of micro/nanoscale surface structures which are critical for enhancing the surface wettability, but the micro/nanostructures are always fragile and easily damaged by external influences during practical use. For example, the mechanical stress will damage the micro/nanoscale surface structures when we extrude the absorbed oil from the sponge-based superwetting materials, which will lead to the loss of the superwettability and restricts the reusability of the material.
- Secondly, quite different from the pure oils with low viscosity used in the reports, the viscosity of crude oil in oil spills is very high. When separating an emulsified crude oil/water mixture using a superhydrophilic–underwater superoleophobic membrane, the crude oil droplets can be easily absorbed on the membrane surface underwater and destroy the superwetting property and separation ability of the membrane. Hence, the oily water in oil spills is more difficult to treat because of the oil-fouling issue.
- Thirdly, in the above reports on the separation of oil/water emulsions, the authors mainly focused on the separation results, and hardly any reports explored the dynamical behaviours of oil or water droplets on the membrane surface during the separation process. The separating mechanism of the emulsified oil/water mixtures is still unclear. How the demulsification process happens on the membrane surface is often

neglected but is very important for building up a theoretical direction for designing high-performance membranes.

- Fourthly, during the separation process of emulsified oil/water mixtures by superwetting membranes, oil droplets will more or less block the membrane pores and lead to the loss of the superwetting property and a quick decline of separation flux and efficiency. Surfactants or other soluble contaminants in the emulsified oily water can be absorbed onto membrane surface and lead to this problem, too. And the higher the separation flux is, the more serious this fouling issue is. At the same time, removing these fouling contaminants during the separation process is a very important barrier we must solve.

We believe the realization process of practical and large-scale treatment of real oily water is only at an early stage. These as-reported superwetting nanomaterials are a good beginning for the realization of mass-producing stable and durable superwetting materials for the large-scale treatment of real oily water.

Future works can be focused on the following major works.

- Firstly, people should develop separation materials which process superwetting property not only on the outer surface but also inside inner part. Or people should develop enough stable and robust surface structures to ensure the durability of the separation materials.
- Secondly, underwater superoleophilic membranes even for high-viscosity oils like crude oil and heavy oil should be developed for the separation of emulsified oily water from oil spills.
- Thirdly, more attention should be focused on how the oil droplets demulsify and coalesce together at membrane surfaces during the separation process.
- Fourthly, superwetting separation membranes or films with self-cleaning ability should be developed to synchronously removing the fouling contaminants including oil droplets and surfactants during the separation process.
- Fifthly, inexpensive and easily scaled-up fabrication methods to obtain these superwetting nanomaterials should be developed for mass production.

References

1. P. Kajitvichyanukul, Y. T. Hung and L. K. Wang, in *Handbook of Environmental Engineering: Membrane and Desalination Technologies*, ed. L. K. Wang, J. P. Chen, Y.-T. Hung and N. K. Shammas, The Humana Press Inc., New York, 2011, ch. 15, vol. 13, pp. 639–668.
2. C. H. Peterson, S. D. Rice, J. W. Short, D. Esler, J. L. Bodkin, B. E. Ballachey and D. B. Irons, *Science*, 2003, **302**, 2082.

3. E. Kintisch, *Science*, 2010, **329**, 735.
4. J. Aurell and B. K. Gullett, *Environ. Sci. Technol.*, 2010, **44**, 9431.
5. B. Wang, W. Liang, Z. Guo and W. Liu, *Chem. Soc. Rev.*, 2015, **44**, 336.
6. M. A. Shannon, P. W. Bohn, M. Elimelech, J. G. Georgiadis, B. J. Mariñas and A. M. Mayes, *Nature*, 2008, **452**, 301.
7. T. Oki and S. Kanae, *Science*, 2006, **313**, 1068.
8. J. Yeston, R. Coontz, J. Smith and C. Ash, *Science*, 2006, **25**, 1067.
9. H. Wang, K.-Y. Lin, B. Jing, G. Krylova, G. E. Sigmon, P. McGinn, Y. Zhu and C. Na, *Water Res.*, 2013, **47**, 4198.
10. T. G. Mason, J. N. Wilking, K. Meleson, C. B. Chang and S. M. Graves, *J. Phys.: Condens. Matter*, 2006, **18**, R635.
11. T. Ichikawa, *Colloids Surf., A*, 2007, **302**, 581.
12. A. B. Nordvik, J. L. Simmons, K. R. Bitting, A. Lewisa and T. S. Kristiansen, *Spill Sci. Technol. Bull.*, 1996, **3**, 107.
13. A. El-Kayar, M. Hussein, A. A. Zatout, A. Y. Hosny and A. A. Amer, *Sep. Technol.*, 1993, **3**, 25.
14. K. Gaaseidnes and J. Turbeville, *Pure Appl. Chem.*, 1999, **71**, 95.
15. A. Y. Hosny, *Sep. Technol.*, 1996, **6**, 9.
16. B. Tansel and J. Regula, *J. Environ. Sci. Health, Part A: Toxic/Hazard. Subst. Environ. Eng.*, 2000, **35**, 1557.
17. T. L. Sun, L. Feng, X. F. Gao and L. Jiang, *Acc. Chem. Res.*, 2005, **38**, 644.
18. X. Yao, Y. L. Song and L. Jiang, *Adv. Mater.*, 2011, **23**, 719.
19. J. Lahann, *Nat. Nanotechnol.*, 2008, **3**, 320.
20. M. M. Pendergast and E. M. V. Hoek, *Energy Environ. Sci.*, 2011, **4**, 1946.
21. L. Feng, Z. Zhang, Z. Mai, Y. Ma, B. Liu, L. Jiang and D. Zhu, *Angew. Chem., Int. Ed.*, 2004, **43**, 2012.
22. Y. Zhu, D. Wang, L. Jiang and J. Jin, *NPG Asia Mater.*, 2014, **6**, e101.
23. J. Y. Lv, Y. L. Song, L. Jiang and J. J. Wang, *ACS Nano*, 2014, **8**, 3152.
24. H. Yang, H. Zhu, M. M. R. M. Hendrix, N. J. H. G. M. Lousberg, G. D. With, C. C. Esteves and J. H. Xin, *Adv. Mater.*, 2013, **25**, 1150.
25. P.-C. Cheng and Z.-K. Xu, *Sci. Rep.*, 2013, **3**, 2776.
26. B. Wang, Y. B. Zhang, W. X. Liang, G. Y. Wang, Z. G. Guo and W. M. Liu, *J. Mater. Chem. A*, 2014, **2**, 7845.
27. J. Ge, Y. D. Ye, H. B. Yao, X. Zhu, X. Wang, L. Wu, J. L. Wang, H. Ding, N. Yong, L. H. He and S. H. Yu, *Angew. Chem., Int. Ed.*, 2014, **53**, 3612.
28. T. Young, *Philos. Trans. R. Soc. London*, 1805, **95**, 65.
29. R. N. Wenzel, *Ind. Eng. Chem.*, 1936, **28**, 988.
30. A. Cassie and S. Baxter, *Trans. Faraday Soc.*, 1994, **40**, 546.
31. A. Marmur, *Langmuir*, 2004, **20**, 3517.
32. Y. Tian, B. Su and L. Jiang, *Adv. Mater.*, 2014, **26**, 6872.
33. A. Lafuma and D. Quéré, *Nat. Mater.*, 2003, **2**, 457.
34. C. Ishino, K. Okumura and D. Quéré, *Europhys. Lett.*, 2004, **68**, 419.
35. M. Reyssat, J. M. Yeomans and D. Quéré, *Europhys. Lett.*, 2008, **81**, 26006.
36. D. Bartolo, F. Bouamrirene, E. Verneuil, A. Buguin, P. Silberzan and S. Moulinet, *Europhys. Lett.*, 2006, **74**, 299.
37. Y. C. Jung and B. Bhushan, *Langmuir*, 2009, **25**, 9208.

38. V. Bahadur and S. V. Garimella, *Langmuir*, 2007, **23**, 4918.
39. J. Bico, C. Tordeux and D. Quéré, *Colloids Surf., A*, 2002, **206**, 41.
40. X. Feng and L. Jiang, *Adv. Mater.*, 2006, **18**, 3063.
41. M. J. Liu, S. T. Wang, Z. X. Wei, Y. L. Song and L. Jiang, *Adv. Mater.*, 2009, **21**, 665.
42. Y. C. Jung and B. Bhushan, *Langmuir*, 2009, **25**, 14165.
43. E. A. Vogler, *Adv. Colloid Interface Sci.*, 1998, **74**, 69.
44. Y. Tian and L. Jiang, *Nat. Mater.*, 2013, **12**, 291.
45. E. F. Hare, E. G. Shafrin and W. A. Zisman, *J. Phys. Chem.*, 1954, **58**, 236.
46. A. K. Kota and A. Tuteja, in *Advances in Fluorine-Containing Polymers*, ed. D. W. Smith, S. T. Iacono, D. J. Boday and S. C. Kettwich, American Chemical Society, Washington, DC, USA, 2012, ch. 11, vol. 1106, ACS Symposium Series, pp. 171–185.
47. A. K. Kota, G. Kwon and A. Tuteja, *NPG Asia Mater.*, 2014, **6**, e109.
48. K. Liu, M. Zhang, J. Zhai, J. Wang and L. Jiang, *Appl. Phys. Lett.*, 2008, **92**, 183103.
49. X. F. Gao and L. Jiang, *Nature*, 2004, **432**, 36.
50. L. Jiang, Y. Zhao and J. Zhai, *Angew. Chem., Int. Ed.*, 2004, **43**, 4338.
51. G. R. J. Artus, S. Jung, J. Zimmermann, H. P. Gautschi, K. Marquardt and S. Seeger, *Adv. Mater.*, 2006, **18**, 2758.
52. J. Zhang and S. Seeger, *Angew. Chem., Int. Ed.*, 2011, **50**, 6652.
53. W. Choi, A. Tuteja, S. Chhatre, J. M. Mabry, R. E. Cohen and G. H. McKinley, *Adv. Mater.*, 2009, **21**, 2190.
54. S. Pan, A. K. Kota, J. M. Mabry and A. Tuteja, *J. Am. Chem. Soc.*, 2013, **135**, 578.
55. A. K. Kota, Y. Li, J. M. Mabry and A. Tuteja, *Adv. Mater.*, 2012, **24**, 5838.
56. K. Golovin, D. H. Lee, J. M. Mabry and A. Tuteja, *Angew. Chem., Int. Ed.*, 2013, **52**, 13007.
57. A. Raza, B. Ding, G. Zainab, M. El-Newehy, S. S. Al-Deyab and J. Y. Yu, *J. Mater. Chem. A*, 2014, **2**, 10137.
58. Y. Dong, J. Li, L. Shi, X. B. Wang, Z. G. Guo and W. M. Liu, *Chem. Commun.*, 2014, **50**, 5586.
59. N. Liu, Y. N. Chen, F. Lu, Y. Z. Cao, Z. X. Xue, K. Li, L. Feng and Y. Wei, *ChemPhysChem*, 2013, **14**, 3489.
60. L.-P. Xu, J. Zhao, B. Su, X. L. Liu, J. T. Peng, Y. B. Liu, H. L. Liu, G. Yang, L. Jiang, Y. Q. Wen, X. J. Zhang and S. T. Wang, *Adv. Mater.*, 2013, **25**, 606.
61. M. J. Liu, S. T. Wang, Z. X. Wei, Y. L. Song and L. Jiang, *Adv. Mater.*, 2009, **21**, 665.
62. Z. X. Xue, S. T. Wang, L. Lin, L. Chen, M. J. Liu, L. Feng and L. Jiang, *Adv. Mater.*, 2011, **23**, 4270.
63. M. Liu, Z. Xue, H. Liu and L. Jiang, *Angew. Chem., Int. Ed.*, 2012, **51**, 8348.
64. P. Calcagnile, D. Fragouli, I. S. Bayer, G. C. Anyfantis, L. Martiradonna, P. D. Cozzoli, R. Cingolani and A. Athanassiou, *ACS Nano*, 2012, **6**, 5413.
65. Q. Zhu, Y. Chu, Z. Wang, N. Chen, L. Lin, F. Liu and Q. Pan, *J. Mater. Chem. A*, 2013, **1**, 5386.

66. J. Li, L. Shi, Y. Chen, Y. Zhang, Z. Guo, B. Su and W. Liu, *J. Mater. Chem.*, 2012, **22**, 9774.
67. C.-F. Wang and S.-J. Lin, *ACS Appl. Mater. Interfaces*, 2013, **5**, 8861.
68. J. Zhang and S. Seeger, *Adv. Funct. Mater.*, 2011, **21**, 4699.
69. B. Wang, J. Li, G. Y. Wang, W. X. Liang, Y. B. Zhang, L. Shi, Z. G. Guo and W. M. Liu, *ACS Appl. Mater. Interfaces*, 2013, **5**, 1827.
70. W. X. Liang and Z. G. Guo, *RSC Adv.*, 2013, **3**, 16469.
71. J. Wu, N. Wang, H. Dong, Y. Zhao and L. Jiang, *ACS Appl. Mater. Interfaces*, 2012, **4**, 3207.
72. B. Deng, R. Cai, Y. Yu, H. Q. Jiang, C. L. Wang, J. Li, L. F. Li, M. Yu, J. Y. Li, L. D. Xie, Q. Huang and C. H. Fan, *Adv. Mater.*, 2010, **22**, 5473.
73. L. Wu, J. P. Zhang, B. C. Li and A. Q. Wang, *Polym. Chem.*, 2014, **5**, 2382.
74. X. Zhou, Z. Zhang, X. Xu, F. Guo, X. Zhu and X. Men, *ACS Appl. Mater. Interfaces*, 2013, **5**, 7208.
75. D. D. Nguyen, N.-H. Tai, S.-B. Lee and W.-S. Kuo, *Energy Environ. Sci.*, 2012, **5**, 7908.
76. A. Li, H. X. Sun, D. Z. Tan, W. J. Fan, S. H. Wen, X. J. Qing, G. X. Li, S. Y. Li and W.-Q. Deng, *Energy Environ. Sci.*, 2011, **4**, 2062.
77. L. Wu, L. Li, B. Li, J. Zhang and A. Wang, *ACS Appl. Mater. Interfaces*, 2015, **7**, 4936.
78. L. Zhang, Z. Zhang and P. Wang, *NPG Asia Mater.*, 2012, **4**, e8.
79. Q. Zhu and Q. Pan, *ACS Nano*, 2014, **8**, 1402.
80. X. Gui, J. Wei, K. Wang, A. Cao, H. Zhu, Y. Jia, Q. Shu and D. Wu, *Adv. Mater.*, 2010, **22**, 617.
81. Z.-Y. Wu, C. Li, H.-W. Liang, Y.-N. Zhang, X. Wang, J.-F. Chen and S.-H. Yu, *Sci. Rep.*, 2014, **4**, 4079.
82. A. K. Kota, Y. X. Li, J. M. Mabry and A. Tuteja, *Adv. Mater.*, 2012, **24**, 5838.
83. D. Tian, X. Zhang, Y. Tian, Y. Wu, X. Wang, J. Zhai and L. Jiang, *J. Mater. Chem.*, 2012, **22**, 19652.
84. G. Kwon, A. K. Kota, Y. Li, A. Li, A. Sohani, J. M. Mabry and A. Tuteja, *Adv. Mater.*, 2012, **24**, 3666.
85. W. B. Zhang, Z. Shi, F. Zhang, X. Liu, J. Jin and L. Jiang, *Adv. Mater.*, 2013, **25**, 2071.
86. X. Gao, L. P. Xu, Z. Xue, L. Feng, J. Peng, Y. Wen, S. Wang and X. Zhang, *Adv. Mater.*, 2014, **26**, 1771.
87. M. Tao, L. Xue, F. Liu and L. Jiang, *Adv. Mater.*, 2014, **26**, 2943.
88. W. B. Zhang, Y. Z. Zhu, X. Liu, D. Wang, J. Y. Li, L. Jiang and J. Jin, *Angew. Chem., Int. Ed.*, 2014, **53**, 856.
89. Z. G. Guo, W. M. Liu and B. L. Su, *Appl. Phys. Lett.*, 2008, **92**, 063104.
90. Z. G. Guo, F. Zhou, J. C. Hao and W. M. Liu, *J. Am. Chem. Soc.*, 2005, **127**, 15670.
91. J. L. Song, S. Huang, K. Hu, Y. Lu, X. Liu and W. Xu, *J. Mater. Chem. A*, 2013, **1**, 14783.
92. S. Sethi and A. Dhinojwala, *Langmuir*, 2009, **25**, 4311.

93. L. B. Zhang, Y. J. Zhong, D. Cha and P. Wang, *Sci. Rep.*, 2013, **3**, 2326.
94. Y. Z. Cao, X. Y. Zhang, L. Tao, K. Li, Z. X. Xue, L. Feng and Y. Wei, *ACS Appl. Mater. Interfaces*, 2013, **5**, 4438.
95. F. Zhang, W. B. Zhang, Z. Shi, D. Wang, J. Jin and L. Jiang, *Adv. Mater.*, 2013, **25**, 4192.
96. C. Gao, Z. Sun, K. Li, Y. Chen, Y. Cao, S. Zhang and L. Feng, *Energy Environ. Sci.*, 2013, **6**, 1147.
97. A. K. Kota, G. Kwon, W. Choi, J. M. Mabry and A. Tuteja, *Nat. Commun.*, 2012, **3**, 1025.
98. H. Li, Y. Cao, J. Qin, X. Jie, T. Wang, J. Liu and Q. Yuan, *J. Membr. Sci.*, 2006, **279**, 328.
99. X. Chen, L. Hong, Y. Xu and Z. W. Ong, *ACS Appl. Mater. Interfaces*, 2012, **4**, 1909.
100. Q. Chang, J. Zhou, Y. Wang, J. Liang, X. Zhang, S. Cerneaux, X. Wang, Z. Zhu and Y. Dong, *J. Membr. Sci.*, 2014, **456**, 128.
101. S. Maphutha, K. Moothi, M. Meyyappan and S. E. Iyuke, *Sci. Rep.*, 2013, **3**, 1509.
102. L. Yang, A. Thongsukmak, K. K. Sirkar, K. B. Gross and G. Mordukhovich, *J. Membr. Sci.*, 2011, **378**, 138.
103. X. S. Yi, S. L. Yu, W. X. Shi, S. Wang, N. Sun, L. M. Jin and C. Ma, *Desalination*, 2013, **319**, 38.
104. D. Rana and T. Matsuura, *Chem. Rev.*, 2010, **110**, 2448.
105. I. Sadeghi, A. Aroujalian, A. Raisi, B. Dabir and M. Fathizadeh, *J. Membr. Sci.*, 2013, **430**, 24.
106. X. Yi, W. Shi, S. Yu, C. Ma, N. Sun, S. Wang, L. Jin and L. Sun, *J. Hazard. Mater.*, 2011, **193**, 37.
107. W. Chen, Y. Su, L. Zheng, L. Wang and Z. Jiang, *J. Membr. Sci.*, 2009, **337**, 98.
108. J. Hyun, H. Jang, K. Kim, K. Na and T. Tak, *J. Membr. Sci.*, 2006, **282**, 52.
109. N. A. Hashim, F. Liu and K. Li, *J. Membr. Sci.*, 2009, **345**, 134.
110. A. Asatekin and A. M. Mayes, *Environ. Sci. Technol.*, 2009, **43**, 4487.
111. Q. Shi, Y. Su, W. Zhao, C. Li, Y. Hu, Z. Jiang and S. Zhu, *J. Membr. Sci.*, 2008, **319**, 271.
112. A. C. Sagle, E. M. V. Wagner, H. Ju, B. D. McCloskey, B. D. Freeman and M. M. Sharma, *J. Membr. Sci.*, 2009, **340**, 92.
113. D. Lin, C. Chang, F. Huang and L. Cheng, *Polymer*, 2003, **44**, 413.
114. M. Yeow, Y. Liu and K. Li, *J. Membr. Sci.*, 2005, **258**, 16.
115. H. C. Yang, J. K. Pi, K. J. Liao, H. Huang, Q. Y. Wu, X. J. Huang and Z. K. Xu, *ACS Appl. Mater. Interfaces*, 2014, **6**, 12566.
116. H.-C. Yang, K.-J. Liao, H. Huang, Q.-Y. Wu, L.-S. Wan and Z.-K. Xu, *J. Mater. Chem. A*, 2014, **2**, 10225.
117. Z.-X. Wang, C.-H. Lau, N.-Q. Zhang, Y.-P. Bai and L. Shao, *J. Mater. Chem. A*, 2015, **3**, 2650.
118. Y. Zhu, F. Zhang, D. Wang, X. F. Pei, W. Zhang and J. Jin, *J. Mater. Chem. A*, 2013, **1**, 5758.

119. X. S. Peng, J. Jin, Y. Nakamura, T. Ohno and I. Ichinose, *Nat. Nanotechnol.*, 2009, **4**, 353.
120. C. C. Striemer, T. R. Gaborski, J. L. McGrath and P. M. Fauchet, *Nature*, 2007, **445**, 749.
121. J. K. Holt, A. Noy, T. Huser, D. Eaglesham and O. Bakajin, *Nano Lett.*, 2004, **4**, 2245.
122. S. Karan, S. Samitsu, X. Peng, K. Kurashima and I. Ichinose, *Science*, 2012, **335**, 444.
123. R. R. Nair, H. A. Wu, P. N. Jayaram, I. V. Grigorieva and A. K. Geim, *Science*, 2012, **335**, 442.
124. H. Li, Z. Song, X. Zhang, Y. Huang, S. Li, Y. Mao, H. J. Ploehn, Y. Bao and M. Yu, *Science*, 2013, **342**, 95.
125. B. Mi, *Science*, 2014, **343**, 740.
126. H. Huang, Z. Song, N. Wei, L. Shi, Y. Mao, Y. Ying, L. Sun, Z. Xu and X. Peng, *Nat. Commun.*, 2013, **4**, 2979.
127. H. Huang, Y. Mao, Y. Ying, Y. Liu, L. Sun and X. Peng, *Chem. Commun.*, 2013, **49**, 5963.
128. K. Huang, G. Liu, Y. Lou, Z. Dong, J. Shen and W. Jin, *Angew. Chem., Int. Ed.*, 2014, **53**, 1.
129. J. Shen, G. Liu, K. Huang, W. Jin, K.-R. Lee and N. Xu, *Angew. Chem., Int. Ed.*, 2015, **54**, 578.
130. L. Sun, Y. Ying, H. Huang, Z. Song, Y. Mao, Z. Xu and X. Peng, *ACS Nano*, 2014, **8**, 6304.
131. Z. Shi, W. Zhang, F. Zhang, X. Liu, D. Wang, J. Jin and L. Jiang, *Adv. Mater.*, 2013, **25**, 2422.
132. S. J. Gao, Z. Shi, W. Zhang, F. Zhang and J. Jin, *ACS Nano*, 2014, **8**, 6344.
133. S. J. Gao, Y. Zhu, F. Zhang and J. Jin, *J. Mater. Chem. A*, 2015, **3**, 2895.
134. S. J. Gao, H. Qin, P. Liu and J. Jin, *J. Mater. Chem. A*, 2015, **3**, 6649.
135. Y. Long, J. F. Hui, P. P. Wang, G. L. Xiang, B. Xu, S. Hu, W. C. Zhu, X. Q. Lu, J. Zhuang and X. Wang, *Sci. Rep.*, 2012, **2**, 612.

CHAPTER 4

Responsive Particle-Stabilized Emulsions: Formation and Applications

MAN-HIN KWOK[a] AND TO NGAI*[a,b]

[a]Department of Chemistry, The Chinese University of Hong Kong, Shatin, N. T., Hong Kong, China; [b]The Chinese University of Hong Kong Shenzhen Research Institute, Shenzhen, 518057, PR China
*E-mail: tongai@cuhk.edu.hk

4.1 Introduction

Emulsions are essential in our modern daily life. The influences and applications of different kinds of emulsions reach pharmaceutical,[1] petroleum,[2] biocatalysis[3,4] and food industries.[5,6] An emulsion is a heterogeneous system which consists of a dispersed liquid in another liquid as a continuous phase.[7] However, the mixing of two or more liquids which have very different polarity is obviously not favourable. As a result, these thermodynamically unstable systems undergo separation in the matter of seconds.

Therefore, one might use surfactants to kinetically stabilize them, slowing the coalescence process of the dispersed droplets and phase separation.[8] Besides amphiphilic molecules of surfactants, there is another category of emulsion stabilizer, which is the solid particle. At the beginning of the last century, Ramsden and Pickering described that colloidal-sized solid particles were also capable of stabilizing emulsion droplets.[9,10] Nowadays, this

RSC Smart Materials No. 20
Smart Materials for Advanced Environmental Applications
Edited by Peng Wang

Published by the Royal Society of Chemistry, www.rsc.org

particle-stabilized emulsion is usually called a "Pickering emulsion". In some senses, particle stabilizers are quite similar to amphiphilic surfactants. For both of them, the relative wettability by the two liquids is a very important factor for one to predict the properties of the resulting emulsion. This relevant wettability is described by the hydrophile–lipophile balance (HLB) and interfacial contact angle for surfactants and particles, respectively. However, there are some important differences between them.[11] For example, the orders of energy of the adsorptions of them are very different. Compared with small surface active molecules, individual colloidal particles are adsorbed to the interface with so much higher energy, which is often in the order of 10^5 k_BT. The high desorption energy of the particle leads to the irreversible adsorption of it and results in outstanding stability for the emulsion droplets.

However, it is not only the formation and stabilization of the emulsions that are important. Also, if one can easily control the stability of the emulsion, it will be much more applicable for different purposes. To manipulate the emulsion stability, obviously the best way is to control the properties of the stabilizers themselves. The changes in the stabilizers' properties are going to affect the properties and stability of the emulsions. It means that the responsiveness of the stabilizer brings responsiveness to the resulting emulsions.[12,13] These responsivenesses are usually referred to thermal, pH, ionic strength, magnetic responsiveness, *etc.*

Therefore, studying and understanding the formation and stabilization of such responsive emulsions help us to prepare the desired product for different innovative applications. Also, different responsive emulsion systems will be discussed and their applications will be emphasized in this chapter.

4.2 Particulate Emulsion Stabilizer

4.2.1 The Stabilization of an Emulsion

Despite the fact that particle stabilizers have been known about for more than one hundred years, increasing interests have been developed since the last decade.[14] It was because the syntheses and characterizations of colloidal particles have become much more accessible. Before different kinds of colloidal particles could be easily prepared with specific properties, the applications of them and the emulsions stabilized by them were greatly limited. As the syntheses of colloidal particles have been extensively studied and reported,[15] particles can be designed and prepared specifically for different situations. Discussion of responsive emulsions is all about the stabilization and destabilization of the emulsion in a controlled manner. Therefore, understanding the fundamental physics of emulsions is very important.

On the other hand, surfactants and amphiphilic molecules with low molecular weight polymers and the stabilization given by them have been well studied. As the stabilization given by them can be quite different from the particulate system, there are some important features which can help us to better understand how the colloid particles behave at the interface.[11]

When two liquids are mixed with each other, the rule of thumb can usually be regarded as "like dissolves like". Therefore, mixing non-polar oil and polar aqueous solution will be thermodynamically not favourable for most of the situation. Also because of this reason, a liquid tends to minimize the area contacting and interacting with another substance. Any attempt that creates extra surface area requires energy. The preparation of an emulsion greatly increases the area of interface between the two liquids. The presence of surfactants can significantly lower the interfacial tension between the two immiscible liquids such that the energy change for the process is less unfavourable. Also, the charges on the surfactant molecules stabilize the emulsion droplets from the coalescence process electrostatically.

Other than thermodynamically stable microemulsions, all emulsions will eventually undergo different processes to achieve separation and the dispersed phase will merge back together. These processes are illustrated in Figure 4.1. They include sedimentation and creaming caused by the action of gravity and the density difference of the two phases; flocculation and aggregation of emulsion droplets bring them to close proximity.[8] However, the thin liquid films between individual droplets are still intact and prevent the droplets from combining. Ostwald ripening originates from the Laplace pressure ΔP. The Young–Laplace equation indicates that the Laplace pressure of the droplet increases as the diameter of it decreases.[7]

$$\Delta P = \frac{2\gamma}{R} \text{ for a spherical droplet} \tag{4.1}$$

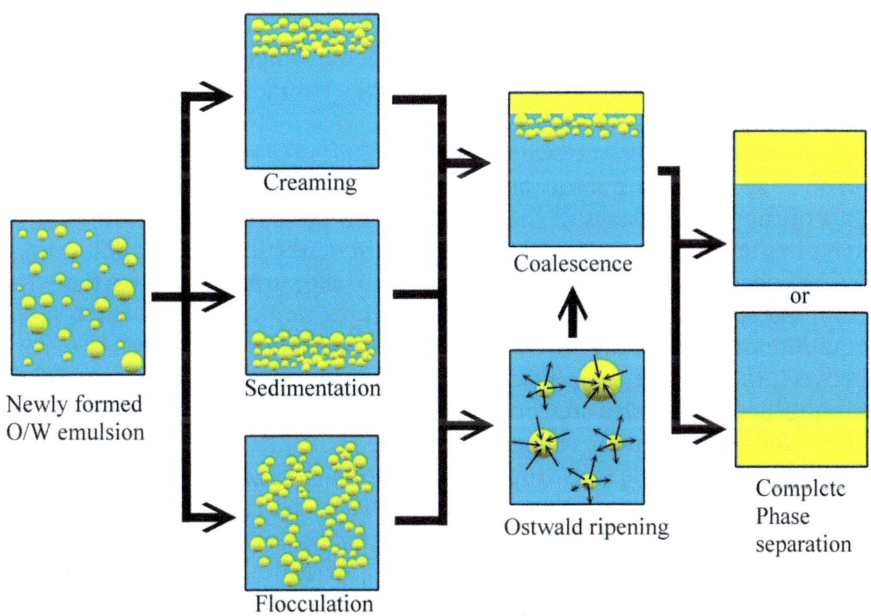

Figure 4.1 Destabilization processes of an emulsion.

The solubility difference between droplets of different sizes causes the large emulsion droplets to slowly absorb the smaller ones, until the difference in Laplace pressure between the droplets is small enough and this process will be essentially stopped. Finally, coalescence is the process where emulsion droplets bump into each other and merge to form a bigger droplet. It eventually leads to phase separation of the emulsion. In addition, Ostwald ripening is not very significant for common oil droplets with diameter larger than a few hundred nanometer.[16,17]

To slow down these processes and stabilize the emulsion, one can lower the interfacial tension by adding amphiphilic molecules or polymers. They are adsorbed to the interface and reduce the interfacial tension significantly. From the Young–Laplace equation, it is found that the Laplace pressure is directly proportional to the interfacial tension. Therefore, surfactants can reduce the Laplace pressure difference between emulsion droplets of different sizes and slow down Ostwald ripening. Besides lowering the interfacial tension, surfactant molecules and amphiphilic polymers which have been adsorbed to the interface also provide electrostatic and steric stabilization to the emulsion. They effectively slow down flocculation and coalescence, which will be discussed later.

4.2.2 Special Features About Particulate Emulsion Stabilizers

Colloidal-sized particles can also be adsorbed to the interface and stabilize emulsions.[18] The surface polarity of the particle usually lies between the polarities of the liquids. It means that these particles can be wetted simultaneously by both the dispersed phase and the continuous phase. This gives the particle interfacial activity. Therefore, they are energetically favourable to be adsorbed onto the interface and reduce the contact area between the two immiscible phases.

Generally speaking, the two immiscible phases in an emulsion consist of an oil phase with low polarity and an aqueous phase with high polarity. Then, part of the particle stays in the oil phase and the other part of the particle stays in the aqueous phase. This equilibrium of wetting areas is determined by the hydrophilicity or the hydrophobicity of the surface of the particle. The resulting special physical quantity is called the contact angle, which is the equilibrated angle of the particle at the interface after the interfacial tensions between the particle, oil phase and aqueous phase are all balanced.[19] First, homogeneous particles are concerned. In Figure 4.2, the three possible situations for a homogeneous spherical particle at the interface are shown. If the surface of the particle is hydrophilic, the interaction with the aqueous phase will be more favourable. It means that the water–particle interfacial tension γ_{wp} is lower than the oil–particle interfacial tension γ_{op}. Figure 4.3 illustrates the Young's equation for a homogeneous spherical particle at an oil–water interface.[19]

$$\gamma_{op} - \gamma_{wp} = \gamma_{ow} \cos\theta \text{ where } \theta \text{ is the contact angle} \tag{4.2}$$

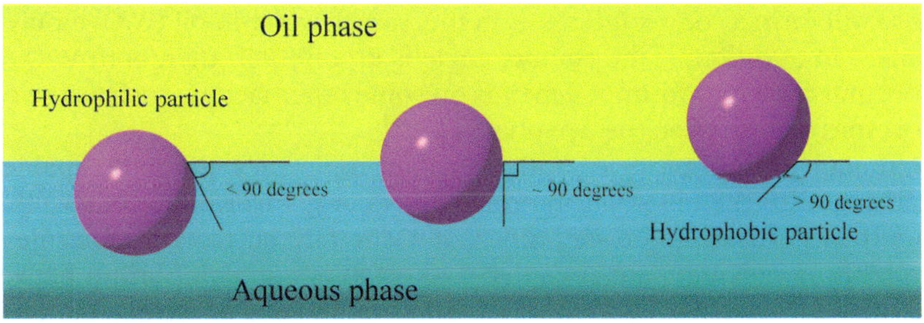

Figure 4.2 Homogeneous spherical particles at oil-water interface. They can be hydrophilic (left), hydrophobic (right) and can have intermediate hydrophilicity (middle). Their contact angles depend on their surface hydrophilicity.

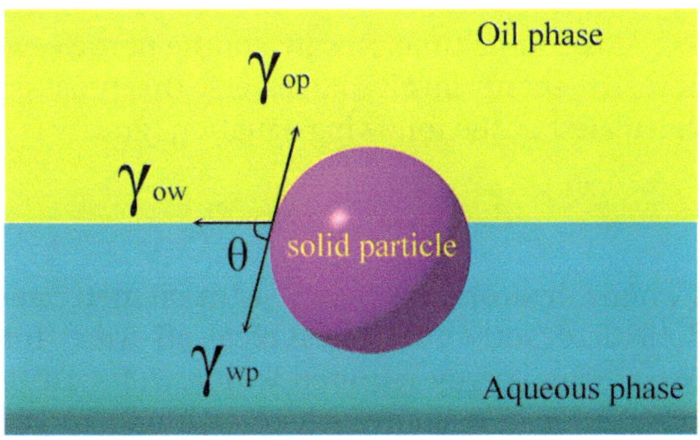

Figure 4.3 Schematic drawing of a homogeneous particle at an oil–water interface. The three interfacial tensions balanced each other according to the Young's equation.

Because $(\gamma_{op} - \gamma_{wp})$ is a positive number for hydrophilic particles, the contact angle will lie between 0° and 90°. As a result, the majority of the particles stay in the aqueous phase when the minority part of it is wetted by the oil phase. On the contrary, hydrophobic particles favour interaction with the oil phase rather than interaction with the aqueous phase, and the majority of the particles remain in contact with the oil phase. The contact angle for a hydrophobic particle at the oil–water interface is then between 90° and 180°. Moreover, if the hydrophilicity of the particle is intermediate, the contact angle of the particle will be very close to 90°. The concept of contact angle can be treated as an analogy of hydrophile–lipophile balance (HLB), which is extensively used for surfactant molecules. It is because the HLB is related to the tendency of curvature of the interface with surfactant molecules.[11] Similar to that of HLB, the contact angle is also related to the tendency of curvature of a particle adsorbed at the interface. If the contact angle of the particle is smaller than 90°, the interface will bend to the oil phase.[14] Then, an oil-in-water (O/W) emulsion forms. However, if the contact angle of the particle is larger than 90°, the aqueous phase will form a convex interface. In this case, a water-in-oil (W/O) emulsion forms.[20] In addition, if the contact angle is around 90°, then both W/O and O/W emulsions may form. It depends on some other factors, such as the type of oil, the ratio between the two phases, *etc.*[21]

Although colloid particles and surfactant molecules are both capable of attaching onto the oil–water interface, the energy involved to remove them from the interface can be very different. Generally, surfactant molecules or amphiphilic polymers are considered to establish an equilibrium between the interface and the bulk solution. The low molecular weight molecules leave the interface easily and molecules in the bulk are adsorbed back simultaneously.[11] However, when the diameters of the colloid particles are larger than around 50 nm, the energy involved in the desorption of individual particles would be very large compared to the thermal energy k_BT.

When a homogeneous spherical particle is adsorbed from the aqueous phase to the interface, it replaces some of the oil–water interface with a particle–oil interface. At the same time, the area of the particle–water interface is reduced. Therefore, the energy involved to reverse this process, the desorption energy, can be calculated by the following equation (gravity is neglected),[22]

$$E_d = \gamma_{ow}\pi r^2(1 - |\cos\theta|)^2 \tag{4.3}$$

For example, when a hydrophilic particle with diameter and contact angle of 100 nm and 45° is desorbed from a typical oil–water interface with γ_{ow} equal to 50 mN m^{-1}, the energy required will be $3.3 \times 10^4\ k_BT$ (T = 298 K). Therefore, the adsorption of colloid particles is usually considered to be irreversible, unless the contact angle is very close to 0° or 180°.[14] The resulting Pickering emulsion usually possesses very high stability. It might not undergo complete phase separation even in the time scale of years, which is normally quite difficult to be achieved by traditional small molecule surfactants. This is a very important feature of a particle-stabilized Pickering emulsion. The strongly adsorbed particles are capable of forming a close-packed layer which surrounds the emulsion droplets. Figure 4.4 shows a SEM image of a colloidosome made by self-assembly in an O/W Pickering emulsion.[23] It is not always a must for the particles to align themselves to a hexagonal-liked packing at the interface. Nevertheless, such packing is often observed, for both O/W and W/O emulsions. This physical barrier effectively protects the droplets from coalescence. It is because a large amount of energy is required to disrupt and deform the particle layer for coalescence to happen. Also, there are interesting interactions between these adsorbed particles, which will be discussed later. The behaviour of this monolayer of particles under compression was studied by Aveyard *et al.* using a Langmuir–Blodgett trough.[24] They found out that the particle monolayer distorted to result in a rhombohedral array. They suggested that particle repulsion and buckling might be general features of such particle monolayers.

Colloidal particle dispersions usually possess very high colloidal stability. The stability of these particles usually originates from the steric stabilization

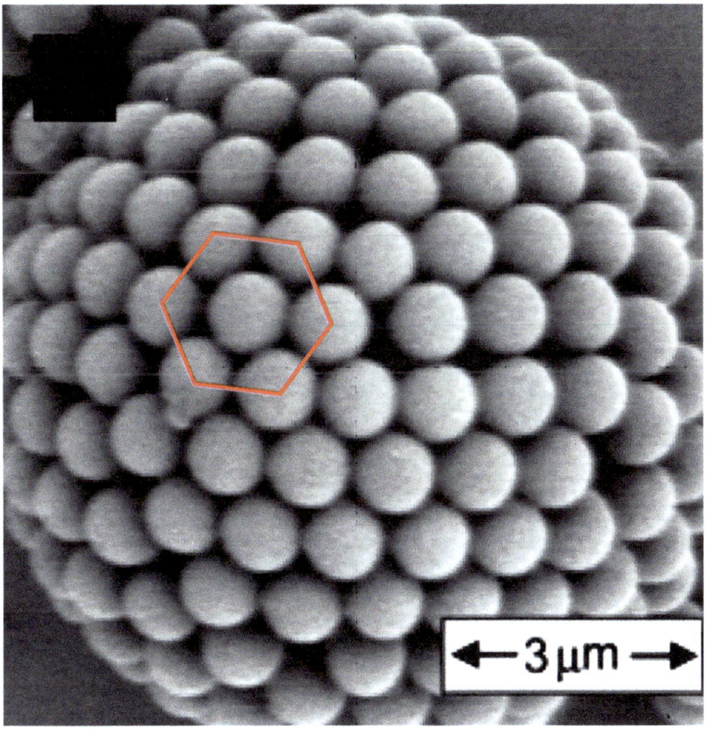

Figure 4.4 SEM image of a colloidosome made by self-assembly of polystyrene particles in O/W Pickering emulsion.[23] Reprinted (adapted) with permission from M. F. Hsu, M. G. Nikolaides, A. D. Dinsmore, A. R. Bausch, V. D. Gordon, X. Chen, J. W. Hutchinson and D. A. Weitz, *Langmuir*, 2005, **21**, 2963–2970. Copyright 2005 American Chemical Society.

and electrostatic stabilization. Obviously, the electrostatic stabilization refers to the repulsion of the charges on the particle surface. Particles can be synthesized or modified so that the surface of the particles possesses ionizable functional groups, such as carboxylic groups, amine groups, sulfonate groups, *etc.*[25] They are ionized in aqueous solution and the surface of the particle will be charged. These charged particles can also stabilize a Pickering emulsion. This electrostatic repulsion of the particles is going to provide the resulting droplets' electrostatic stabilization as well. In Figure 4.5, two droplets covered by charged particles are illustrated. When they approach each other, the repulsion between the charges on the particles repels the droplets.

For steric stabilization, it is given by the dangling chains on the particle surface (Figure 4.5). These dangling chains are solvated by the continuous phase. When the emulsion droplets are very close to each other, the dangling chains layer may be overlapped. As it happens, the osmotic pressure of the overlapping region increases. Then, the continuous phase will be drawn to the overlapping region and prevent the droplets from getting closer. This steric stabilization is quite common for polymeric colloidal particles, such as

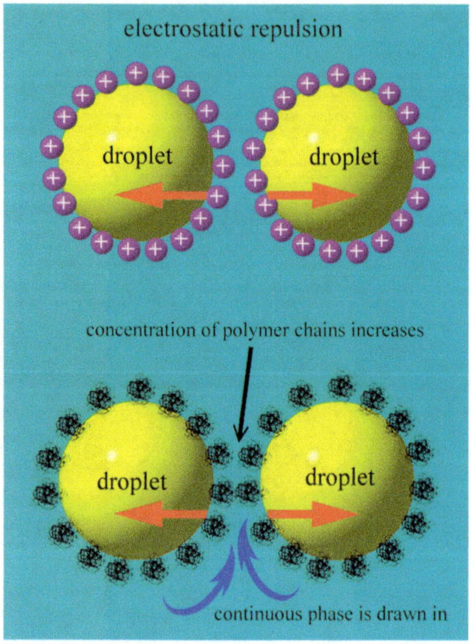

Figure 4.5 Illustration of electrostatic and steric stabilization given by microgel particles.

poly(*N*-isopropylacrylamide) microgel particles. Similar to the case of electrostatic stabilization of the particles, steric repulsion of the particles also contributes in enhancing the stability of the emulsion droplets in a Pickering emulsion. They stabilize the emulsion by slowing down the flocculation and coalescence process as they avoid the droplets from getting to close proximity.

Particle–particle interactions are not limited to the particles on different droplets. They also exist between the particles on the same droplet. Kralchevsky and Nagayama discussed and reviewed different capillary interactions between particles at interfaces.[26,27] Figure 4.6 shows some of these interactions. Normally flotation interactions, which are driven by the gravity, are negligible for colloid particles. However, the immersion interactions, which are driven by the wetting of particles, are usually large compared to the thermal energy k_BT.[26] This happens even if the diameter of the particle is just a few nanometers. Therefore, Brownian force might not be able to counteract the capillary interactions between the particles. Then, they are attracted by each other and aggregate at the interface. At the same time, the interactions originated on the particles themselves, which are mentioned above, balance this capillary attraction. These complicated particle–particle interactions affect the packing of the adsorbed particles at the interface. Thus the emulsion stability of the resulting Pickering emulsion is also dependent on them.

Another important feature of a Pickering emulsion is called maximum capillary pressure, P_{max}.[28] First, two colliding emulsion droplets are considered in an O/W Pickering emulsion. Each of the oil droplets is protected by a

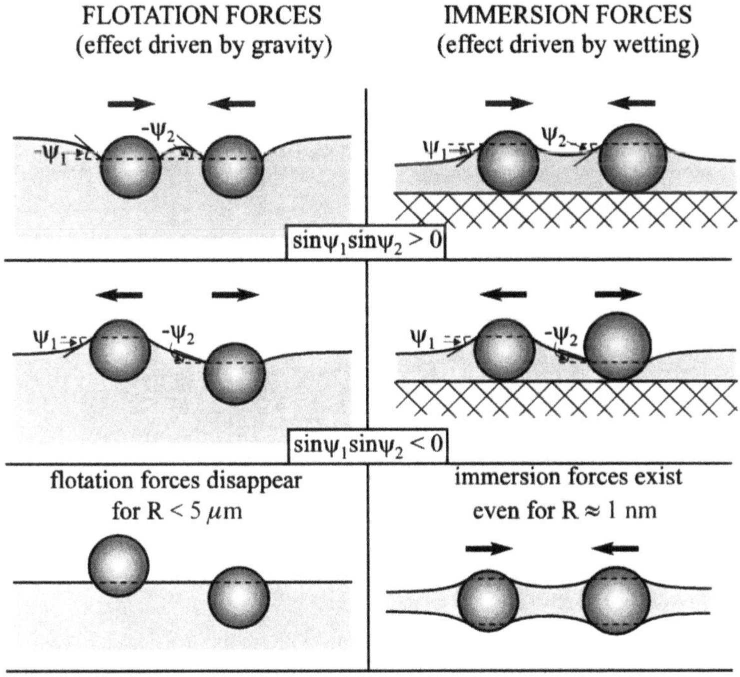

Figure 4.6 Different capillary forces acting on solid particles at an interface.[26] Reprinted from P. A. Kralchevsky and K. Nagayama, Capillary interactions between particles bound to interfaces, liquid films and biomembranes, *Adv. Colloid Interface Sci.*, 2000, **85**, 145–192, Copyright 2000, with permission from Elsevier.

single layer of particles. When they are pushed together, each particle shell is deformed and rearranged. A bridged interface with single layer of particles between the two oil droplets will then be formed (see Figure 4.7).[29] Then, there will be a thin aqueous film located right between the oil droplets. If these two touching droplets are further pushed together, it causes the aqueous interfilm to drain. Finally, the interfilm is completely drained, and the droplets will touch and combine. However, as the particles here are hydrophilic, there will be a capillary pressure which counteracts the interfilm drainage. The capillary pressure increases as the interfilm drainage goes on, until the pressure pushing the two droplets together exceeds the maximum capillary pressure of the system. At this point, the aqueous interfilm between them is ruptured and coalescence of the droplets occurs.

Denkov and co-workers[28] have introduced this idea as a stabilization mechanism of particle-stabilized emulsions. Later, this study was also studied by other groups.[30] The maximum capillary pressure of a particle layer between two oil droplets can be calculated by the follow equation:

$$P_{\text{max}} = \frac{2P\gamma_{\text{ow}}\left|\cos\theta\right|}{r} \tag{4.4}$$

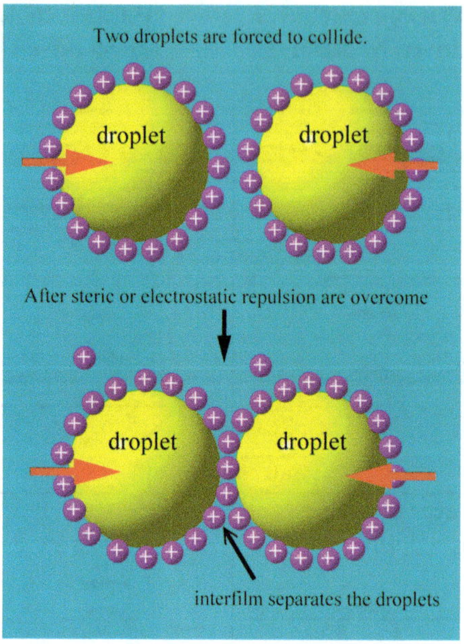

Figure 4.7 Schematic drawing of two colliding oil droplets. Eventually, an aqueous film will be formed in between the droplets. Further pushing causes the film to drain and a capillary pressure is built up.

where *P* in the equation is called theoretical packing parameter and *r* is the radius of the particles. In another study, which was done by Tcholakova *et al.*,[31] the maximum capillary pressure was studied with β-lactoglobulin protein by the film trapping technique. Besides that, they also measured the maximum osmotic pressure of the emulsion before coalescence by putting the emulsion into a centrifuge. They found out that this maximum osmotic pressure was a complete analogy of the maximum capillary pressure. It was because centrifugation applied a pressure to the aqueous film and drained it by the density difference of the two phases. Nevertheless, measuring the maximum osmotic pressure by centrifugation was very convenient. It was also applied by other researchers for comparing emulsion stability.[32] Maximum capillary pressure or maximum osmotic pressure of the emulsion gives us a general and convenient way to compare and study the emulsion stability of a Pickering emulsion. Note that there is an interesting relationship between the desorption energy and the maximum capillary pressure. In eqn (4.3), desorption energy of the particle increases as the contact angle approaches 90°. On the other hand, eqn (4.4) indicates that the maximum capillary pressure of the interfilm with the particles increases as the contact angle approaches 0° (or 180° for W/O Pickering emulsion). For a contact angle approaching 90°, the particle is likely to be adsorbed to the interface more strongly for both O/W and W/O emulsion. However, the particle would not be preferentially wetted by

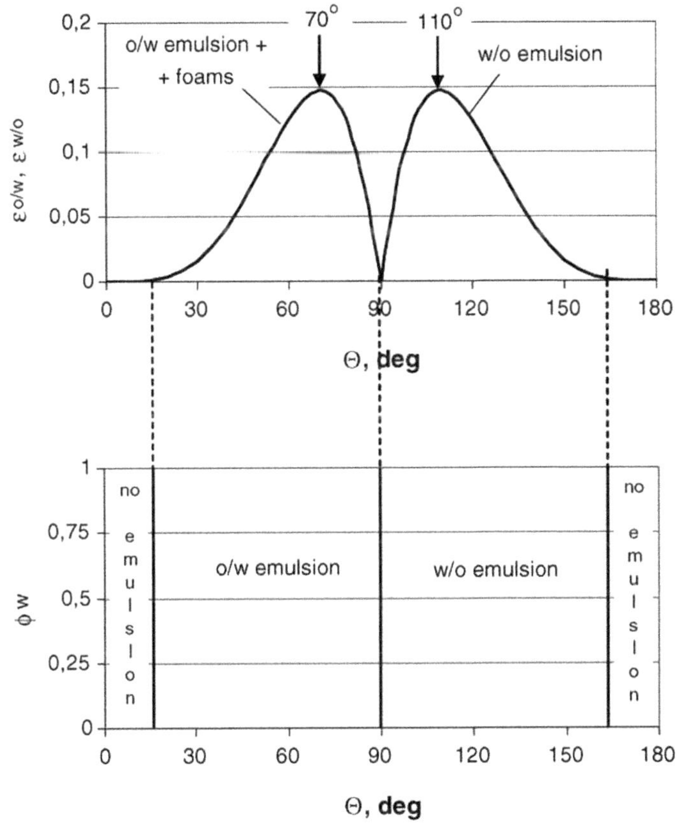

Figure 4.8 Combined effect of desorption energy and maximum capillary pressure, which is considered energetically.[30] Reprinted from G. Kaptay, On the equation of the maximum capillary pressure induced by solid particles to stabilize emulsions and foams and on the emulsion stability diagrams, *Colloids Surf., A*, 2006, **282**, 387–401, Copyright 2006, with permission from Elsevier.

the continuous phase significantly. The capillary pressure which draws in the continuous phase and maintains the interfilm would be much lower. Therefore, Kapkay *et al.* evaluated the combined effect of them.[30] It was predicted that the best contact angle for stabilizing a Pickering emulsion would be 70° (or 110° for W/O Pickering emulsion), which is shown in Figure 4.8.

Therefore, to control the stability of a particle-stabilized Pickering emulsion, the above factors are to be altered by designing specific particles and changing the external conditions. This means that the emulsion can be prepared with excellent stability, yet it can be destabilized in a controlled manner when needed. These responsivenesses will be discussed in the next section.

4.3 Categories of Particles

4.3.1 Inorganic Particles

At first, inorganic particles are investigated in the field of Pickering emulsions. When Pickering investigated and prepared those solid particle stabilized emulsions, a number of different metal oxides and metal hydroxides were used as the "insoluble emulsifiers".[9] For example, magnesium carbonate, calcium hydroxide, basic copper sulfate, *etc.* They were used because they can be prepared conveniently. Many kinds of fine particles of transition metal hydroxides were prepared by mixing lime solution (calcium hydroxide) with the corresponding metal sulfate and the stability of the resulting Pickering emulsions was compared. As they are ionic salts, their contact angles are usually smaller than 90°.[33] They are hydrophilic and capable of stabilizing O/W Pickering emulsions. Besides insoluble salts,[34] clays,[35,36] carbon black[37] and silica[38] also are very popular choices for inorganic particulate emulsion stabilizers. Nonomura *et al.* used mica particles to stabilize both O/W and W/O Pickering emulsions (Figure 4.9).[21] The phase inversion of the emulsions was also studied at different particle/oil/water ratios as well.

Pickering emulsions stabilized by LAPONITE® clay particles were studied by Ashby and Binks.[35] These disc-like, non-spherical particles stabilized toluene-in-water Pickering emulsions with great stability. However, these inorganic particles usually do not show any responsiveness to changes in surrounding conditions such as temperature or pH value. To use them as responsive particulate emulsion stabilizers, surface modification should be done. Iron oxides nanoparticles are not considered to be responsive. Nevertheless, in Lan's work, bilayer of oleic acid was coated on their Fe_3O_4

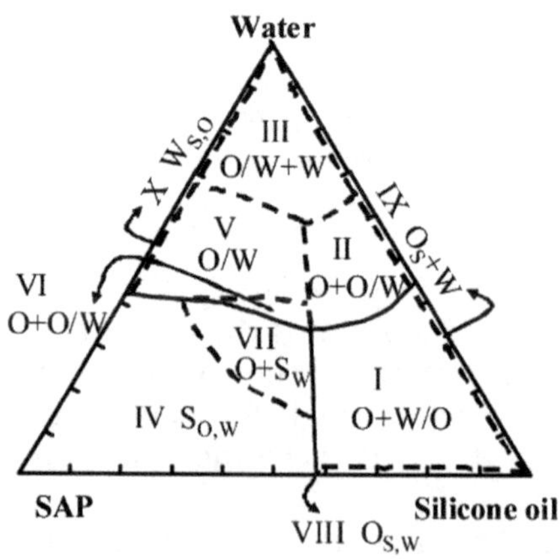

Figure 4.9 The types of emulsion formed in different surface active mica particle (SAP)/water/silicone oil ratios.[21] Reprinted from Y. Nonomura and N. Kobayashi, Phase inversion of the Pickering emulsions stabilized by plate-shaped clay particles, *J. Colloid Interface Sci.*, 2009, **330**, 463–466, Copyright 2009, with permission from Elsevier.

nanoparticles.[39] So that the contact angle of the Fe_3O_4 nanoparticle changed in different pH values, which will be discussed more in the next section.

Certainly, silica particles should be the most representative one. It is because the syntheses of silica particles have been well studied.[40] The design and preparation of the desired silica particles are more accessible compared to other types of inorganic particles. Many methods can be applied to produce such particles. One common method is ion exchange reaction of sodium silicate in acidic condition. To control the properties of the resulting silica particles, one can alter the parameters used in the synthesis, for examples, the duration of the reaction, temperature or the pH value of the reaction mixture. Other novel methods were also developed to prepare silica particles. Rao *et al.* applied ultrasonication during the hydrolysis of tetraethyl orthosilicate.[41] The effects of different experimental parameters on the resulting particle diameters were extensively studied. They found out that the hydrolysis of tetraethyl orthosilicate yielded silica particles with diameters between 20 nm and 460 nm. Figure 4.10 shows the particle diameter dependence on temperature and water to tetraethyl orthosilicate ratio.

Silica particles are naturally hydrophilic because the surface of silica particles are covered by hydrophilic siloxane groups and silanol groups.[42] Actually, this is the biggest advantage of using silica particles because surface modification done by reacting these silanol groups is very accessible. The hydrophilicity of the particles can be controlled by replacing different amounts of silanol groups on the surface. It can be achieved by the reaction between surface silanol groups and dimethyldichlorosilane to different extents.[20] Silica particles with a high percentage of silanol are likely to be

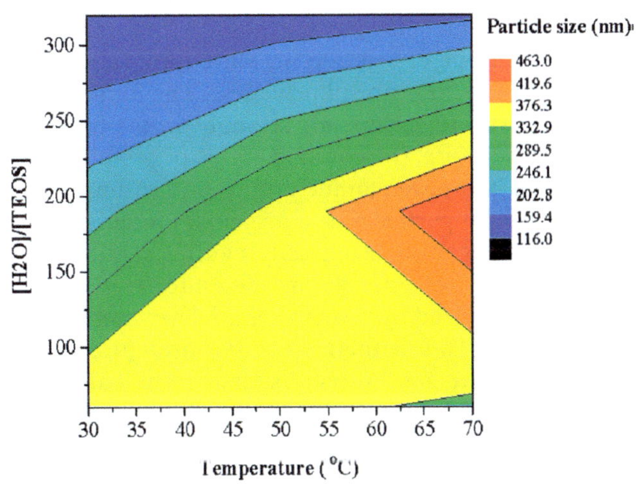

Figure 4.10 Particle diameter dependence on temperature and water to tetraethyl orthosilicate ratio.[41] Reprinted from K. S. Rao, K. ElHami, T. Kodaki, K. Matsushige and K. Makino, A novel method for synthesis of silica nanoparticles, *J. Colloid Interface Sci.*, 2005, **289**, 125–131, Copyright 2005, with permission from Elsevier.

more hydrophilic and stabilize O/W emulsions. On the contrary, particles with a low percentage of silanol indicate that most of the silanol groups are replaced by hydrophobic substituents. They probably favour stabilizing W/O emulsions. Moreover, as the surface of silica particles is covered by silanol groups, they can link to other functional groups such as bromo groups. Then, polymers can be polymerized on these particles through atom transfer radical polymerization (ATRP). Perruchot *et al.* have prepared such silica particles covered by different polymers.[43]

4.3.2 Biological Particles

To prepare solid-stabilized emulsions, the particles chosen are synthetic particles for most of the cases. However, biological particles which are present naturally are also interesting. These biological particles include proteins,[31,44] bacteria,[45,46] viruses,[47,48] polysaccharide particles,[49,50] *etc.* For synthetic nanoparticles, it often requires so much effort and resources to increase the monodispersity of the product. However, most of these bio-nanoparticles are exactly monodisperse. Also, some proteins and polysaccharide particles are actually common chemicals which appear in our daily life, such as starches and celluloses. Therefore, they are non-toxic and capable of being used in food processing.

Proteins are important biological molecules which are essential for every organism on the planet. They control and catalyse chemical reactions in organisms. Proteins mainly consist of polypeptide chains which are linear polymers of amino acids. The properties of the resulting proteins depend on the number of amino acids, the proportion of different amino acids and also the sequence of them. The size, wettability, conformation and charge of different proteins can be very different. More importantly, most of these bio-nanoparticles possess multi-responsiveness. The folding of polypeptide, which forms the final protein, depends strongly on the interactions on the polypeptide chain. Therefore, the conformations of proteins are sensitive to temperature changes. Moreover, amine groups and carboxylic groups introduce special electrical properties to the proteins. Proteins can be positively charged, negatively charged or electrically neutral at different pH values, which depend on their isoelectric points. de Folter *et al.* studied the O/W Pickering emulsion stabilized by protein zein.[51] They prepared the particles simply by precipitation. The diameters of their protein zein particles were around 100 nm with quite good monodispersity. The particles were stable in acidic and alkaline conditions. However, they aggregated at intermediate pH values because the electrostatic repulsion were weak when the pH was close to the isoelectric point of zein, which was around 6.5. Figure 4.11 shows the Pickering emulsions prepared by them at different pH values. At these pH values, the zein particles were all stable colloidal dispersions and yet the stability of their emulsions was different. The emulsion was not very stable at pH 2.5. At pH 8.5 and pH 9.8 zein particles stabilized the emulsion with good emulsion stability. Our group also reported a novel high internal phase emulsion (HIPE) stabilized by bovine serum albumin (BSA) protein nanoparticles (see Figure 4.12).[52] The excess BSA proteins were also capable of forming a network in the continuous phase.

Polysaccharide particles are also choices for bio-nanoparticle stabilizers. They are natural polymers of carbohydrates, such as starch and cellulose. They are abundant in natural resources, especially in plants. Starch consists of two major polymers of glucose. They are linear amylose and highly branched amylopectin. However, compared to proteins, starch is much simpler and starch is considered to be not responsive at all. Therefore, to obtain responsive starch particles, surface modifications are usually required. Tan *et al.* successfully prepared responsive starch-based nanospheres by using acetic

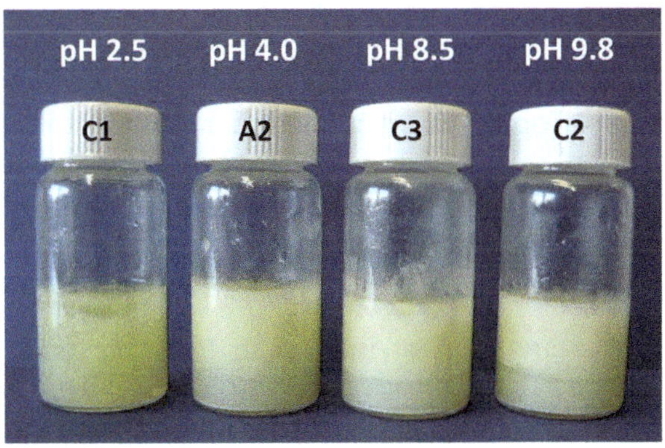

Figure 4.11 Emulsions stabilized by protein zein particles and the pH dependence of the emulsions.[51] Reproduced from ref. 51 with permission from The Royal Society of Chemistry.

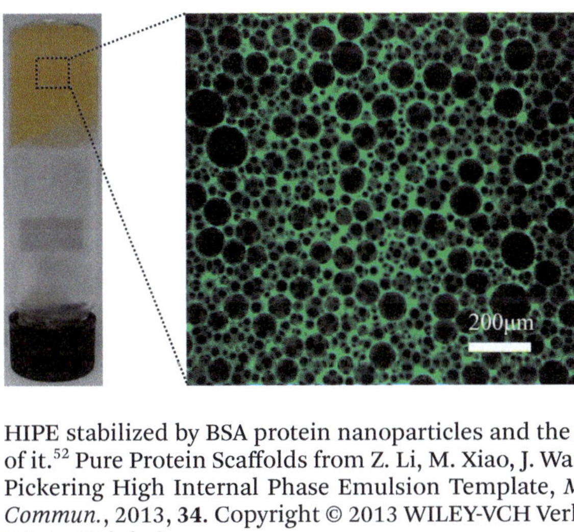

Figure 4.12 HIPE stabilized by BSA protein nanoparticles and the confocal image of it.[52] Pure Protein Scaffolds from Z. Li, M. Xiao, J. Wang and To Ngai, Pickering High Internal Phase Emulsion Template, *Macromol. Rapid Commun.*, 2013, **34**. Copyright © 2013 WILEY-VCH Verlag GmbH & Co. KGaA, Weinheim.

anhydride and phthalic anhydride.[53] Hydroxyl groups are present all over the starch surface. Then, carboxylic acid groups were added to the surface of the starch particles and they were used to stabilize Pickering emulsions. In addition, Yusoff *et al.* modified the surface of starch granules partially with octenyl succinic anhydride.[54] These hydrophobic molecules enhanced the hydrophobicity of the starch particles. The emulsions stabilized by their modified particles were still O/W. Nevertheless, the ability of particle adsorption was enhanced. Cellulose nanoparticles can also be applied in stabilizing Pickering emulsions. Kalashnikova *et al.* reported emulsions stabilized by cellulose nanocrystals.[55] The cellulose nanocrystals were prepared by hydrolysis of bacterial cellulose and they were called bacterial cellulose nanocrystals (BCNs). Similar to that of starch particles, cellulose particles are also covered by hydroxyl groups and are usually hydrophilic. Therefore, surface modifications of cellulose particles are conveniently accessible. Thermo-responsive cellulose nanocrystals were synthesized by Zoppe *et al.*[56] Thermo-responsive polymers of poly(*N*-isopropylacrylamide) brushes were grafted onto the cellulose nanocrystals.

4.3.3 Polymeric Particles (Synthetic) and Microgel Dispersions

Polymeric particles are of great interest in Pickering emulsions. This is because polymeric particles can be easily synthesized and modified to numerous characteristics. They can be hydrophilic particles such as poly(*N*-isopropylacrylamide) (PNIPAM). They can also be hydrophobic polystyrene particles (PS).[33] Colloidal-sized polymeric particles are usually synthesized by emulsion polymerization or suspension polymerization.[57–59] This is radical polymerization of unsaturated carbon compounds, which form colloidal-sized polymeric particles. However, surfactants are used in the syntheses. The stability of the particles at the early stage of the reaction is not high. Surfactants stabilize them from uncontrolled aggregation so that monodispersed particles can be obtained. Consequently, the surfactants stay with the final particles. Complete removal of these surfactant molecules from the particles is very difficult and time consuming, and yet these surfactants may sometimes be undesirable.[60] Therefore, surfactant-free emulsion polymerization (SFEP), also known as precipitation polymerization, has been developed. Polymeric particles prepared by precipitation polymerization do not suffer from the problem of surfactant removal. One well-known example of particles prepared by precipitation polymerization is PNIPAM microgel particles.[61] PNIPAM microgels are cross-linked gel particles, where their diameters are in micron or submicron range. They were first prepared by Pelton *et al.* in 1986.[62] The reaction was very simple. Only the monomer NIPAM, cross-linker *N*,*N*′-methylenebisacrylamide (MBA) and initiator (potassium persulfate, KPS) were added to the reaction mixture and then the temperature was maintained at around 70 °C. The diameters of the resulting particles were around 1 μm in the swollen state. At that time, the diameters of the swollen and dried particles were just compared under TEM. Indeed, PNIPAM microgel dispersion systems have been so popular in the recent decades because PNIPAM microgel particles undergo volume-phase transition at increasing temperature. Figure 4.13 shows a typical diameter dependence on temperature of a PNIPAM microgel sample.

When the temperature is at room temperature, which is lower than the lower critical solution temperature (LCST), the PNIPAM microgel is very hydrophilic. It is at its highly swollen state, which may have water content above 90%.[63] Then, the temperature is increased and the diameters of the microgel particles decrease gradually. However, when the temperature reaches the LCST of PNIPAM, the diameter is reduced sharply in a matter of a few degrees Celsius. This is because the particles become much less hydrophilic and they expel their water content.[64] The microgel shrinks because of the formation of intra-molecular hydrogen bonds and the loss of hydration with the methyl groups in the polymeric network. Lai *et al.* studied the mechanism of this volume-phase transition of the microgel by using infrared spectroscopy.[65] Amazingly, this transition often introduces a ten-fold shrinkage in volume of the particle.[62] Besides the changes in volume and surface area of the particles, the dangling chains also turn into globular conformation. Steric repulsion given by them will be greatly reduced. The contact angle of PNIPAM is also a function of temperature as it becomes much less hydrophilic at temperatures above its LCST (Figure 4.14).[66] Therefore, the sharp changes in microgel properties at LCST allow microgel dispersions to be a great choice of responsive particles, which can be used in preparing responsive Pickering emulsions.

The thermo-responsive property is definitely an important advantage of microgel dispersions. Nevertheless, the pH-responsive property of PNIPAM-based microgel gels is even more interesting. Originally, PNIPAM

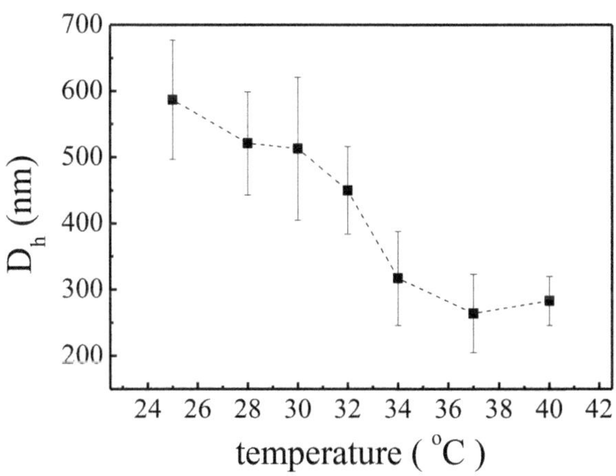

Figure 4.13 Dependence of hydrodynamic diameter on temperature for a PNIPAM microgel.

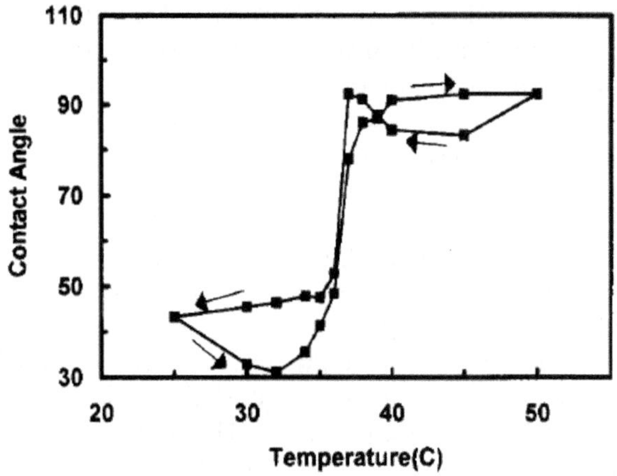

Figure 4.14 The relationship between contact angle of water on PNIPAM gel and temperature.[66] Reprinted (adapted) with permission from J. Jhang, R. Pelton and Y. L. Deng, *Langmuir*, 1995, **11**, 2301–2302. Copyright 1995 American Chemical Society.

polymers possess amide groups, which are not pH responsive at all. But an olefin monomer with ionizable function groups can easily co-polymerize with NIPAM. For examples, they are acrylic acid (AA), methacrylic acid (MAA),[60,67] vinylacetic acid (VAA),[68] 2-aminoethyl-methacrylate hydrochloride (AEM),[69] *etc.* After the co-polymerization, these ionizable functional groups are linked to the polymeric networks of the microgels. They usually are carboxylic acid or amine groups. Just like ordinary organic acids (or bases), they are weak acids (or bases). The degree of dissociation α depends on the pH value. At low pH values, the carboxylic groups are protonated and they deprotonate to carboxylate at high pH values. When they are deprotonated, counterions are introduced to the polymeric network. Therefore, the microgel particles become even more hydrophilic and the osmotic pressure increased by the ions draws in more water.

Figure 4.15 shows the pH dependence of pH-responsive microgel samples with different morphologies. The micron-sized microgels were prepared by semi-batched precipitation polymerization with temperature programmed technique.[60] This demonstrated the flexibility of polymeric particles syntheses. Figure 4.16 shows the pH and temperature dependence of a typical PNIPAM-*co*-MAA microgel. From Figure 4.16, it can be discovered that the thermo-responsive property is also a function of pH if carboxylic groups are co-polymerized.[70] Other than PNIPAM microgel particles, there are many more examples of responsive polymeric colloidal particles. They can be responsive polystyrene particles,[71,72] poly(methyl methacrylate) (PMMA) particles,[73] poly(vinylamine) microgels (which are highly swollen at low pH values rather than alkaline conditions).[74] The diversity and speciality of polymeric particles are just too many to be included in a few paragraphs.

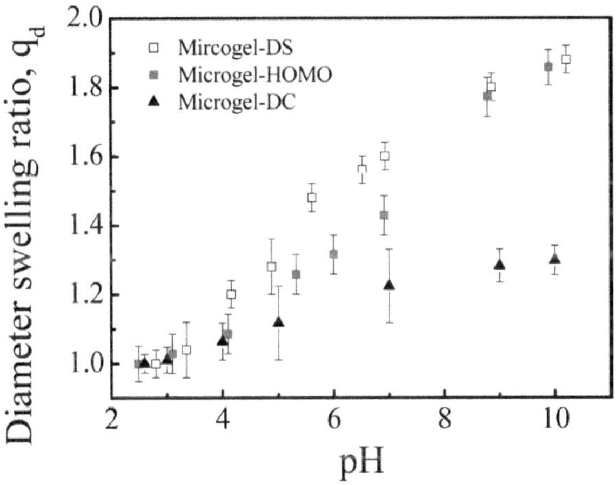

Figure 4.15 pH dependence of PNIPAM-*co*-MAA microgel samples with different morphologies.[60] Reprinted (adapted) with permission from M. H. Kwok, Z. F. Li and T. Ngai, *Langmuir*, 2013, **29**, 9581–9591. Copyright 2013 American Chemical Society.

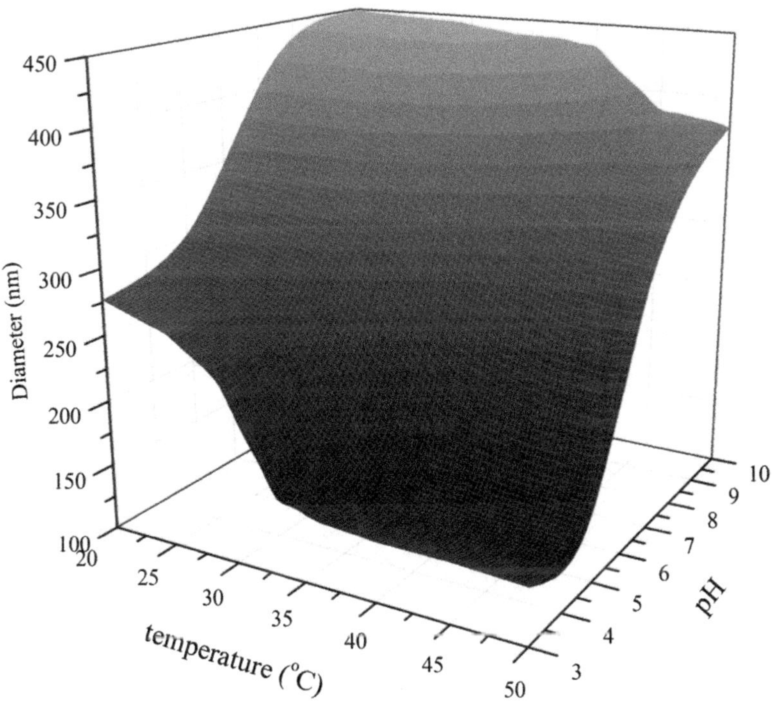

Figure 4.16 The pH and temperature dependence of a typical PNIPAM-*co*-MAA microgel. As the degree of ionization of the carboxylic groups increases, the LCST of the particle also increases.

4.3.4 Janus Particles

In recent years, Janus particles have been extensively studied and reported.[75,76] Janus particles are non-centrosymmetric and are different from homogeneous particles. Their surfaces can be divided into different parts with different chemical components, structures or polarity. The simplest case of a Janus particle can be considered to be a spherical particle divided into two equal parts. One of them is hydrophilic and the other part is hydrophobic. The wetting behaviours of these parts are highly diversified. They are also not limited by spherical shape. Certainly, the interfacial properties of Janus particles also are of great interest. This is because the attachment energy of Janus particles is usually much higher than that of homogeneous particles. The hydrophobic section and hydrophilic section of Janus particles are highly wetted by the oil phase and the aqueous phase respectively. One report even suggested that Janus particles may be capable of stabilizing thermodynamically stable emulsions.[77] The extraordinary high attachment energy might compensate the energy required to generate the emulsions. Therefore the free energy change of the whole process might be negative.

Figure 4.17 shows Pickering emulsions stabilized by Janus particles prepared by Jiang *et al.*[78] Their Janus particle samples were modified so that they have different fractions that are hydrophilic. The unmodified bare silica

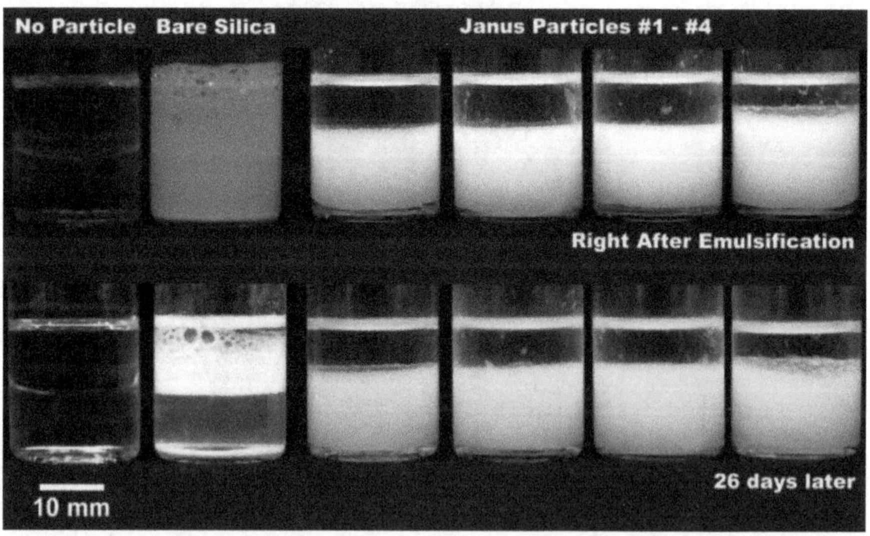

Figure 4.17 Comparison of stability of Pickering emulsions stabilized by Janus particles with different fractions of the hydrophilic part. The emulsion stabilized by the unmodified bare silica was also compared.[78] S. Jiang, Q. Chen, M. Tripathy, E. Luijten, K. S. Schweizer and S. Granickm, Janus Particle Synthesis and Assembly, *Adv. Mater.*, 2010, **22**. Copyright © 2010 WILEY-VCH Verlag GmbH & Co. KGaA, Weinheim.

particles were also used as particulate stabilizers and their emulsion stability was compared. After the emulsions were prepared for 26 days, the fraction of free oil, aqueous phase and the emulsion were analysed. Janus particles provided much better emulsion stability compared to the untreated bare silica particles. Also, Janus particles with more balanced hydrophilic and hydrophobic parts were better Pickering stabilizers.

Just like the biological and polymeric particles which were mentioned above, responsive Janus particles can also be synthesized. One interesting Janus particle dispersion was prepared by Tanaka *et al.*,[79] in which they prepared a mushroom-shaped Janus particle with both thermo- and pH-responsiveness (Figure 4.18). There were three types of polymer contained in one Janus particle. First, poly(methyl methacrylate)/poly(styrene-2-(2-bromoisobutyryloxy) ethyl methacrylate) (PMMA/PS-BIEM) Janus particles were prepared by common solvent evaporation. Then, the surface with PS-BIEM underwent surface-initiated atom transfer radical polymerization (ATRP) and 2-(dimethylamino)ethyl methacrylate (DM) was grafted onto the surface to form the mushroom head. The responsive properties of these interesting Janus particles will be discussed in the next section.

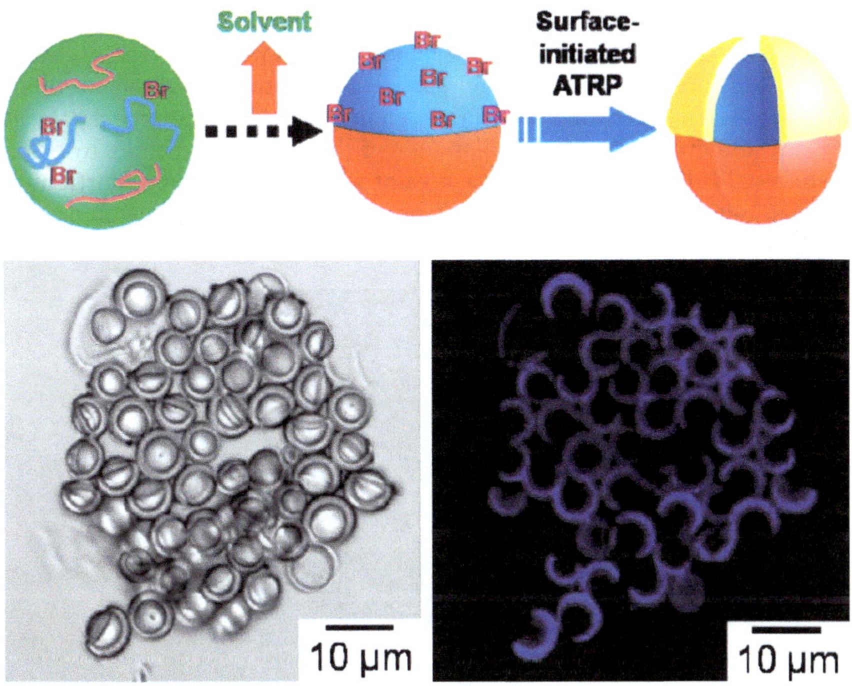

Figure 4.18 Images of mushroom-shaped Janus particles prepared by T. Tanaka *et al.*[79] Reprinted (adapted) with permission from T. Tanaka, M. Okayama, H. Minami and M. Okubo, *Langmuir*, 2010, **26**, 11732–11736. Copyright 2010 American Chemical Society.

4.4 Responsiveness of Emulsions

4.4.1 Thermal Stimulation

Responsive Pickering emulsions have drawn more and more attention since the report of a multi-responsive Pickering emulsion.[13] This was stabilized by pH and thermal responsive microgel particles. The first category of responsive emulsion in our discussion is thermo-responsive Pickering emulsions, which can be obtained by using thermo-responsive particulate stabilizers. The properties of these particles are very sensitive to changes in temperature. Therefore, their ability to stabilize emulsions also changes with the temperature.

In the previous sections, thermo-responsive particles were discussed. They are capable of being particulate stabilizers. Figure 4.19 shows emulsions stabilized by one of our microgel samples at different temperatures. The microgel particles were made of PNIPAM and 3% wt/wt *N*,*N*′-methylenebisacrylamide (MBA) was used as the cross-linking agent. The emulsion was stable at temperatures below the LCST of the PNIPAM polymer and did not undergo significant phase separation in at least six months, even though creaming occurred right after the agitation. However, the emulsion was unstable when the temperature was increased above the LCST. The mechanism of this thermo-responsiveness was studied by many different researchers.[12,80] First, the microgel particles expel theirs water content and shrink at high temperature.[64] Surface area provided by the microgel particles is reduced significantly. If the diameter of the microgel shrinks to half of the original value, then the surface area of microgel is just one-fourth of the original. It leads to an obvious consequence. Some of the emulsion droplets might be exposed instead of being covered by the particles. As a result, the

Figure 4.19 A typical thermo-responsive Pickering emulsion stabilized by PNIPAM microgel particles.

droplets will be vulnerable to coalescence when the droplets meet each other. Destribats *et al.* studied the relationship between the deformability of microgel particles and the resulting emulsion stability.[81] They found out that soft, more deformable microgel samples, which possessed less cross-linker content, are better stabilizers. Combining their observation under the cryogenic scanning electron microscopy (Cryo-SEM), they attributed the emulsion stability of microgel-stabilized Pickering emulsions to the lateral overlapping and interfacial elasticity. Therefore, it may also explain the reduction in stability of microgel-stabilized emulsion when the temperature is increased above the LCST. As the microgel particles shrink at increased temperature, the polymer chains in the polymeric network hold each other strongly and resist any deformation. Hashmi *et al.* reported that the fully shrunk microgel particle could be ten times stiffer than the fully swollen microgel particle.[82]

Eqn (4.3) in Section 4.2.2 predicts that the desorption energy of each particle decreases as the radius of it is reduced and increases as the contact angle approaches 90°. Nevertheless, changes in desorption energy might not be that significant in the discussion. This is because the desorption energy of individual particles is still in a way larger than the thermal energy. However, there is one more major factor which causes particle desorption – the aggregation of microgel particles. As the temperature increases above the LCST of PNIPAM, polymer–polymer interactions would be more favourable compared to polymer–solvent interactions. Then the microgel particles may coagulate and even form macroscopic aggregates. This actually disables the stabilization given by the microgel dispersion. Figure 4.20 shows a PNIPAM microgel dispersion at different temperatures. The microgel was a stable pale light dispersion at room temperature. When the temperature was increased to just above the LCST, the scattering intensity of the solution increased. As the temperature was further increased, the microgel particles finally aggregated to form observable aggregates.

Besides microgel particles, thermo-responsive composite particle dispersions are also popular choices for preparing responsive Pickering

Figure 4.20 The thermo-responsiveness of a typical PNIPAM microgel dispersion.

emulsions. Tsuji and Kawaguchi studied the thermo-sensitive Pickering emulsions with microgel particles and PNIPAM-carrying polystyrene particles.[83] Stability of emulsions prepared by different oils was compared at different temperatures. Their microgel particles were just synthesized by conventional precipitation polymerization with no special features. Nevertheless, they grafted PNIPAM polymer onto polystyrene particles. The hydrodynamic diameters of the two particles were similar and around 800 nm. It was discovered that the hairy responsive particles showed better thermo-responsiveness than the conventional microgel particles. Therefore, the emulsions stabilized by hairy responsive particles also demonstrated better thermo-responsiveness. The emulsions stabilized by conventional microgel particles only showed partial coalescence of droplets at elevated temperature. Figure 4.21 shows the thermo-responsiveness of emulsions stabilized by their hairy particles. The responsiveness of emulsions prepared by different oils was different. Nevertheless, all of them demonstrated the thermal-induced destabilization when the temperature was increased above the LCST of PNIPAM.

As mentioned in the previous section, polysaccharide, which is not thermo-responsive, can also be modified so that it can be used as a responsive particulate stabilizer. Zoppe *et al.* prepared cellulose nanocrystals grafted with thermo-responsive polymer brushes [poly(NIPAM)-*g*-CNCs].[56] Their cellulose nanocrystals were prepared from ramie fibres. The emulsions stabilized by these particles can be stable for more than four months, if the temperature was not raised above the LCST of PNIPAM. However, when the Pickering emulsions were heated to temperature above 35 °C for one minute, phase separation was observed for their emulsions. Grafted nanoparticle aggregation was also observed by them. Therefore, they attributed the destabilization to the desorption and aggregation of the modified nanoparticles.

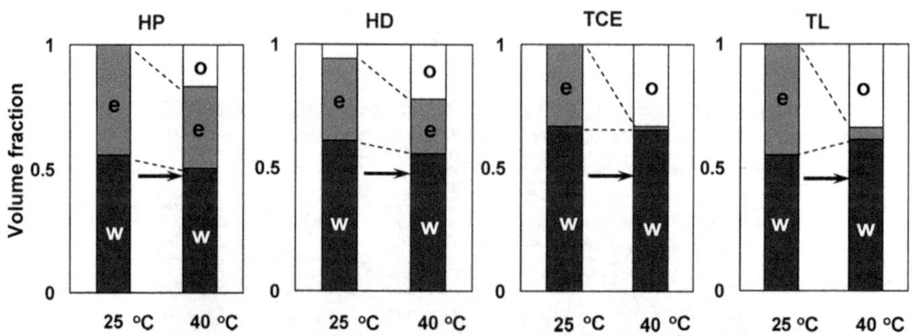

Figure 4.21 Thermo-responsive emulsions prepared by Tsuji and Kawaguchi.[83] Hairy PNIPAM grafted polystyrene particles were the particulate stabilizers. Different types of oil were used in their experiment. HP, HD, TCE and TL referred to heptane, hexadecane, trichloroethylene and toluene. Reprinted (adapted) with permission from S. Tsuji and H. Kawaguchi, *Langmuir*, 2008, **24**, 3300–3305. Copyright 2008 American Chemical Society.

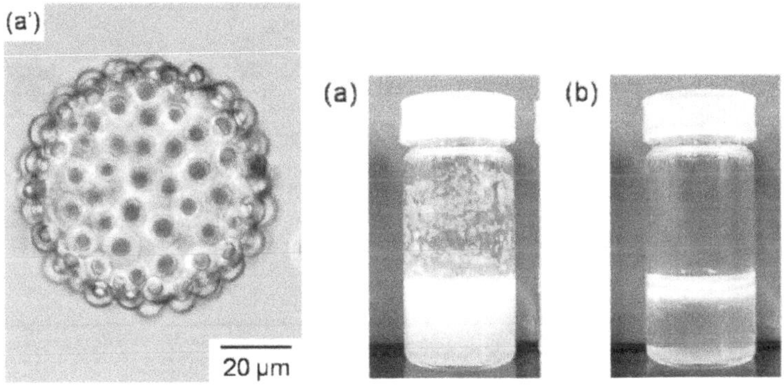

Figure 4.22 (a′) Stable 1-octanol/water Pickering emulsion droplet prepared by Tanaka *et al.* The particles were mushroom-shaped responsive Janus particles. (a) Emulsion formed at pH 7.2 and 25 °C. (b) Destabilization of emulsion at pH 7.2 and 60 °C.[79] Reprinted (adapted) with permission from T. Tanaka, M. Okayama, H. Minami and M. Okubo, *Langmuir*, 2010, **26**, 11732–11736. Copyright 2010 American Chemical Society.

In the previous section, Janus particles prepared by Tanaka *et al.* were discussed.[79] In Figure 4.22a, the microscope image of a 1-octanol droplet is shown. It was observed that the hydrophilic PDM grafted polymer faced the aqueous phase. At pH 7.2, a stable emulsion could be formed at room temperature. However, when the temperature was increased to 60 °C, the Janus particles desorbed and entered the oil phase. The emulsion underwent complete phase separation.

4.4.2 pH Stimulation

When the thermo-responsiveness of Pickering emulsions is discussed, pH responsiveness is often also included in the emulsions. This is because pH-responsive functional groups can be conveniently co-polymerized with the thermo-responsive PNIPAM. The pH responsiveness of microgel-stabilized emulsions is even more complicated. As the degree of ionization increases, the size, surface charge, hydrophilicity and deformability are also changed. Nevertheless, empirical results help us to understand more about this complicated problem.

Indeed, the responsive Pickering emulsions which were first reported by Ngai *et al.* responded to both pH and temperature changes.[13] It was discovered that octanol-in-water Pickering emulsions are stable in alkaline condition. Highly swollen charged microgel particles seemed to be a better particulate stabilizer for octanol-in-water Pickering emulsions. Figure 4.23 shows the qualitative relationship between temperature, pH and emulsion stability. When the pH value was under 3, the emulsion underwent phase separation with microgel particles likely left in the oil phase. This spontaneous phase separation of

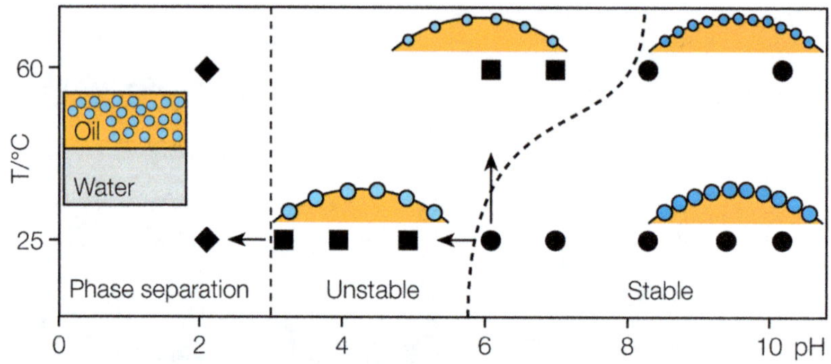

Figure 4.23 Stability of Pickering emulsion stabilized PNIPAM based multi-responsive microgel particles at different pH and temperatures.[13] Reproduced from ref. 13 with permission from The Royal Society of Chemistry.

the emulsion was a direct result of particle desorption into the octanol. More interestingly, at intermediate pH values, the temperature dependence of emulsion stability changed with the pH value. For example, at pH 6, the emulsion became unstable when the temperature was increased to around 35 °C. At pH 8, the emulsion was remained stable until the temperature was above 60 °C. These trends in emulsion stability were consistent with the properties of the microgel particles. In the discussion in Section 4.3, it was found that the LCST of multi-responsive PNIPAM-based microgel particles was dependent on the degree of ionization of the ionizable functional groups. If the degree of ionization of the carboxylic groups is higher, the microgel particles will collapse and shrink at a higher temperature. Moreover, the deprotonation of carboxylic acid groups also suppressed the thermo-responsiveness of the microgel. Therefore, the Pickering emulsions were stable at pH higher than 8, even when the temperature was increased. Therefore, this destabilization presumably originated from the shrinkage of particles.

The choice of the oil phase is also a crucial factor for responsive Pickering emulsions. In the above work done by Ngai *et al.*, a more polar oil phase was used. In some of our studies, emulsions consisting of a simple alkane, such as decane, were investigated. In addition, PNIPAM-*co*-MAA microgel particles with different morphologies were prepared for the study. Semi-batch syntheses were applied in order to obtain a loose core–dense shell (DS) and a loose shell–dense core (DC) microgel.[60,84] In acidic conditions, it was discovered that both DS and DC microgel particles did not enter the decane oil phase. Both of them were able to stabilize decane-in-water Pickering emulsions. Interestingly, emulsions could be formed in both acidic and alkaline conditions. Figure 4.24 shows the emulsions formed by DS and DC microgel particles at different pH values. It was found that the appearance of the alkaline DS microgel-stabilized emulsion was different from the others. Therefore, the maximum osmotic pressures of these emulsions were determined by centrifugation (Figure 4.24). The maximum osmotic pressure of the alkaline DS

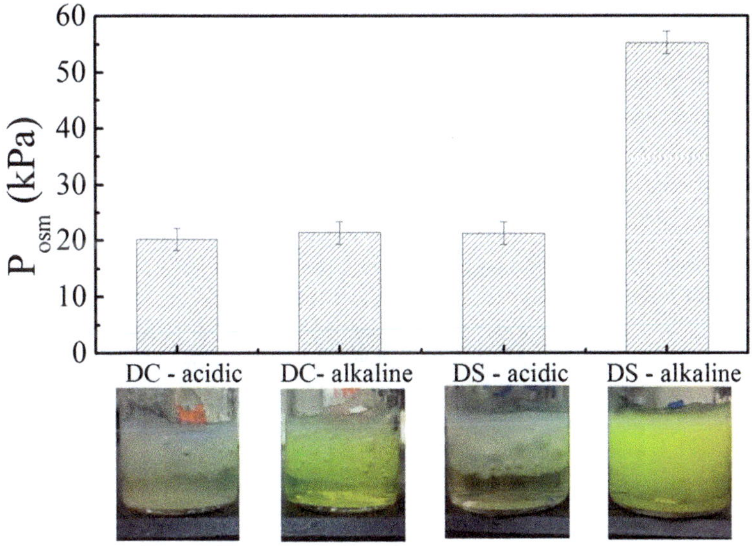

Figure 4.24 The appearances and the maximum osmotic pressures of emulsions stabilized by loose core–dense shell (DS) and a loose shell–dense core (DC) microgel at different pH values (pH 3 for acidic conditions and pH 10 for alkaline conditions).

microgel-stabilized emulsion was significantly higher than the others. This phenomenon should be the result of the charge distribution of the microgel particles. However, more detail is to be studied in the future.

Besides PNIPAM-based polymeric particles, there still are many choices for preparing pH-responsive Pickering emulsions. In addition, it seems that the demulsification mechanism of these emulsions would be quite different from the microgel-stabilized emulsions. Morse *et al.* have prepared novel pH-responsive poly(*tert*-butylaminoethyl methacrylate) (PTBAEMA) latexes.[85] They co-polymerized poly(ethylene glycol) methacrylate (PEGMA) macromonomer with the PTBAEMA particles. In contrast to PNIPAM-*co*-MAA microgel particles, these amine group-possessing particles swelled at low pH values and were relatively shrunk at alkaline conditions. This was because the amine groups were protonated and positively charged at reduced pH values. Co-polymerization with PEGMA significantly increased the steric stability of the resulting particles. They determined the dependence of zeta potential of the particles on pH values. At pH around 9, the zeta potential of the particles was close to zero. The electrostatic stabilization of the particles was very weak at this pH value. Therefore, co-polymerized PEGMA was expected to provide extra steric stabilization for their PTBAEMA particles. Both PEGMA-PTBAEMA and PTBAEMA particles were used to prepare Pickering emulsions with *n*-dodecane at pH around 10. Figure 4.25 shows the Pickering emulsions stabilized by different concentrations of PTBAEMA particles. At pH 10, stable emulsions could be formed at particle concentration as

Figure 4.25 Pickering emulsions stabilized by PTBAEMA particles at different particle concentrations. The destabilization of the emulsions was also demonstrated by acidification by hydrochloric acid.[85] Reprinted (adapted) with permission from A. J. Morse, D. Dupin, K. L. Thompson, S. Armes, K. Ouzineb, P. Mills and R. Swart, *Langmuir*, 2012, **28**, 11733–11744. Copyright 2012 American Chemical Society.

low as 0.2% wt/wt. However, after the addition of concentrated hydrochloric acid, immediate phase separation was observed. They attributed the destabilization of the emulsions to the desorption of charged, swollen PTBAEMA latexes. As the amine groups were protonated in acidic condition, the contact angle of the particles became too low. Therefore, the unprotected emulsion droplets coalesced immediately after the acidification. This phenomenon was not usually found in microgel dispersion system. At pH values causing the microgel to be in a swollen state (alkaline conditions for PNIPAM-*co*-MAA microgel), microgel particles were still able to be adsorbed to the interface and effectively stabilized a Pickering emulsion.

Charged-induced particle desorption seems to be a direct and fast method for emulsion destabilization in such pH-responsive Pickering emulsions.[87] Therefore, it is not difficult to find other studies which are based on this principle. Fujii *et al.* focused on silica nanocomposite microgel particles.[86] They prepared the nanocomposite particles by lightly cross-linking poly(4-vinylpyridine) (P4VP) and silica. P4VP was also a weak base which could be protonated in acidic conditions. Just like the previous example, this change in hydrophilicity caused particle desorption at low pH values. Consequently, the uncovered emulsion droplets underwent coalescence. Figure 4.26 shows the pH-triggered destabilization. Three different oils were used in the experiment and all of them showed great emulsion responsiveness to pH changes.

In another study which was also done by Morse *et al.*,[88] they prepared other acid-swelling latexes with poly[2-(diethylamino)ethyl methacrylate] (PDEA). The mechanism of the destabilization of this Pickering emulsion was just the same as the previous examples. Nevertheless, not only conventional acid and alkali were used to adjust the pH of the system; they also used CO_2/N_2 gas to demonstrate the pH changing cycles. In Figure 4.27b, the responsiveness was almost completely reversible when CO_2/N_2 gas were used. In contrast, the

Figure 4.26 pH-responsive Pickering emulsions prepared by Fujii *et al.*[86] Reprinted (adapted) with permission from S. Fujii, S. P. Armes, B. P. Binks and R. Murakami, *Langmuir*, 2006, **22**, 6818–6825. Copyright 2006 American Chemical Society.

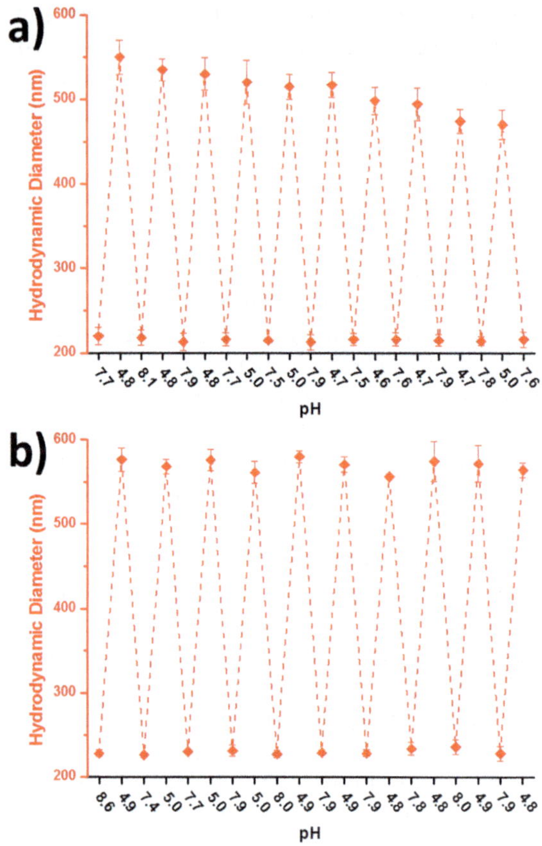

Figure 4.27 pH responsiveness of PEGMA–PDEA particles.[88] In (a), the pH was adjusted by HCl and KOH. In (b), the pH was changed by purging with CO_2 and N_2 gas. Reprinted (adapted) with permission from A. Morse, S. Armes, K. Thompson, D. Dupin, L. Fielding, P. Mills and R. Swart, *Langmuir*, 2013, **29**, 5466–5475. Copyright 2013 American Chemical Society.

acid-swollen diameters of the particles decreased after several cycles. This was attributed to the fact that the salt formed in the neutralization increased the ionic strength of the system. The reversibility of the responsive emulsion system is also an important factor in applications.

Besides destabilizing the emulsions, the type of emulsion can be controlled by changing the pH of the system. Lan *et al.* synthesized Fe_3O_4 nanoparticles which were chemically modified with oleic acid.[89] These modified Fe_3O_4 nanoparticles were capable of adsorbing another oleic acid layer at certain pH values. Therefore, the wettability of the modified Fe_3O_4 nanoparticles they prepared was a strong function of the pH value. In Figure 4.28, their stability measurements of the resulting emulsions are shown. At relatively low pH values, the oleic acid was protonated and the particles had a higher contact angle. The formation of W/O emulsions was

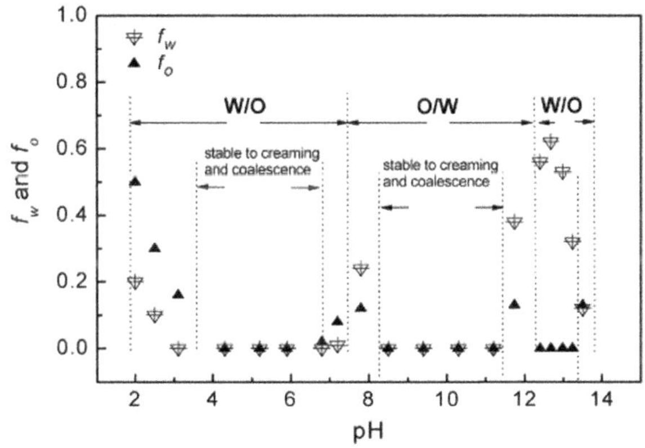

Figure 4.28 Fraction of free water and paraffin oil released three months after the emulsification. The emulsions were stabilized by modified Fe_2O_3 nanoparticles.[89] Reprinted from Q. Lan, C. Liu, F. Yang, S. Liu, J. Xu and D. Sun, Synthesis of bilayer oleic acid-coated Fe_3O_4 nanoparticles and their application in pH-responsive Pickering emulsions, *J. Colloid Interface Sci.*, 2007, **310**, 260–269, Copyright 2007, with permission from Elsevier.

favoured. At intermediate pH values, the oleic acid was ionized and the more hydrophilic particles favoured the stabilization of O/W emulsions. Their Pickering emulsions showed excellent emulsion stability even three months after the preparation. Li and Stöver also reported an interesting doubly pH-responsive Pickering emulsion stabilized by commercially available alumina-coated silica nanoparticles and hydrogen phthalate.[90] As the hydrogen phthalate was protonated, forming the mono-anion and di-anion at different pH values, it interacted with the silica particles differently. In their study, only when the phthalate was in its mono-anion form, could a stable emulsion be obtained.

4.4.3 Magnetic Stimulation

For industrial applications, sometimes it is not desirable to control the emulsion stability by temperature or pH. This is because this may take large amounts of chemicals or energy to achieve. Therefore, magnetic responsive emulsion stabilizers are introduced. The control of the Pickering emulsion can be easily achieved by turning on and off the magnetic field. As the syntheses of different kinds of particles have been well studied, the preparation of magnetic particles are also much more accessible than before. These magnetic particles can be modified, just like other types of particles, to form composite stabilizers.

Zhou *et al.* prepared a magnetic Pickering emulsion using superparamagnetic Fe_3O_4 nanoparticles.[91] It seems that Fe_3O_4 particles are a popular choice for magnetic nanoparticles. Fe_3O_4 was not an expensive material and the syntheses of such magnetic nanoparticles were very accessible. They prepared the Fe_3O_4 nanoparticles by conventional co-precipitation of iron(II) and iron(III) salts in alkaline conditions. The diameter distribution of the resulting nanoparticles was not very monodispersed. Nevertheless, these particles were still capable of stabilizing Pickering emulsions. As mentioned in the previous section, inorganic salts were usually hydrophilic; this was also true for the Fe_3O_4 nanoparticles. They investigated O/W Pickering emulsions with different kinds of oil. The Fe_3O_4 nanoparticles could stabilize O/W Pickering emulsions well when the oil was less polar. However, if the polarity of the oil was high, stable emulsions could not be formed using their Fe_3O_4 nanoparticles. In Figure 4.29, it shows that their particles were not effective in stabilizing butyl butyrate and decanol. The three phase contact angles were also studied for different oils. When oil with high polarity was considered, the contact angle was quite small. Therefore, stable O/W emulsions could not be formed. However, the magnetic properties of their magnetic Pickering emulsions were not described in detail in the study.

In other research done by Melle, Lask and Fuller, carbonyl iron particles were used to prepare magnetic responsive Pickering emulsions.[92] The diameters of the particles were around a few micrometres. In Figure 4.30, the effects of a static magnetic field are clearly shown. When the magnetic field was turned on, the oil droplets were attracted to the bottom of the vial. They

Figure 4.29 Emulsions prepared by Zhou *et al.* Different oil fractions were studied.[91] Also four oils were used in their experiment. They are (a) dodecane, (b) PDMS, (c) butyl butyrate and (d) decanol. Reprinted (adapted) with permission from J. Zhou, X. Qiao, B. P. Binks, K. Sun, M. Bai, Y. Li and Y. Liu, *Langmuir*, 2011, 27, 3308–3316. Copyright 2011 American Chemical Society.

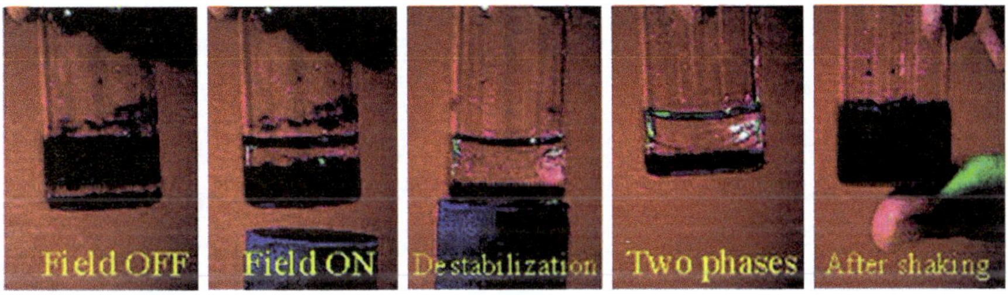

Figure 4.30 A magnetic responsive Pickering emulsion prepared by Melle *et al.*[92] By controlling the strength of the magnetic field, the emulsion droplets could be attracted and broken. Reprinted (adapted) with permission from S. Melle, M. Lask and G. G. Fuller, *Langmuir*, 2005, **21**, 2158–2162. Copyright 2005 American Chemical Society.

found that the critical field strength was a function of particle concentration. As the concentration increased, the magnetic field required to move the oil droplets was reduced. In addition, they demonstrated at stronger magnetic fields that the Pickering emulsion could be destabilized and undergo complete phase separation. In this case, the critical magnetic field strength was independent of the particle concentration. This can be understandable because the force required to pull the particles away from the interface should be the same for different particle concentrations. Moreover, they proved that this stabilization destabilization cycle was reversible by agitation. Therefore, we can see that the stabilization and destabilization of emulsions, separation and recycling of the magnetic particles could be easily controlled by a static magnetic field. These features are very desirable in industrial applications.

Brugger and Richtering have reported their stimuli-responsive particle emulsifiers.[93] Their stimuli-responsive particles were sensitive to magnetic and thermal stimuli. NIPAM, MAA and 2-hydroxyethylmethylacrylate (HEMA) were used to prepare the primary microgel. Then, iron(II,III) oxide magnetic nanoparticles were synthesized inside the microgel particles, which was previously reported by Zhang *et al.*[94] Many small magnetic nanoparticles with a diameter of 20 nm were formed within a larger microgel particle. In their study, the direct destabilization was not performed by applying a strong magnetic field. On the contrary, a moderate magnetic field was applied so that the emulsion droplets could be attracted and gathered by the field. More importantly, the field strength was not strong enough to cause coalescence and phase separation in their work.

In Figure 4.31, the capture of the emulsion droplets by a static magnetic field is demonstrated. When the static magnetic field was removed, the oil droplets were released and the density difference caused them to rise again. The size distribution of the Pickering emulsion was compared before and after the magnetic treatment. It was almost unchanged. This was because they would like to demonstrate the thermo-induced destabilization by induction heating rather than using a static magnetic field. A high frequency magnetic field was applied to the Pickering emulsion. Induction heating of the

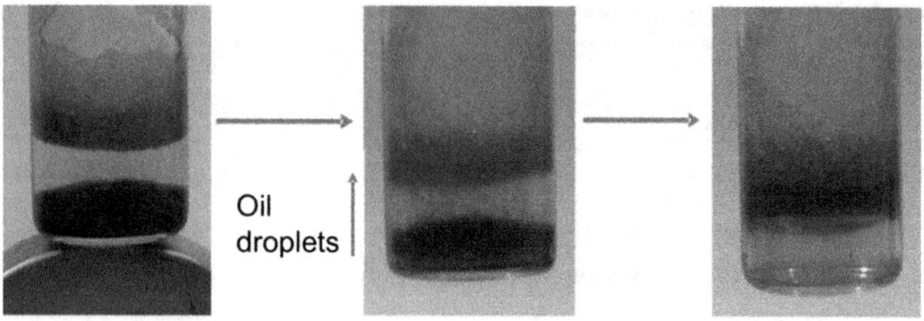

Figure 4.31 Static magnetic field could attract and gather the emulsion droplets without breaking the emulsion. After the static magnetic field was removed, the oil droplets were released.[93] B. Brugger and W. Richtering, Magnetic, Thermosensitive Microgels as Stimuli-Responsive Emulsifiers Allowing for Remote Control of Separability and Stability of Oil in Water-Emulsions, *Adv. Mater.*, 2007, **19**. Copyright © 2007 WILEY-VCH Verlag GmbH & Co. KGaA, Weinheim.

magnetic particles in the microgel increased the temperature of the emulsion. As the microgel stabilizers were thermo-responsive, the destabilization of the emulsion could be done by this remote controlled localized heating technique. It was a great example for combining the features of a thermo-responsive microgel with magnetic particles.

4.4.4 Other Stimulations

Besides the above common examples of responsive Pickering emulsions, some other innovative responsive systems have been reported in recent years. One of them consisted of an ion-specific polymer, poly(2-(methacryloyloxy)-ethyl-trimethyl-ammonium chloride) (PMETAC) and it was responsive specifically to perchlorate ions (ClO_4^-).[95]

Tan *et al.* have modified the surface of silica nanoparticles with this polymer.[96] Surface-initiated atom-transfer radical polymerization (ATRP) was applied to prepare the composite nanoparticle. The nanoparticles and their PMETAC polymer were well characterized by different methods, such as transmission electron microscopy (TEM), dynamic light scattering (DLS) and ellipsometry. The presence of perchlorate ions normally caused a hydrophobic-induced collapse of the PMETAC polymer brush and an increase in contact angle. In their study, when the concentration perchlorate was increased to above around 20 mM, particle aggregation was observed and it was dependent on the polymer chain length. Then, they used the nanoparticles as particulate stabilizers to prepare ion-responsive Pickering emulsions. Figure 4.32 shows their results of stabilizing dodecane-in-water Pickering emulsions with their composite particles. No stable emulsion were formed when the concentration of perchlorate ions was lower then 20 mM. Different

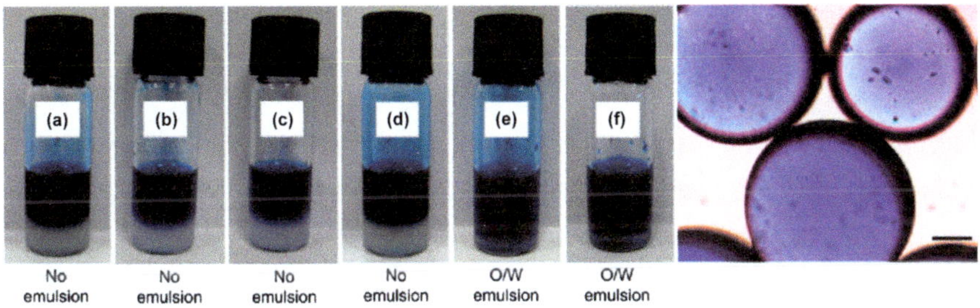

Figure 4.32 Pickering emulsions prepared by Tan *et al.*[96] using ion-responsive PMETAC-grafted silica nanoparticles. Particles were in (a) deionized water, (b) 18.5 mM NaCl, (c) 100 mM NaCl, (d) 10 mM $NaClO_4$, (e) 40 mM $NaClO_4$, (f) 40 mM $NaClO_4$, but the photo was taken after 60 h. Reprinted (adapted) with permission from K. Y. Tan, J. E. Gautrot and W. T. Huck, *Langmuir*, 2011, **27**, 1251–1259. Copyright 2011 American Chemical Society.

concentrations of NaCl were also tested in order to show that the particles were not quite responsive to Cl^- anions. When the concentration of perchlorate was 40 mM, shown in Figure 4.32e and f, a stable emulsion could be formed. After 30 h, they found that the emulsion was still stable. They measured that there was a minor decrease in emulsion volume fraction 60 h after the agitation. Therefore, one can expect that dilution of the originally stable emulsion may cause it to be destabilized.

Similar to magnetic-responsive emulsions, photo-responsive emulsions can also be controlled conveniently. This is because no chemicals are required to be added to the system. Takahashi *et al.* successfully demonstrated a photo-induced demulsification of emulsions.[98] They called their surfactant "Photo-responsive gemini surfactant". There was a famous photo-sensitive azobenzene centre in their surfactant. Therefore, UV light was capable of flipping the azobenzene from *trans* to *cis*. This UV light-induced flipping caused the stable O/W emulsion to be destabilized and undergo phase separation. Although this example does not concern a Pickering emulsion, the photo-responsive property was very interesting and future designs of particulate stabilizers can also refer to such a system.

On the other hand, Chen *et al.* have reported the light-controlled inversion of Pickering emulsions.[97] The wettability of their nanoparticles was controlled by spiropyran. Spiropyran is an interesting type of organic compound because it is photochromic. UV light can change the hydrophobic spiropyran to hydrophilic merocyanine and visible light is capable of reversing the change. Also, lanthanide-doped up-conversion nanoparticles (UCNPs) were used in their work. UV light emission of a large anti-Stokes shift can be obtained when they are excited by near-infrared radiation. All of these features were combined on silica nanoparticles. Therefore, instead of irradiating the system with UV light, near-infrared could also be used to trigger the Pickering emulsion inversion. In Figure 4.33, it could be found that the particles were hydrophobic and the contact angle was slightly higher

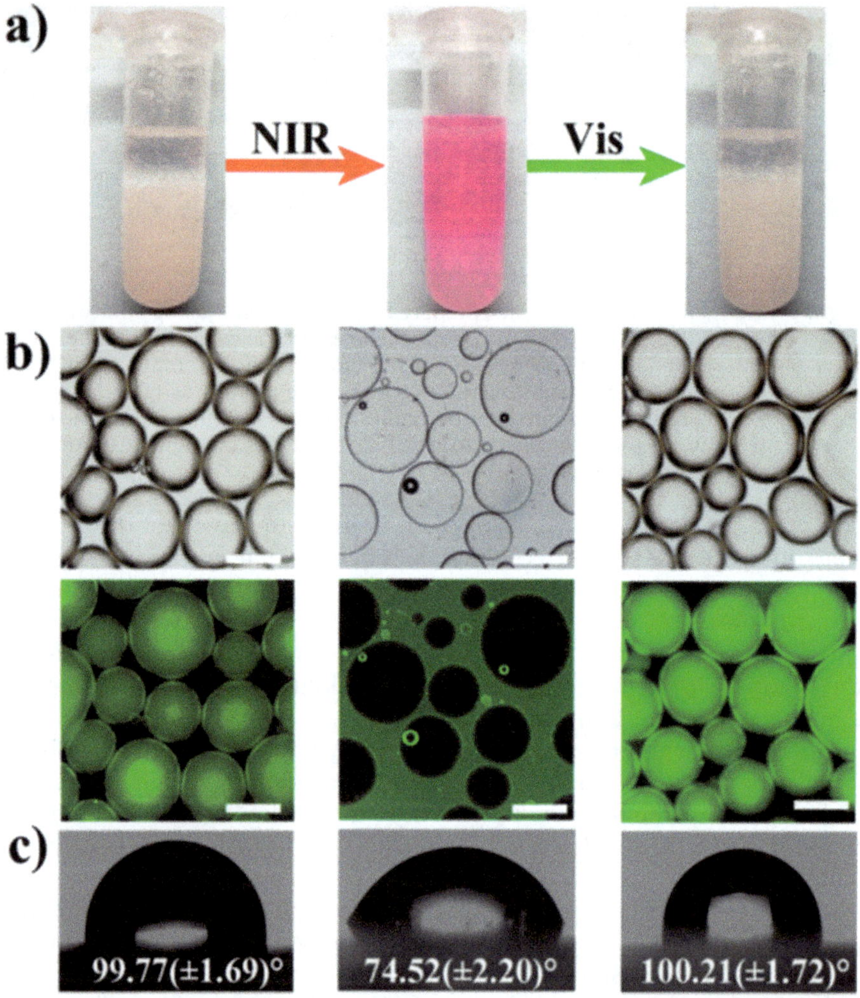

Figure 4.33 Photo-responsive Pickering emulsions stabilized by light-sensitive nanoparticles. The wettability of the nanoparticles could be changed by near-infrared and visible light.[97] Reprinted (adapted) with permission from Z. Chen, L. Zhou, W. Bing, Z. Zhang, Z. Li, J. Ren and X. Qu, *J. Am. Chem. Soc.*, 2014, **136**, 7498–7504. Copyright 2014 American Chemical Society.

than 90°. After exposure to near-infrared radiation, the colour and the wettability of the particles changed. In this situation, O/W Pickering emulsions were formed instead. After that, the merocyanine could be changed back to spiropyran by visible light. The application of such a responsive Pickering emulsion will be discussed in the next section. Although there are not many studies about photo-responsive Pickering emulsions reported, the potential offered by such systems should not be overlooked.

4.5 Applications

4.5.1 Pharmaceutical Applications

Pickering emulsion systems have attracted much attention in recent years. Pickering emulsions show advantages when they are compared with conventional emulsions stabilized by low molecular weight surfactants or amphiphilic polymers. For example, the stability of Pickering emulsions is usually much higher, even at high ionic strength. Also, the solid particles surrounding the emulsion droplets act as a physical barrier, which can be used to control the transfer of substances.[99] Haase *et al.* have demonstrated the encapsulation of amphoteric substances using responsive Pickering emulsions.[100] O/W Pickering emulsions were stabilized by their silica particles together with 8-hydroxyquinoline (8-HQ). Stable emulsions could only be prepared in a small pH range, which controlled the protonation states of 8-HQ. They also characterized the release kinetics of 8-HQ from the emulsions. It was found that the kinetics depended on the pH value. A faster release was obtained at pH 4.5 when compared to that at pH 5.9. They attributed the difference in the kinetics to the available interfacial area and the partition of 8-HQ in the two phases at different pH values.

To demonstrate the controlled release of encapsulated substances, Frasch-Melnik *et al.* prepared interesting fat-crystal stabilized W/O responsive Pickering emulsions.[101] They used a mixture of monoglyceride and triglyceride crystals to stabilize their Pickering emulsions. Sodium chloride was used by them to demonstrate the controlled release of encapsulated materials. Differential scanning calorimetry (DSC) was also utilized to study the thermal behaviour of the fat crystal. The fat crystals melted in a large temperature range from around 15 °C to 50 °C. As the fat crystal melted, the stability of the Pickering emulsion decreased. The solid particles were important in stabilizing the emulsion droplets at the interface. Therefore, the rate of salt release was increased at the higher temperature. They also observed a rapid destabilization of the emulsion at 30 °C, which was near the melting-peak temperature of the fat crystal. The rates of salt release were determined at different temperatures. Their results were summarized in Table 4.1. At 5 °C, it took a very long period of time to release 5% of the

Table 4.1 Time required to release 5% of total salt from the emulsion prepared by Frasch-Melnik *et al.* at different temperatures.[101] 0.5% monoglyceride and 1% tripalmitin were used as the stabilizers. Reprinted from S. Frasch-Melnik, I. T. Norton and F. Spyropoulos, Fat-crystal stabilised w/o emulsions for controlled salt release, *J. Food Eng.*, **98**, 437–442, Copyright 2010, with permission from Elsevier.

Temperature (°C)	Time
5	>100 days
22	70 days
30	<1 h
50	20 s

salt. At this temperature, the salt was securely encapsulated in the Pickering emulsion. When temperature was increased to 22 °C, the encapsulation was still quite secure. However, when the temperature was increased to 30 °C, the rate of salt release increased rapidly. It took less than 1 hour to release 5% of the salt. Finally, when the temperature was increased to 55 °C, complete release of salt could be finished in around 200 seconds.

4.5.2 Petroleum Industry

In the petroleum industry, emulsion science is always related to the processing of oil. For example, this includes reservoir flooding, oil recovery, oil transportation, *etc.* Some surface-active molecules like naphthenic acids and asphaltenes exist naturally in crude oil.[102] They are capable of forming very stable water-in-oil emulsions. There are ionic salts dissolved in the water-in-oil emulsions, which damage the facilities for oil processing. On the other hand, reservoir flooding processes also rely on emulsion science. Conventionally, surfactants are added to the water for reservoir flooding processes. Surfactants can effectively lower the interfacial tension between the flooding solution and crude oil.[103] The efficiency of crude oil extraction from a reservoir can be enhanced. However, after extracting the oil from the ground, the removal and recycling of these surfactants is not easy and it requires a lot of effort. During the oil recovery process, the emulsions are required to be collected and broken. By considering these features, it seems that responsive Pickering emulsion systems will be suitable for such applications.

Wang and co-workers have developed a recyclable pH-responsive magnetic nanoparticle system.[104] For their particles, the preliminary magnetic cores were surrounded by silica shells. Then, the pH-responsive polymer of poly(dimethylaminoethyl methacrylate) (PDMAEMA) was grown on the silica surface. These particles were used in an oil–water separation process. Figure 4.34 shows the isolation of diesel oil from a diesel-in-water emulsion. First, their hybrid magnetic nanoparticles were added to a diesel-in-water emulsion. Then, the magnetic nanoparticles were adsorbed onto the oil–water interface. They could be collected by a static magnetic field. Therefore, the clear aqueous phase was removed separately. In addition, they recycled the magnetic nanoparticles and repeated this separation six times. This demonstration suggested that such a Pickering emulsion system was capable of being applied in oil recovery processes in petroleum industry.

Besides crude oil extraction and recovery, emulsions can also be applied in heavy crude oil transportation.[105] In order to transport crude oil by pipe lines, the viscosity of the crude oil must be lowered under a certain value. The formation of crude oil-in-water emulsions can achieve this objective. However, properties of such emulsions are also very important. For example, the stability of the emulsions is required to be high in both dynamic and static conditions. In addition, after the transportation, the emulsions can be conveniently broken before entering the refinery facilities. Therefore, the excellent stability of Pickering emulsions makes it a nice choice for this

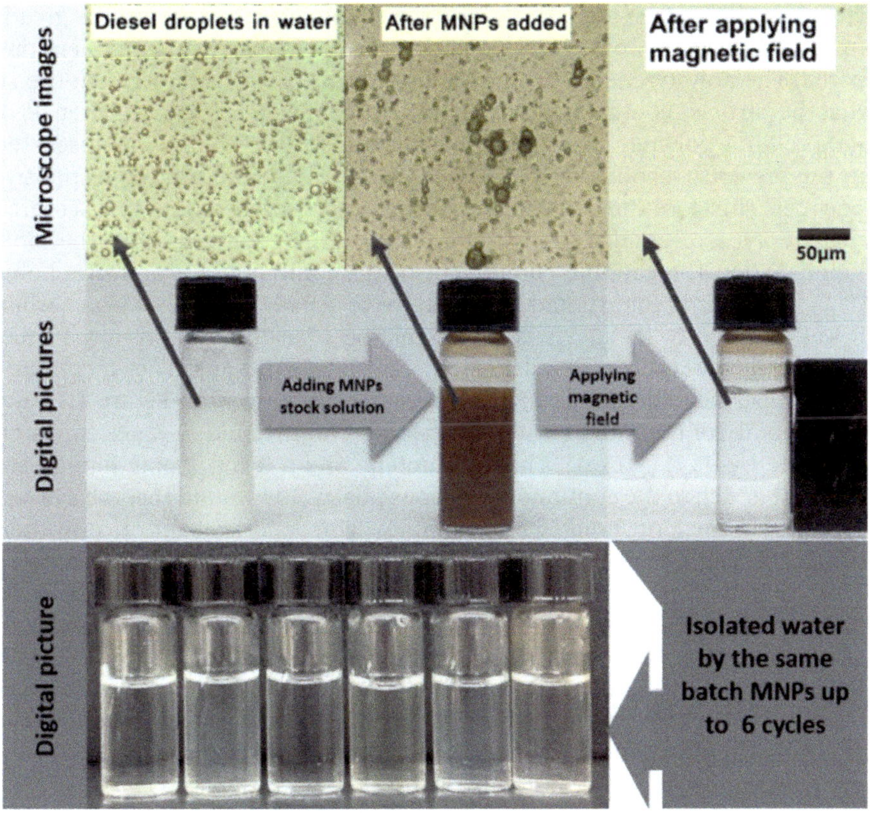

Figure 4.34 Schematic illustration of the oil–water separation process using magnetic nanoparticles prepared by Wang *et al.*[104] Reprinted from X. Wang, Y. Shi, R. W. Graff, D. Lee and H. Gao, Developing recyclable pH-responsive magnetic nanoparticles for oil-water separation, *Polymer*, 2015, online published, Copyright 2015, with permission from Elsevier.

application. Moreover, the destabilization of the responsive Pickering emulsions can easily be triggered after the transportation.

4.5.3 Extraction

Responsive Pickering emulsions are gaining attractiveness to be used in extraction. This is because they have excellent stability compared to conventional emulsions and can be demulsified on demand. The extraction content can be conveniently collected by destabilizing the emulsion while the remaining stabilizers can be reused.

Monteillet *et al.* have developed a novel system which applies ionic liquid emulsions in extraction.[106] Ionic liquids are often considered to be green extraction solvents for extracting proteins, lipids and other molecules. The use of emulsions is desirable in such applications because emulsions offer a very large specific surface area. Therefore, the extraction efficiency can be greatly improved. In their work, PNIPAM/polystyrene composite particles were prepared. The polystyrene (PS) cores were synthesized with a fluorescent dye so that the particles could be easily seen under a confocal microscope. Then, PNIPAM-*co*-MAA polymer shell was polymerized onto the PS core. As discussed in the previous sections, PNIPAM-*co*-MAA was a pH- and thermo-responsive polymer. This gave the composite particles responsiveness. It was discovered that the type of emulsion was dependent on the composition of the ionic liquid. When the cationic component of the ionic liquid was hydrophobic $[P^{+}_{6,6,6,14}]$, water-in-ionic liquid emulsions were usually formed. However, when it was replaced by $[N^{+}_{1,8-10,8-10,8-10}]$, the microgel particles stayed in the aqueous phase. The resulting emulsion was ionic liquid-in-water. They mentioned that the permeability of the interface was very important. In Figure 4.35, the permeability of the interface was tested by an ionic liquid-in-water emulsion prepared by $[P_{6,6,6,14}][DCA]$. Then, β-carotene and a hydrophobic fluorescent dye were added to the continuous aqueous phase. They found that the dye was extracted to the ionic liquid within minutes. This was also confirmed by their confocal images, which are shown in Figure 4.35. The red dye was extracted into the ionic liquid even when the interface was fully covered by the composite microgel particles. In some of the previous examples, the collecting of emulsion droplets was demonstrated by applying a mild static magnetic field. In this study, instead of using magnetic particles, the ionic liquid itself was chosen to be paramagnetic, where $[P_{6,6,6,14}][FeCl_4]$ was used. Therefore, the emulsion droplets could be simply collected by a state magnetic field. Finally, the temperature-induced destabilization of the emulsions was also successfully demonstrated by them. When the temperature was increased above the LCST of PNIPAM, the emulsion was broken as expected.

Another example was reported by Chen *et al.*[107] The particles prepared by them were core–shell particles with magnetic Fe_3O_4 cores and PNIPAM shells. Their responsive Pickering emulsion system was designed for effective oil harvesting. In Figure 4.36, the responsiveness of the emulsions is shown. The core–shell particles were capable of stabilizing toluene-in-water emulsions. These oil droplets could be collected by a magnetic, just like the other examples which were discussed previously. Then, the destabilization of the emulsions was achieved by increasing the temperature above the LCST of PNIPAM. Their system showed potential applications in oil spill accidents and as a treatment for industrial oily wastewater.

4.5.4 Catalysis

In the previous section, a photo-responsive Pickering emulsion prepared by Chen *et al.* was discussed.[97] However, they also performed a biphasic enantioselective biocatalysis with their emulsion. They converted (*R*,*S*)-mandelonitrile to *R*-(−)-mandelic acid by *A. faecalis* ATCC 8750. (*R*,*S*)-Mandelonitrile was more soluble in the oil phase while *R*-(−)-mandelic acid and *A. faecalis* ATCC 8750 remained in the aqueous phase. Conventionally, monophasic or biphasic approaches were often considered

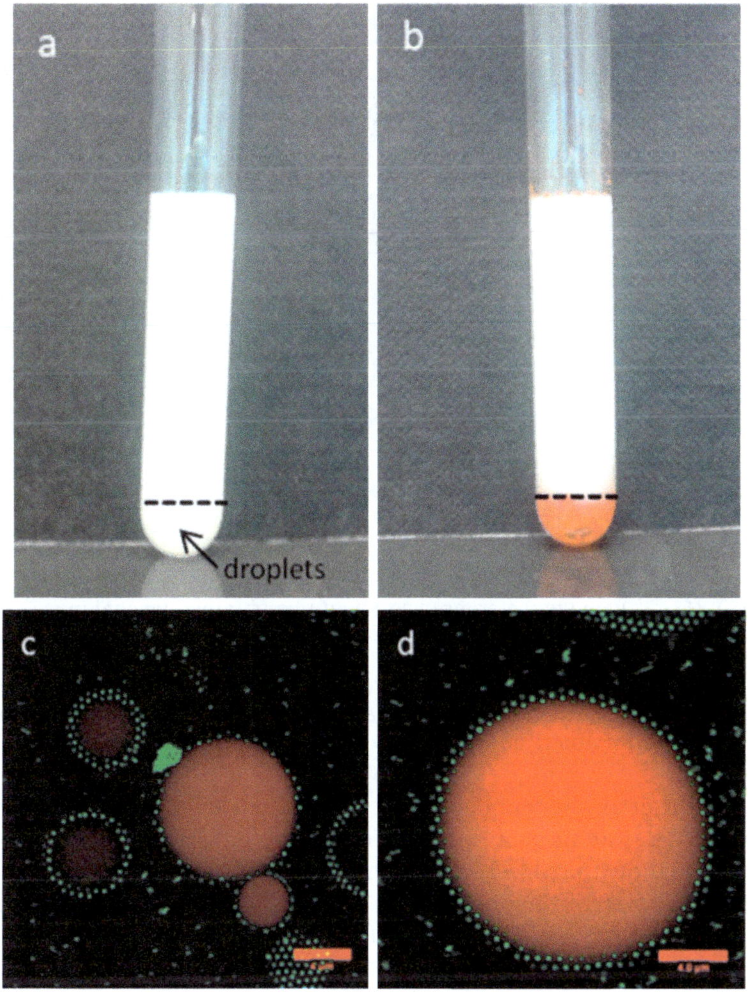

Figure 4.35 (a) Photos of $[P_{6,6,6,14}]$[DCA] in a water emulsion stabilized by composite microgel particles.[106] (b) β-Carotene was adsorbed by the ionic liquid. (c) and (d) were the confocal images of the emulsion droplets. Reproduced from ref. 106 with permission from The Royal Society of Chemistry.

for the reaction. However, there were disadvantages for both monophasic and biphasic reactions. In monophasic reactions, the concentrations of the reactants were limited because of the low solubility. On the other hand, the reaction rate was limited in biphasic reactions as the catalyst and the reactants stayed in two different phases. The transportation of materials to different phases limited the rate of reaction.

Therefore, Chen and co-workers applied their responsive Pickering emulsions to the hydrolysis reaction. For this biphasic system, the interfacial area was greatly increased because of the formation of emulsion droplets. Figure 4.37 compares the rates of conversion in different conditions. It was

Figure 4.36 Toluene-in-water emulsion could be obtained by shaking. The emulsion droplets could then be collected by a magnet. A thermal triggered demulsification was also demonstrated.[107] Reprinted (adapted) with permission from Y. Chen, Y. Bai, S. Chen, J. Ju, Y. Li, T. Wang and Q. Wang, *ACS Appl. Mater. Interfaces*, 2014, **6**, 13334–13338. Copyright 2014 American Chemical Society.

discovered that at low mandelonitrile concentration (50 mM), monophasic and emulsion systems shared similar efficiency. However, when the concentration of mandelonitrile was increased to 100 mM and 300 mM, the Pickering emulsion system was much more desirable than the other two systems for the hydrolysis. The rate of conversion could be greatly improved. Furthermore, the inversion capability of the responsive emulsion benefited the product separation after the hydrolysis. After the reaction, near-infrared radiation could change the emulsions from W/O to O/W. The Pickering emulsion stabilizers and biocatalyst could be recovered. In addition, the use of near-infrared instead of UV radiation for the inversion did less damage to the biocatalyst. Therefore, the efficiency of the biocatalyst did not deteriorate quickly. The application of responsive Pickering emulsions in catalysis was also reported by many different researchers.[4,108,109]

4.5.5 Pickering Emulsion Polymerization

Emulsion polymerization is a well-known technique for preparing polymeric colloidal particles.[110] Conventionally, surfactant molecules are added to the reaction mixture in order to provide stabilization to the newly formed colloidal particles. The application of Pickering emulsions in emulsion polymerization has attracted more and more attention. These surfactant-free

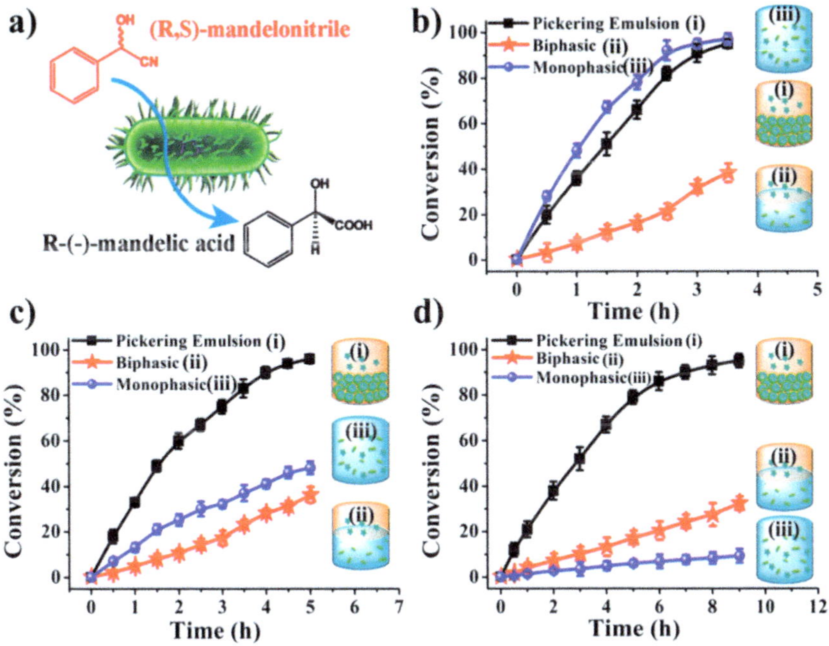

Figure 4.37 The rate of conversion of (*R*,*S*)-mandelonitrile by *A. faecalis* ATCC 8750 at different mandelonitrile concentrations. (a) shows the schematic illustration of the reaction. The concentrations in (b)–(d) were 50 mM, 100 mM and 300 mM respectively.[97] Reprinted (adapted) with permission from Z. Chen, L. Zhou, W. Bing, Z. Zhang, Z. Li, J. Ren and X. Qu, *J. Am. Chem. Soc.*, 2014, **136**, 7498–7504. Copyright 2014 American Chemical Society.

emulsions are considered to be more environmentally friendly as they are less toxic and the stabilizing particles are easily recyclable. Colloidal particles can often be used as an antifoaming agent.[111] Foam reduction is also an advantage of using particulate stabilizers.

Wei *et al.* used lignin extracted from furfural residues in emulsion polymerization of polystyrene.[112] Figure 4.38 showed the cycle of using lignin particles as particulate stabilizers. First, lignin was dissolved in an alkaline solution. Then, lignin colloidal particles could be formed by addition of acid. These particles act as the emulsion stabilizers in the emulsion polymerization of styrene. After the polymerization, alkali was added and the lignin became soluble again and the whole process could be repeated for next batch of styrene. The recycling of stabilizer is not easy for a conventional surfactant approach.

4.6 Concluding Remarks

Finally, a long time after the discovery of Pickering emulsions, more and more attention has been placed on the development of innovative Pickering emulsion systems. Certainly, responsive Pickering emulsions

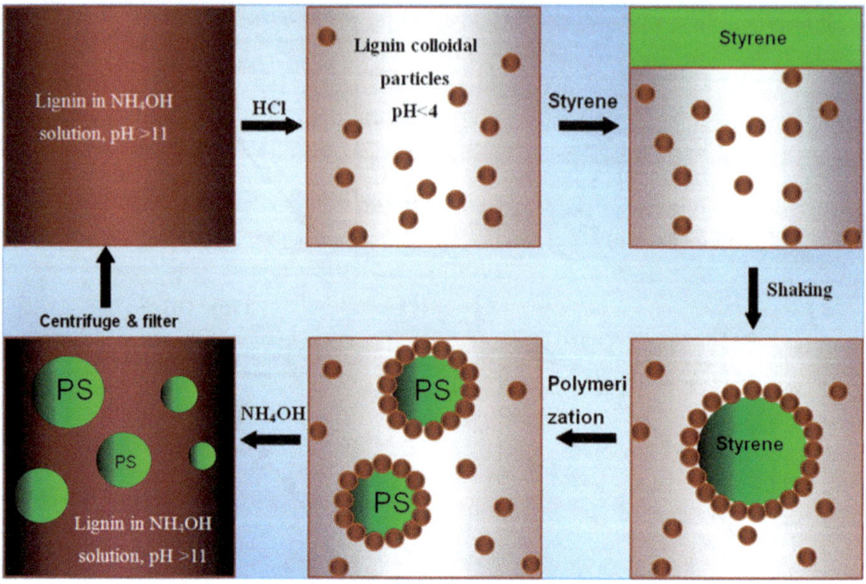

Figure 4.38 Schematic illustration of the polymerization of PS particles using lignin colloidal particles as emulsion stabilizers.[112] Reproduced from ref. 112 with permission from The Royal Society of Chemistry.

are a most interesting system. The potential applications of such emulsions can extensively reach different areas, such as drug delivery systems, crude oil transportation, *etc.* However, there might be problems which are required to be solved beforehand. For example, the large industrial-scale production of particulate stabilizers with good qualities is very important for them to be used in different applications. Nevertheless, as more resources are being poured into the related research, the future developments of responsive Pickering emulsions are exciting. Their potential cannot be overlooked.

Acknowledgements

The financial support of this work by the National Nature Science Foundation of China (B040606, 21374091) is gratefully acknowledged.

References

1. T. Takino, K. Konishi, Y. Takakura and M. Hashida, *Biol. Pharm. Bull.*, 1994, **17**, 121–125.
2. M. Santander, R. T. Rodrigues and J. Rubio, *Colloids Surf., A*, 2011, **375**, 237–244.
3. K. Hofstetter, J. Lutz, I. Lang, B. Witholt and A. Schmid, *Angew. Chem., Int. Ed.*, 2004, **43**, 2163–2166.

4. S. Wiese, A. C. Spiess and W. Richtering, *Angew. Chem., Int. Ed.*, 2013, **52**, 576–579.
5. D. G. Dalgleish, *Food Hydrocolloids*, 2006, **20**, 415–422.
6. D. G. Dalgleish, *Trends Food Sci. Technol.*, 1997, **8**, 1–6.
7. *Physical Chemistry of Surfaces*, ed. A. W. Adamson and A. P. Gast, Wiley-Interscience, Canada, 1997.
8. B. P. Binks, *Modern Aspects of Emulsion Science*, The Royal Society of Chemistry, Cambridge, 1990.
9. S. U. Pickering, *J. Chem. Soc., Trans.*, 1907, **91**, 2001–2021.
10. W. Ramsden, *Proc. R. Soc. London*, 1903, **72**, 156–164.
11. B. P. Binks, *Curr. Opin. Colloid Interface Sci.*, 2002, **7**, 21–41.
12. T. Ngai, H. Auweter and S. H. Behrens, *Macromolecules*, 2006, **39**, 8171–8177.
13. T. Ngai, S. H. Behrens and H. Auweter, *Chemical communications*, Cambridge, England, 2005, pp. 331–333.
14. T. N. Hunter, R. J. Pugh, G. V. Franks and G. J. Jameson, *Adv. Colloid Interface Sci.*, 2008, **137**, 57–81.
15. S.-M. Yang, S.-H. Kim, J.-M. Lim and G.-R. Yi, *J. Mater. Chem.*, 2008, **18**, 2177–2190.
16. S. Simovic and C. A. Prestidge, *Langmuir*, 2004, **20**, 8357–8365.
17. R. J. Hunter, *Foundations of Colloids Science*, Oxford University Press, 1987.
18. Z. Li, K. Geisel, W. Richtering and T. Ngai, *Soft Matter*, 2013, **9**, 9939–9946.
19. B. P. Binks and J. H. Clint, *Langmuir*, 2002, **18**, 1270–1273.
20. B. P. Binks and S. O. Lumsdon, *Langmuir*, 2000, **16**, 8622–8631.
21. Y. Nonomura and N. Kobayashi, *J. Colloid Interface Sci.*, 2009, **330**, 463–466.
22. S. A. F. Bon, in *Particle-Stabilized Emulsions and Colloids: Formation and Application*, ed. T. Ngai and S. A. F. Bon, The Royal Society of Chemistry, 2014, pp. 1–7.
23. M. F. Hsu, M. G. Nikolaides, A. D. Dinsmore, A. R. Bausch, V. D. Gordon, X. Chen, J. W. Hutchinson and D. A. Weitz, *Langmuir*, 2005, **21**, 2963–2970.
24. R. Aveyard, J. H. Clint, D. Nees and N. Quirke, *Langmuir*, 2000, **16**, 8820–8828.
25. C. E. Reese, C. D. Guerrero, J. M. Weissman, K. Lee and S. A. Asher, *J. Colloid Interface Sci.*, 2000, **232**, 76–80.
26. P. A. Kralchevsky and K. Nagayama, *Adv. Colloid Interface Sci.*, 2000, **85**, 145–192.
27. P. A. Kralchevsky, N. D. Denkov and K. D. Danov, *Langmuir*, 2001, **17**, 7694–7705.
28. N. D. Denkov, I. B. Ivanov, P. A. Kralchevsky and D. T. Wasan, *J. Colloid Interface Sci.*, 1992, **150**, 589–593.
29. T. S. Horozov, R. Aveyard, J. H. Clint and B. Neumann, *Langmuir*, 2005, **21**, 2330–2341.
30. G. Kaptay, *Colloids Surf., A*, 2006, **282**, 387–401.
31. S. Tcholakova, N. D. Denkov, I. B. Ivanov and B. Campbell, *Langmuir*, 2002, **18**, 8960–8971.

32. A. San-Miguel and S. H. Behrens, *Langmuir*, 2012, **28**, 12038–12043.
33. R. Aveyard, B. P. Binks and J. H. Clint, *Adv. Colloid Interface Sci.*, 2003, **100**, 503–546.
34. J. H. Schulman and J. Leja, *Trans. Faraday Soc.*, 1954, **50**, 598–605.
35. N. P. Ashby and B. P. Binks, *Phys. Chem. Chem. Phys.*, 2000, **2**, 5640–5646.
36. S. Guillot, F. Bergaya, C. de Azevedo, F. Warmont and J.-F. Tranchant, *J. Colloid Interface Sci.*, 2009, **333**, 563–569.
37. A. Schlaepfer, *J. Chem. Soc.*, 1918, **113**, 522.
38. J. Frelichowska, M.-A. Bolzinger and Y. Chevalier, *Colloids Surf., A*, 2009, **343**, 70–74.
39. Q. Lan, C. Liu, F. Yang, S. Liu, J. Xu and D. Sun, *J. Colloid Interface Sci.*, 2007, **310**, 260–269.
40. R. I. Nooney, D. Thirunavukkarasu, Y. M. Chen, R. Josephs and A. E. Ostafin, *Chem. Mater.*, 2002, **14**, 4721–4728.
41. K. S. Rao, K. El-Hami, T. Kodaki, K. Matsushige and K. Makino, *J. Colloid Interface Sci.*, 2005, **289**, 125–131.
42. I. Blute, R. J. Pugh, J. van de Pas and I. Callaghan, *J. Colloid Interface Sci.*, 2007, **313**, 645–655.
43. C. Perruchot, M. A. Khan, A. Kamitsi, S. P. Armes, T. von Werne and T. E. Patten, *Langmuir*, 2001, **17**, 4479–4481.
44. S. Fujii, A. Aichi, M. Muraoka, N. Kishimoto, K. Iwahori, Y. Nakamura and I. Yamashita, *J. Colloid Interface Sci.*, 2009, **338**, 222–228.
45. P. Wongkongkatep, K. Manopwisedjaroen, P. Tiposoth, S. Archakunakorn, T. Pongtharangkul, M. Suphantharika, K. Honda, I. Hamachi and J. Wongkongkatep, *Langmuir*, 2012, **28**, 5729–5736.
46. L. S. Dorobantu, A. K. Yeung, J. M. Foght and M. R. Gray, *Appl. Environ. Microbiol.*, 2004, **70**, 6333–6336.
47. J. T. Russell, Y. Lin, A. Böker, L. Su, P. Carl, H. Zettl, J. He, K. Sill, R. Tangirala and T. Emrick, *Angew. Chem., Int. Ed.*, 2005, **44**, 2420–2426.
48. G. Kaur, J. He, J. Xu, S. Pingali, G. Jutz, A. Böker, Z. Niu, T. Li, D. Rawlinson and T. Emrick, *Langmuir*, 2009, **25**, 5168–5176.
49. M. V. Tzoumaki, T. Moschakis, V. Kiosseoglou and C. G. Biliaderis, *Food Hydrocolloids*, 2011, **25**, 1521–1529.
50. A. Timgren, M. Rayner, M. Sjöö and P. Dejmek, *Procedia Food Sci.*, 2011, **1**, 95–103.
51. J. W. de Folter, M. W. van Ruijven and K. P. Velikov, *Soft Matter*, 2012, **8**, 6807–6815.
52. Z. Li, M. Xiao, J. Wang and T. Ngai, *Macromol. Rapid Commun.*, 2013, **34**, 169–174.
53. Y. Tan, K. Xu, C. Liu, Y. Li, C. Lu and P. Wang, *Carbohydr. Polym.*, 2012, **88**, 1358–1363.
54. A. Yusoff and B. S. Murray, *Food Hydrocolloids*, 2011, **25**, 42–55.
55. I. Kalashnikova, H. Bizot, B. Cathala and I. Capron, *Langmuir*, 2011, **27**, 7471–7479.
56. J. O. Zoppe, R. A. Venditti and O. J. Rojas, *J. Colloid Interface Sci.*, 2012, **369**, 202–209.
57. W. V. Smith and R. H. Ewart, *J. Chem. Phys.*, 1948, **16**, 592.

58. C. S. Chern, *Prog. Polym. Sci.*, 2006, **31**, 443–486.
59. C. K. Ober, K. P. Lok and M. L. Hair, *J. Polym. Sci., Part C: Polym. Lett.*, 1985, **23**, 103–108.
60. M. H. Kwok, Z. F. Li and T. Ngai, *Langmuir*, 2013, **29**, 9581–9591.
61. B. R. Saunders and B. Vincent, *Adv. Colloid Interface Sci.*, 1999, **80**, 1–25.
62. R. H. Pelton and P. Chibante, *Colloids Surf.*, 1986, **20**, 247–256.
63. R. Pelton, *Macromol. Symp.*, 2004, **207**, 57–65.
64. R. Pelton, *Adv. Colloid Interface Sci.*, 2000, **85**, 1–33.
65. H. Lai and P. Wu, *Polymer*, 2010, **51**, 1404–1412.
66. J. Jhang, R. Pelton and Y. L. Deng, *Langmuir*, 1995, **11**, 2301–2302.
67. Z. Li, M.-H. Kwok and T. Ngai, *Macromol. Rapid Commun.*, 2012, **33**, 419–425.
68. T. Hoare and R. Pelton, *Macromolecules*, 2004, **37**, 2544–2550.
69. F. Meunier, A. Elaissari and C. Pichot, *Polym. Adv. Technol.*, 1995, **6**, 489–496.
70. J. D. Debord and L. A. Lyon, *Langmuir*, 2003, **19**, 7662–7664.
71. J. I. Amalvy, S. P. Armes, B. P. Binks, J. A. Rodrigues and G. F. Unali, *Chem. Commun.*, 2003, 1826–1827.
72. J. I. Amalvy, G. F. Unali, Y. Li, S. Granger-Bevan, S. P. Armes, B. P. Binks, J. A. Rodrigues and C. P. Whitby, *Langmuir*, 2004, **20**, 4345–4354.
73. B. H. Tan, K. C. Tam, Y. C. Lam and C. B. Tan, *Polymer*, 2005, **46**, 10066–10076.
74. S. Thaiboonrod, C. Berkland, A. H. Milani, R. Ulijn and B. R. Saunders, *Soft Matter*, 2013, **9**, 3920–3930.
75. A. Walther and A. H. Müller, *Soft Matter*, 2008, **4**, 663–668.
76. N. Glaser, D. J. Adams, A. Böker and G. Krausch, *Langmuir*, 2006, **22**, 5227–5229.
77. R. Aveyard, *Soft Matter*, 2012, **8**, 5233–5240.
78. S. Jiang, Q. Chen, M. Tripathy, E. Luijten, K. S. Schweizer and S. Granick, *Adv. Mater.*, 2010, **22**, 1060–1071.
79. T. Tanaka, M. Okayama, H. Minami and M. Okubo, *Langmuir*, 2010, **26**, 11732–11736.
80. B. Brugger, B. A. Rosen and W. Richtering, *Langmuir*, 2008, **24**, 12202–12208.
81. M. Destribats, V. Lapeyre, M. Wolfs, E. Sellier, F. Leal-Calderon, V. Ravaine and V. Schmitt, *Soft Matter*, 2011, **7**, 7689–7698.
82. M. H. Sara and R. D. Eric, *Soft Matter*, 2009, **5**, 3682–3688.
83. S. Tsuji and H. Kawaguchi, *Langmuir*, 2008, **24**, 3300–3305.
84. P. Sheikholeslami, C. M. Ewaschuk, S. U. Ahmed, B. A. Greenlay and T. Hoare, *Colloid Polym. Sci.*, 2012, 1181–1192.
85. A. J. Morse, D. Dupin, K. L. Thompson, S. Armes, K. Ouzineb, P. Mills and R. Swart, *Langmuir*, 2012, **28**, 11733–11744.
86. S. Fujii, S. P. Armes, B. P. Binks and R. Murakami, *Langmuir*, 2006, **22**, 6818–6825.
87. J. Amalvy, S. Armes, B. Binks, J. Rodrigues and G. Unali, *Chem. Commun.*, 2003, 1826–1827.
88. A. Morse, S. Armes, K. Thompson, D. Dupin, L. Fielding, P. Mills and R. Swart, *Langmuir*, 2013, **29**, 5466–5475.

89. Q. Lan, C. Liu, F. Yang, S. Liu, J. Xu and D. Sun, *J. Colloid Interface Sci.*, 2007, **310**, 260–269.
90. J. Li and H. D. Stöver, *Langmuir*, 2008, **24**, 13237–13240.
91. J. Zhou, X. Qiao, B. P. Binks, K. Sun, M. Bai, Y. Li and Y. Liu, *Langmuir*, 2011, **27**, 3308–3316.
92. S. Melle, M. Lask and G. G. Fuller, *Langmuir*, 2005, **21**, 2158–2162.
93. B. Brugger and W. Richtering, *Adv. Mater.*, 2007, **19**, 2973–2978.
94. J. Zhang, S. Xu and E. Kumacheva, *J. Am. Chem. Soc.*, 2004, **126**, 7908–7914.
95. O. Azzaroni, A. A. Brown and W. T. Huck, *Adv. Mater.*, 2007, **19**, 151–154.
96. K. Y. Tan, J. E. Gautrot and W. T. Huck, *Langmuir*, 2010, **27**, 1251–1259.
97. Z. Chen, L. Zhou, W. Bing, Z. Zhang, Z. Li, J. Ren and X. Qu, *J. Am. Chem. Soc.*, 2014, **136**, 7498–7504.
98. Y. Takahashi, K. Fukuyasu, T. Horiuchi, Y. Kondo and P. Stroeve, *Langmuir*, 2013, **30**, 41–47.
99. J. Frelichowska, M.-A. Bolzinger, J.-P. Valour, H. Mouaziz, J. Pelletier and Y. Chevalier, *Int. J. Pharm.*, 2009, **368**, 7–15.
100. M. F. Haase, D. Grigoriev, H. Moehwald, B. Tiersch and D. G. Shchukin, *J. Phys. Chem. C*, 2010, **114**, 17304–17310.
101. S. Frasch-Melnik, I. T. Norton and F. Spyropoulos, *J. Food Eng.*, 2010, **98**, 437–442.
102. L. Y. Zhang, R. Lopetinsky, Z. Xu and J. H. Masliyah, *Energy Fuels*, 2005, **19**, 1330–1336.
103. J. Tang, P. J. Quinlan and K. C. Tam, *Soft Matter*, 2015, **11**, 3512–3529.
104. X. Wang, Y. Shi, R. W. Graff, D. Lee and H. Gao, *Polymer*, 2015, **72**, 361–367.
105. D. Langevin, S. Poteau, I. Hénaut and J. Argillier, *Oil Gas Sci. Technol.*, 2004, **59**, 511–521.
106. H. Monteillet, M. Workamp, X. Li, B. Schuur, J. M. Kleijn, F. A. Leermakers and J. Sprakel, *Chem. Commun.*, 2014, **50**, 12197–12200.
107. Y. Chen, Y. Bai, S. Chen, J. Ju, Y. Li, T. Wang and Q. Wang, *ACS Appl. Mater. Interfaces*, 2014, **6**, 13334–13338.
108. Z. Fang, D. Yang, Y. Gao and H. Li, *Colloid Polym. Sci.*, 2015, **293**, 1505–1513.
109. H. Yang, T. Zhou and W. Zhang, *Angew. Chem.*, 2013, **125**, 7603–7607.
110. C. Chern, *Prog. Polym. Sci.*, 2006, **31**, 443–486.
111. G. C. Frye and J. C. Berg, *J. Colloid Interface Sci.*, 1989, **127**, 222–238.
112. Z. Wei, Y. Yang, R. Yang and C. Wang, *Green Chem.*, 2012, **14**, 3230–3236.

CHAPTER 5

Intrinsic Self-Healing Polymeric Materials for Engineering and Environmental Applications

LIN LI[a], JINGSI CHEN[a], BIN YAN[a], AND HONGBO ZENG*[a]

[a]Department of Chemical and Materials Engineering, University of Alberta, Edmonton, AB T6G 2V4, Canada
*E-mail: hongbo.zeng@ualberta.ca

5.1 Introduction

The ability of self-healing and regeneration of function upon inflicted damage, such as the healing of bone fractures and the closure of injured blood vessels, are pervasive in biological systems while rare in man-made materials.[1] During the last decade self-healing has enjoyed great popularity in materials science because it can provide reduced material damage during general usage, reduced replacement costs, and improved product safety, especially for applications located in poorly accessible areas that demand long-term reliability. Polymers are by far the mostly studied material class in the context of self-healing behavior due to the facile functionalization and modification of polymeric systems. Generally the self-healing polymeric materials can be classified into two kinds: to achieve healing extrinsically based on external healing components or intrinsically by reversible bond formation.[2] As an example, the self-healing process can be achieved extrinsically by pre-embedded micro- or nanocapsules in the polymer matrix. Upon crack intrusion, the

RSC Smart Materials No. 20
Smart Materials for Advanced Environmental Applications
Edited by Peng Wang

Published by the Royal Society of Chemistry, www.rsc.org

healing agents are released from the embedded capsules and polymerized to bond the crack faces.[3] However, the depletion of the local healing agent after damage-triggered release results in only one singular healing event, suggesting an inability to repeatedly heal the polymer at the same location.[2] This limitation has motivated the development of vascular self-healing materials, mimicking the capillary network of the dermis layer in animals. Healing agents are contained in hollow channels in the polymer matrix and could be refilled from an interconnected undamaged region of the vasculature or from an external source by human intervention.[4] Despite the considerable progress gained in extrinsic self-healing polymeric materials, developing intrinsic self-healing polymeric materials has become increasingly appealing as they require no additional healing agents and the healing event can be repeatedly conducted even after multiple damage.[5]

This chapter reviews recent advances on the intrinsic self-healing polymeric materials, starting with a general introduction of various intrinsic self-healing mechanisms in Section 5.2. Section 5.3 gives a brief review on the self-healing polymeric materials inspired by mussel threads, and case studies of using self-healing polymeric materials for environmental applications are summarized in Section 5.4.

5.2 Self-Healing Polymeric Materials *via* Reversible Bond Formation

Unlike "extrinsic" self-healing based on a pre-embedded healing agent, "intrinsic" self-healing is achieved through reversible bonds, either covalent or non-covalent. These polymeric materials with various functionalities have the ability to repair themselves autonomously or upon an external stimulus *via* bond cleavage and reformation.[6] In this section, several selected examples of dynamic covalent bonding are highlighted in the first part, and the second part focuses on the supramolecular chemistry mechanisms, which include hydrogen bonding, ionic interactions, metal–ligand coordination, and π–π stacking.[7]

5.2.1 Self-Healing Polymeric Materials *via* Dynamic Covalent Bonding

Due to their high bonding strength, covalent polymers compose the majority of polymeric structural engineering materials.[8] Generally, covalent bonds are irreversible, and traditional covalent polymers are permanently damaged by the formation and propagation of cracks.[9] Recently, reversible covalent linkages have been exploited to achieve self-healing by reforming new bonds around the damaged zone.[10] Various self-healing polymeric materials have been synthesized *via* dynamic covalent bonds,[11] including Diels–Alder (DA) reactions, disulfide bonds, reversible radical reactions, and photocycloaddition.

The Diels–Alder (DA) reaction is a convenient route for cross-linking polymers through the formation of C–C bonds.[12,13] The thermo-reversibility of the DA reaction offers opportunity for polymers to achieve repeated healing.[14] Wudl *et al.* developed highly cross-linked polymeric materials with furan (diene) and maleimide (dienophile) *via* a DA reaction.[15,16] As shown in Figure 5.1(A), the "intermonomer" linkages could be cleaved (retro-DA reaction) upon heating or stress, and at lower temperatures the covalent bonds could be reconstructed to repair the crack.[17] This process was fully reversible under mild conditions, and the cross-linked polymers were tough solids with mechanical properties comparable to commercial epoxy resins for engineering applications. The concept of a reversible DA reaction has been used to develop several self-healing polymers, such as polyepoxides,[18] polylamides,[19] and polyacrylates.[20] Besides the homogeneous bulk polymers, Peterson *et al.* utilized this concept to build reversible interfaces in composites.[21] The DA reaction between a furan-functionalized epoxy-amine matrix and a maleimide-functionalized glass fiber was used to afford healing capacity at the polymer–glass interface, extending the fatigue life of glass fiber-reinforced composites for potential applications.[22]

Disulfide bonds can undergo exchange reactions, where two neighboring S–S bonds are disrupted and reformed at moderate temperatures,[23] as depicted in Figure 5.1(B-1), leading to the reconnection of cross-links across

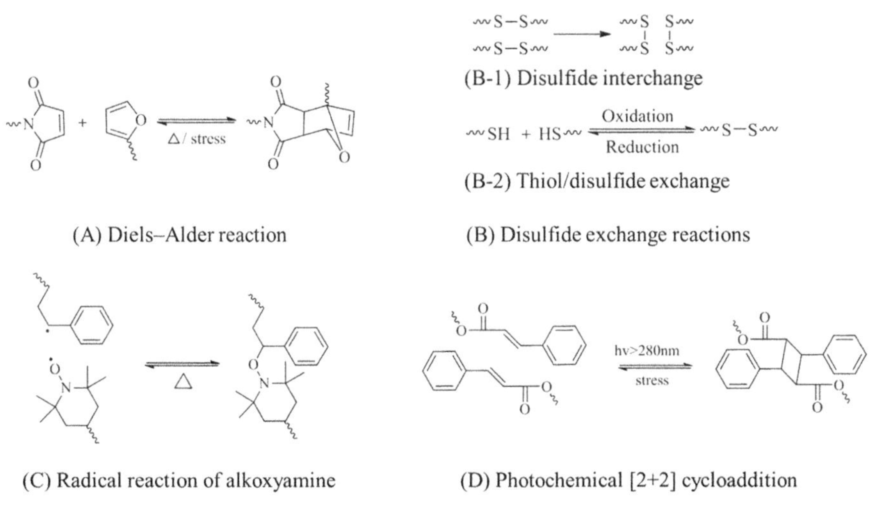

Figure 5.1 Self-healing reactions *via* dynamic covalent bonding. (A) Diel–Alder reaction.[15] (B) Disulfide exchange reactions.[24,29] ((B-1), reproduced with permission from ref. 24, copyright 2011 American Chemical Society; (B-2), reproduced with permission from ref. 29, copyright 2011 American Chemical Society). (C) Reversible radical reaction of alkoxyamine.[31] (Reproduced with permission from ref. 31, copyright 2011 American Chemical Society.) (D) Photochemical cycloaddition of TCE.[33] (Reproduced with permission from ref. 33, copyright 2004 American Chemical Society.)

the damaged area. Canadell *et al.* incorporated disulfide linkages into epoxy resins to construct a self-healing rubber, which was able to recover the mechanical properties at 60 °C.[24] This approach is applicable to other polymers with low glass transition temperature (T_g), such as polyurethanes[25] and polyesters,[26] since chain mobility is essential for the exchange reaction to take place. This approach is also ideal for developing self-healing coatings with corresponding applications of the polymers. As illustrated in Figure 5.1(B-2), previous studies have suggested that S–S bonds can be cleaved by reduction to form two thiol (S–H) groups and reconnected by oxidation.[27,28] Yoon and co-workers prepared polymer gels from star polymers with reversible disulfide cross-links at the periphery of branches.[29] The cross-linked star polymer films showed a rapidly spontaneous self-healing behavior from cuts micromachined with the atomic force microscopy (AFM) tip at room temperature, as studied by AFM imaging. The self-healing ability was attributed to the regeneration of S–S bonds *via* thiol/disulfide exchange reactions and the healing efficiency was strongly dependent on the film thickness and the width of the damaged area.

Free radicals are commonly the most reactive species resulting from bond cleavage.[30] As shown in Figure 5.1(C), dynamically reversible C–ON bonds in alkoxyamine moieties were employed to cross-link polystyrene (PS) backbones, where covalent bond dissociation and radical recombination took place synchronously upon heating.[31] Self-healing of cracks was achieved with efficiencies of 65–76%, without losing integrity and load bearing ability to meet practical application requirements. Moreover, the healing temperature could be modulated within a wide range by tuning molecular structure of alkoxyamines.

In addition to thermally reversible reactions, photo-reversible reactions are also important and commonly used to realize self-healing of polymers, which would be more advantageous over other approaches since light is a remote stimulus and can be controlled with great ease of operation.[32] Photochemical [2 + 2] cycloaddition of cinnamate monomer 1,1,1-tris-(cinnamoyloxymethyl)ethane (TCE) was utilized to create self-healing polymers *via* reversible conversion between cyclobutane and original cinnamoyl groups.[33] As shown in Figure 5.1(D), TCE can cross-link to form a hard solid upon ultraviolet (UV) irradiation at $\lambda > 280$ nm. When the polymer suffers an impact, C–C bond cleavage of cyclobutane rings results in the formation of cinnamoyl groups, and the healing process occurs due to the recovery of the cross-linked networks upon UV exposure. Similarly, coumarin was introduced into polyurethane as cross-links because of its reversible photodimerization and photocleavage characteristics.[34] In this case, the cleavage of coumarin dimers can be achieved by mechanical damage or 254 nm radiation, while reconnection requires UV irradiation at 350 nm.

Self-healing covalent polymers are not limited to those quoted above, which are mainly designed for structural materials. To achieve the aim of practical application, further efforts are needed to deal with remaining challenges, such as extending healing ability to common engineering materials, increasing healing efficiency, and reducing dependency of the healing on external stimuli.[11]

5.2.2 Self-Healing Polymeric Materials *via* Supramolecular Chemistry

Self-healing processes with reversible covalent bonds generally require relatively complex treatment, *e.g.* high temperatures or UV exposure, to break and reform the bonds. To overcome this limitation, supramolecular polymers are becoming an attractive approach as the networks can be remodeled rapidly and reversibly.[30] Supramolecular chemistry focuses on polymers assembled by non-covalent bonds, including hydrogen bonding,[35] ionic interactions,[36,37] metal–ligand coordination,[38–40] and π–π stacking.[41,42] As these interactions have relatively lower strength as compared to the covalent bonds aforementioned, they are highly reversible and ideal to be incorporated into polymer structures to achieve self-healing.

5.2.2.1 Hydrogen Bonding

Hydrogen bonding interaction is one of the most developed approaches to generate supramolecular polymers because of its moderate strength, directionality and affinity.[43] As a single hydrogen bond is generally weak with bonding energy in the range of 1–5 kcal mol^{-1},[44] multiple hydrogen bonding groups, which can form double, triple, or quadruple hydrogen bonds within two complementary moieties, are favorable due to their relatively high strength, excellent reversibility, and complementary recognition.[45] The association strength between multiple hydrogen bonding units depends on the nature of donor/acceptor, solvent, number of hydrogen bonds, and arrangements. Figure 5.2 shows examples of several multiple hydrogen bonding groups and their association constants (K_a) in $CDCl_3$.[46–49] Under proper circumstances, these hydrogen bonding interactions could be disrupted and reformed dynamically, and have the potential to heal cracks.

Several kinds of self-healing polymers have been developed by incorporating hydrogen bonding functionalities. In 2007, Leibler and his co-workers designed thermo-reversible rubbers consisting of oligomers with multiple hydrogen bonding units.[35] These polymers were obtained using a two-step synthetic route, where fatty acids from renewable resources were condensed with an excess of diethylenetriamine, followed by reactions with urea. It resulted in the formation of a hydrogen bonded network with a mixture of oligomers containing a variety of different hydrogen bonding residues, as depicted in Figure 5.3. When the hydrogen bonded material was plasticized by dodecane and cut with a razor, it could be repaired by bringing the fractured surfaces together at room temperature. This is because the non-associated –C=O groups of amide and amine-functionalized ends on the

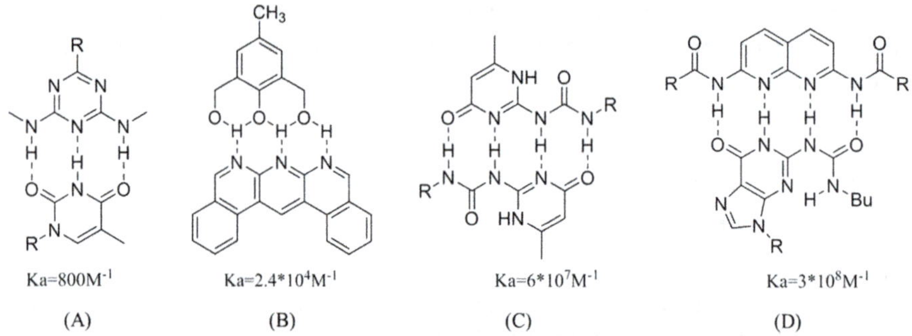

Figure 5.2 Examples of supramolecular hydrogen bonding groups and their association constants in $CDCl_3$ previously reported by (A) Binder *et al.*[46] (reproduced with permission from ref. 46, copyright 2005 WILEY-VCH), (B) Djurdjevic *et al.*[47] (reproduced with permission from ref. 47, copyright 2007 American Chemical Society), (C) Meijer *et al.*[48] (reproduced with permission from ref. 48, copyright 1998 WILEY-VCH), (D) Zimmerman *et al.*[49] (reproduced with permission from ref. 49, copyright 2005 American Chemical Society).

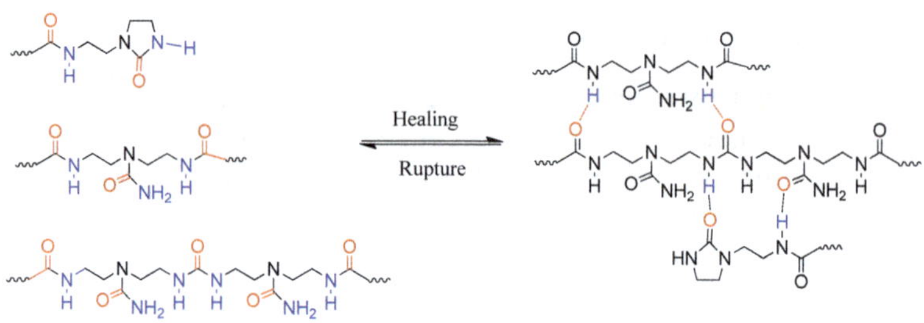

Figure 5.3 The hydrogen bonding network components of a self-healing rubber reported by Leibler *et al.*[35] Reproduced with permission from ref. 35, copyright 2008 Nature Publishing Group.

fractured surfaces tend to link with each other when brought into contact. The investigation of self-healing kinetics indicates that self-healing reactions compete with external conditions.[50,51] When freshly cut specimens were exposed to heat or moisture, the self-healing efficiency was significantly diminished due to the redistribution of active groups on each surface. Quadruple hydrogen bonding unit 2-ureido-4[1*H*]-pyrimidinone (UPy) has become a preeminent motif in supramolecular chemistry since it was introduced by Meijer and his co-workers,[52] because it is self-complementary with high dimerization constant and readily accessible in a one-step reaction from inexpensive starting materials. As the dimerization of UPy groups is reversible and highly thermally responsive,[53,54] a number of UPy-based healable materials have been reported either by applying a heat trigger or contact pressure trigger.[55] The functional hydrogen bonding groups can be employed as telechelic ends, chain-extended segments, or pedant groups to form diverse supramolecular architectures.[56] The introduction of UPy groups to a series of polymers has led to corresponding thermally healable materials manufactured by Suprapolix, including polysiloxane, polyethers, and polyesters.[57] At elevated temperatures, rearrangement of hydrogen bonding interactions could be achieved due to reduced viscosity and enhanced chain mobility of polymers, while autonomous self-healing at room temperature could occur for low T_g materials, which showed higher level of dynamics of the polymer chains.[57] Taking advantage of high segmental mobility of polymer matrix, UPy groups were incorporated as side chains into soft poly(*n*-butyl acrylate). They showed dynamic dissociation/dimerization at room temperature, and the lifetime was about 1.2 seconds.[58]

To construct matcrials with high modulus, Guan and co-workers produced multiphase supramolecular thermoplastic elastomers with a hydrogen bonding brush polymer consisting of a polystyrene backbone (D.P. ≈ 114) with approximately 11 polyacrylic amide side chains (D.P. ≈ 186).[59] These polymers could self-assemble into a two-phase nanostructure due to the immiscibility of polystyrene and polar side groups. The polystyrene backbones collapsed as discrete "hard" domains, which were connected by "soft" polyacrylic amid pendant chains with multiple hydrogen bonding sites. The low T_g (≈ 5 °C) of the polymer matrix facilitated the regeneration of hydrogen bonds at room temperature with healing efficiency of 92% after 24 hours. Another heterogeneous strategy to achieve self-healing was designed by Hentschel with poly(*n*-butyl acrylate)-*b*-polystyrene (PBA-PS) end-functionalized with UPy groups.[60] These multiphase supramolecular polymers retained the hard (PS)/soft (PBA) two-phase morphology with reversible hydrogen bonding moieties in polymer matrix, which combined stiffness of the thermoplastic elastomers with dynamic healing capabilities. This concept can be applicable to a wide range of multi-phase polymers and various types of supramolecular motifs for the design of stiff, strong, and tough self-healing polymers.

Plasticizing generally reduces the glass transition temperature of a polymer and increases its chain mobility.[35] Hydrogel systems can be viewed as highly plasticized systems with water being the plasticizer, and have the potential to show autonomous self-healing at room temperature. Phadke reported permanently cross-linked hydrogels with self-healing properties in an aqueous environment.[61] The feature was achieved by arming the hydrogel with acryloyl-6-aminocaproic acid (PA6ACA) side chains to form hydrogen bonds across the two hydrogel interfaces through the amide and carboxylic functional groups. The healing process could be switched on and off by changing pH due to protonation/deprotonation of carboxyl groups, correlating with association/dissociation of hydrogen bonds. As illustrated in Figure 5.4(A), when pH < 3, two separated PA6ACA hydrogel pieces could heal rapidly and the healed hydrogel was strong enough to sustain repeated stretch. However, the healed pieces could be separated again when exposed to high pH (pH > 9) solution, preventing the formation of hydrogen bonds by electrical repulsion. Cui *et al.* reported a thermally regulated self-healing random copolymer comprising 2-(dimethylamino)-ethyl methacrylate (DMAEMA)

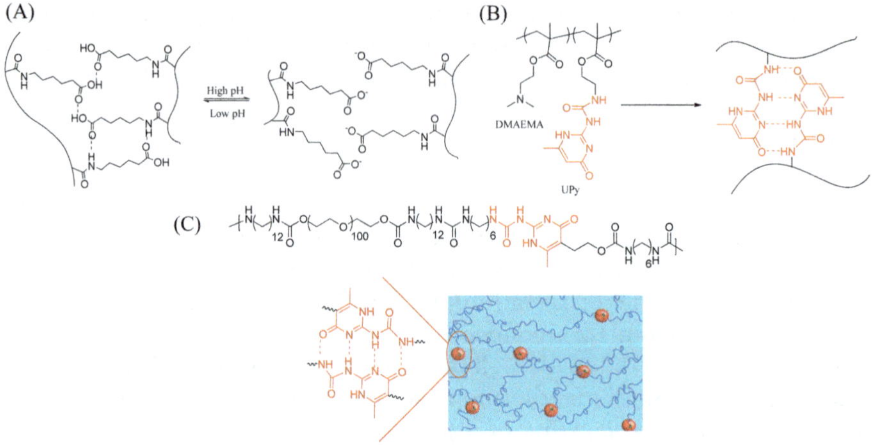

Figure 5.4 Hydrogen bond-based self-healing hydrogels. (A) pH-regulated self-healing mechanism of PA6ACA hydrogels[61] (reproduced from ref. 61, copyright 2012 National Academy of Sciences, USA), (B) chemical structure of poly(DMAEMA-*co*-UPy) and the dynamic dimerization of UPy moiety[62] (reproduced with permission from ref. 62, copyright 2012 Royal Society of Chemistry). (C) Chemical structural of PEG–UPy chain-extended polymers and illustrative depiction of hydrogel morphology[65] (reproduced with permission from ref. 65, copyright 2014 American Chemical Society).

and UPy-based monomers, as shown in Figure 5.4(B).[62] The reversibility of multivalent UPy groups imparted self-healing ability to the hydrogel. The gelation of the polymer solution was induced when the pH increased to 8, due to the dimerization of the UPy units. Self-healing of this sample occurred automatically within 5 minutes and could be controlled by the thermal response of the polymer matrix. Above low critical solution temperature (LCST), the aggregation of PDMAEMA restricted the diffusion of polymer chains and undermined the healing process. However, below LCST, the self-healing ability was regained. Similar self-healing systems could be obtained by using different monomers copolymerized with UPy moieties, such as 2-hydroxyethyl methacrylate (HEMA), 2-(2-methoxyethoxy) ethyl methacrylate (MEO_2MA) or *N*-isopropyl-acrylamide (NIPAAm).[63,64] A tough stimuli-responsive supramolecular hydrogel was obtained by adding Upy groups into a poly(ethylene glycol) (PEG) main chain.[65] As depicted in Figure 5.4(C), the bulk materials contain nanoscopic physical cross-links composed of UPy–UPy dimers. The hydrogen bonding groups were embedded in segregated hydrophobic domains dispersed within the PEG matrix both in the dry state and in water-swollen hydrogels. As UPy groups were shielded by alkyl spacers, they formed strong hydrogen bonded arrays offering the gel elasticity and recovery of mechanical properties after deformation. Up to now, some applications of self-healing gels include coatings and sealants in industrial fields, as well as tissue adhesives, drug delivery, and cell therapy in the biomedical field.[66] However, most self-healing gels have not found practical applications, mainly due to their relatively poor mechanical properties.[67] Developing tough self-healing hydrogels to address this problem is desperately needed. It is also promising to develop self-healing gels combining with multifunctional systems, which would be potentially applied for artificial electronic skin, soft actuator, and artificial muscle.[68]

5.2.2.2 Ionic Interactions

Ionomers are a class of polymers containing up to 15 mol% ionic groups along the polymer backbone.[69] The ion pairs aggregate into discrete regions, which act as crosslinks exerting influence on the mechanical and physical properties of ionomers.[70] They are commonly used for commercial applications such as molded coatings, packaging films, and membranes in fuel cells.[71] The reversible nature of ionic interactions leads to rearrangement of the clustered regions and contributes to the self-healing behavior, which was demonstrated by projectile testing of poly(ethylene-*co*-methacrylic acid) (EMAA)-based ionomers.[32,33] Following ballistic puncture, low-density polyethylene (LDPE) without ion pairs did not show healing ability, while EMAA films healed the bullet hole leaving only a small scar on the surface. The self-healing property was attributed to the ionic functionalities and the polar acid groups in EMAA polymers. Based on the effect of temperature and ionic content on self-healing efficiency, a healing mechanism was proposed by Kalista *et al.*[72] The heat generated by the friction during the high-energy impact leads to a localized melting of the damage area, and the surfaces fuse together due to interdiffusion of the melt, followed by the reformation of thermally responsive ionic cross-linking as well as hydrogen bonding. Although the existing EMAA ionomers are not suitable for applications at elevated temperatures, it is promising to develop applicable self-healing ionomers, as they are able to recover severe damage autonomously within a short time and at low cost.[73]

5.2.2.3 Metal–Ligand Coordination

Responsive, supramolecular materials utilizing metal–ligand interactions are particularly attractive because reversibility and physical properties of the coordination complexes could be facilely tuned by incorporating different metal ions and ligands.[74] In 2005, Rowan *et al.* reported a class of organic/inorganic hybrid materials self-assembled from telechelic poly(tetrahydrofuran) end-capped with 2,6-bis(benzimidazolyl)-4-oxypyridine derivatives. The polymer chain was extended by forming complexes with a range of metal ions (*e.g.*, Fe^{II}, Co^{II}, Zn^{II}, or Cd^{II}).[75] This concept of metallo-supramolecular polymers was also demonstrated by Schubert, using terpyridine-functionalized poly(*N*-iso-propylacrylamide) and divalent metal ions to prepare switchable polymers *via* dynamic metal–ligand interactions.[76]

The first examples of optically healable materials with metal–ligand interactions were developed by Rowan and co-workers.[38] The polymers comprised a low molecular mass poly(ethylene-*co*-butylene) core with 2,6-bis(1′-methylbenzimidazolyl)pyridine ligands at the termini that were coordinated with Zn^{2+} or La^{3+}. The formation of metal ion binding between rubbery poly(ethylene-*co*-butylene) macromonomers revealed microphase-separated lamellar morphologies, where the "hard phase" consisted of metal–ligand complexes and the "soft" domains were formed by the telechelic macromonomer cores.[77] As opposed to direct thermal stimulation, the metallosupramolecular polymers damaged by scratching could undergo autonomous healing upon exposure to UV light. It is because the metal–ligand motifs are electronically excited under UV radiation, converting the adsorbed energy into heat. The resultant temporary disengagement of the reversible bonds decreases the polymer's viscosity and induces the following healing process. Since the rubbery blocks of the polymers are poor thermal conductors, selective heating of the material occurs only around areas that are directly exposed to the light, which offers an opportunity for the materials to be healed under load. Mechanical analysis of fully healed samples indicated that the healing efficiency could be up to 100%. Polyethylenimine–copper (C_2H_5N–Cu) supramolecular networks capable of UV-induced self-repairs were developed *via* Cu–N coordination bonds.[78] It is assumed that mechanical damage cleaves Cu–N bonds to the uncomplexed state, but upon UV exposure, the metal–ligand complex undergoes square planar to tetrahedral transition, which leads to network remodeling and is responsible for self-healing.

Self-healing gels based on reversible metal–ligand coordination have drawn much attention in recent years.[79] Self-healing poly(acrylic acid) hydrogels were synthesized *via* facial free radical polymerization containing both covalently and ionically cross-linked networks.[80] The covalent cross-linking of polymers provided a mechanical support to sustain the shape of the hydrogels, while the autonomous self-healing of the hydrogel was achieved through the dynamic ionic bonding between Fe^{3+} ions and carboxylic groups. Banerjee's group[81] discovered a series of self-healing metallo-hydrogels based on the interaction between amphiphilic tyrosine and Ni^{2+}, which showed stimuli responses upon mechanical shaking, heat, pH, and external chemicals. The remarkable self-healing property of the metallo-hydrogels could be nicely modulated by changing the chain length of the amphiphiles. The coordination between Fe^{3+} and catechol ligands inspired by mussel adhesive proteins has also been widely investigated due to its uniqueness.[39] The degree of polymer cross-linking was pH sensitive and the polymers showed self-healing properties through restoration of broken catechol–Fe^{3+} crosslinks.[40] The self-healing properties and mechanism of polymers based on catechol–Fe^{3+} coordination will be thoroughly illustrated in Section 5.3. The self-healing hydrogel materials are promising for applications in smart coatings and biomedical devices. However, for biomedical applications, the high concentration of metal ions may not be desirable[80] and special attention should be also paid to the water solubility of certain ligands for specific applications.

5.2.2.4 π–π Stacking

π–π stacking interactions are the association between electronically complementary aromatic π systerms,[82] and play an important role in supramolecular chemistry. Computational studies on model compounds indicated that a "chain-folding" motif comprising two π-electron-deficient diimide units separated by a triethylenedioxy linking unit provides an ideal stereoelectronic center for binding complementary π-electron-rich aromatic groups.[83,84]

As shown in Figure 5.5(A), Burattini synthesized a series of polymer blends consisting of a chain-folding polydiimide with π-electron-deficient diimide units and telechelic polysiloxane[42]/polyamide[85] with π-electron-rich pyrenyl groups. The proposed self-healing mechanism is illustrated in Figure 5.5(B). Upon heating, the π–π stacking interactions are partially disrupted, and the disengagement of pyrenyl groups from polydiimide enables the blend to flow. With reduced temperature, repair of damage and recovery of mechanical strength will occur due to the increasing levels of π–π stacking. The visual demonstration of healing characteristics of the polymer blend is given in Figure 5.9(C). A fracture up to 70 μm wide could be rapidly healed in seconds by heating the material above 80 °C. Comparing the tensile modulus of the pristine sample with the values measured after healing, 100% recovery was achieved after 5 minutes healing at 50 °C.

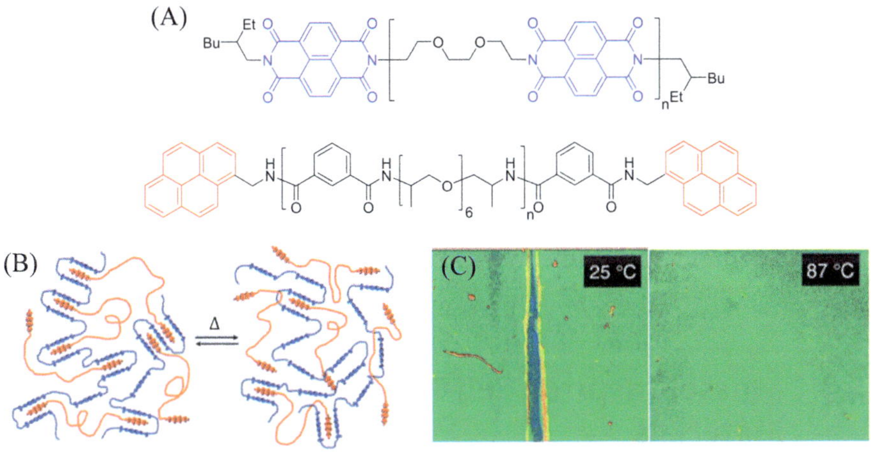

Figure 5.5 Self-healing supramolecular polymer blend *via* π–π stacking interactions.[42,85] (A) Polydiimide containing multiple π-electron-deficient chain-folds and the telechelic polyamide end-capped with π-electron-rich pyrenyl residues. (B) Schematic representation of the thermo-triggered self-healing for the polymer blend (reproduced with permission from ref. 42, copyright 2009 Royal Society of Chemistry). (C) Environmental scanning electron microscopy images of a fracture zone of the self-healing polymer annealing at 25 °C and 87 °C (reproduced with permission from ref. 85, copyright 2009 Royal Society of Chemistry).

Another approach to produce self-healing materials is to combine two distinct types of supramolecular interactions, such as π–π stacking and hydrogen bonding, within one system. By blending a chain-folding polyimide with a low molecular weight polyurethane with pyrenyl end-groups, a tough, elastomeric material was produced.[41] The polymer network contained both hydrogen bonding moieties (ureas and urethanes) and complementary π–π stacking residues. This polymer showed significant enhancement of tensile modulus (3×10^5 Pa) over the purely hydrogen bonded rubber reported by Leibler *et al.*[35] (1×10^4 Pa), and its mechanical properties could be repeatedly regained at elevated temperature. Xu *et al.* prepared cholesteryl (Chol) derivatives containing nitrobenzoxadiazole (NBD), which could form a gel with rapid self-healing ability in the mixture solution of pyridine–methanol.[86] Hydrogen bonding and π–π stacking were the main driving forces to promote the gelation of the system, and the length of the spacer-connecting NBD and Chol units played a critical role in healing properties. Taking advantage of the superior mechanical strength and self-healing ability, the gels can act as candidates for vibration-resistant materials and biomedical materials. Besides, the fluorescence of the NBD moiety also brings convenience for real-life applications.

The universal design of self-healing supramolecular systems is based on the scission and reformation of weak supramolecular interactions rather than stronger covalent bonds. This concept provides an upper limit to the strength of these materials as there is a dilemma between bonding strength and self-healing ability.[59] Up to now, most healable supramolecular materials are soft elastomers with relatively low glass transition temperature (T_g). Supramolecular nano-composites may offer a promising approach to enhancing the strength of the material by utilizing nanoscale particles[68] or high aspect-ratio fillers[87] as the strengthening phase. This class of supramolecular composites keeps the self-healing properties offered by supramolecular polymer matrix, and has the potential to generate materials with useful engineering and environmental applications, especially for high-modulus structural materials.

5.3 Mussel-Inspired Self-Healing Polymeric Materials

Mussels can obtain secure underwater attachment to various surfaces including rocks, metals, wood structures, and concretes in tidal waves through the formation of mussel byssus.[88,89] During a typical attachment process, the mussel foot stretches out of the shell to conduct a "tiptoeing" exploration on the substratum to locate the suitable spot before it starts a surface preparation. Then several kinds of liquid proteins are secreted from the mussel foot to form a bundle of byssus consisting of byssal threads which connect the mussel shells and byssal plaques which hold fast to the substratum (Figure 5.6).[90–92] Besides the remarkable wet adhesion capability, mussel

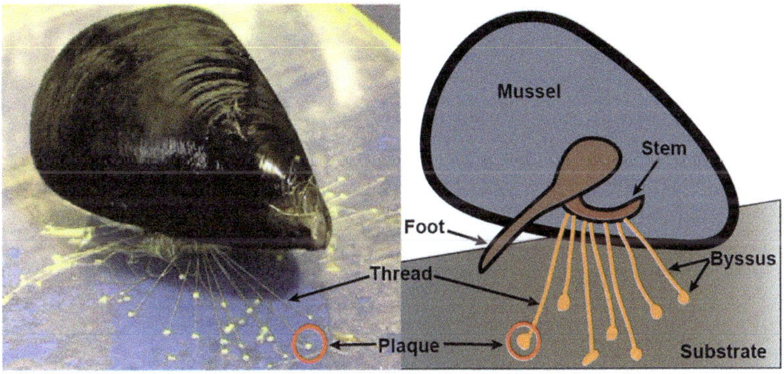

Figure 5.6 Photograph (left) and schematic internal structure (right) of a typical adult mussel with byssus attaching to a substrate.

threads also bear an inherent property of self-healing which allows themselves to autonomously recover the initial length and modulus after yield or plastic deformation, confirmed by both natural observation and laboratory cyclic stress–strain tests.[93] It is proposed that some reversible covalent and non-covalent intermolecular cross-links existing in the mussel threads could serve as breakable sacrificial bonds which rupture at elevated strain levels and reform when the load is removed. Through years of endeavors, various interactions contributing to the self-healing behaviors of mussel threads have been found, including catechol–metal coordination, histidine–metal coordination, and catechol-mediated hydrogen bonding and aromatic interactions, which will be introduced in detail in the following sections.

5.3.1 Catechol-Mediated Interactions

Formed by post-translational modification of tyrosine, the catecholic amino acid 3,4-dihydroxyphenyl-L-alanine (DOPA) plays a crucial role in achieving the remarkable underwater adhesion by participating in various types of interactions including hydrogen bonding, metal–catechol coordination, electrostatic interaction, cation–π interaction and π–π aromatic interactions.[90,94–98] Meanwhile DOPA makes up a major component in mussel foot protein-1 (mfp-1) which is found exclusively in the cuticle of byssal threads and plaques. Further studies demonstrate that DOPA also accounts for the self-healing behavior of mussel threads, through various reversible interactions including catechol–metal coordination, catechol-mediated hydrogen bonding, and aromatic interactions.

5.3.1.1 Catechol–Fe^{3+} Non-Covalent Coordination

Catechol–Fe^{3+} coordination has been the most widely applied mechanism for preparing mussel-inspired self-healing polymeric materials. It was proposed that mussel foot proteins pre-bound Fe^{3+} in mono-DOPA-Fe^{3+} complexes in

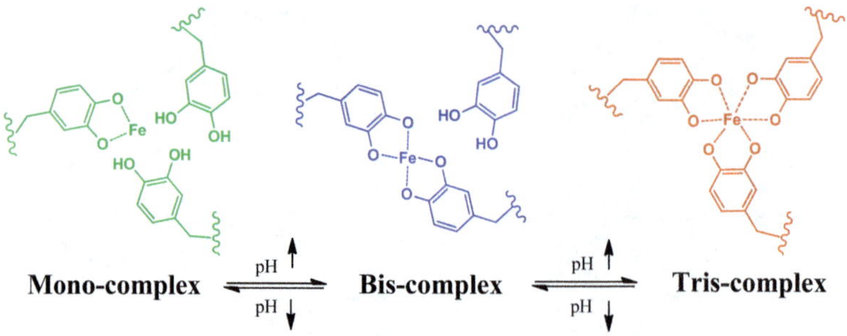

Figure 5.7 Cross-linking mechanism of pH-dependent catechol–Fe^{3+} coordination.

secretory granules at pH ≤ 5, while when released to sea water (pH ~8) the thread material would undergo spontaneous cross-linking *via* bis- and/or tris-dopa-Fe^{3+} complexes due to the considerable pH increase. It was further testified that the catechol–Fe^{3+} interpolymer cross-linking was readily pH-controlled, with the mono complex dominating at pH < 5.6, the bis complex at 5.6 < pH < 9.1 and the tris complex at pH > 9.1 (Figure 5.7), and the bis and tris complexes can act as breakable sacrificial bonds for constructing self-healing materials.[40]

Based on this mechanism, DOPA-containing recombinant mussel adhesive proteins (MAPs), which were mass produced by *Escherichia coli* and modified by mushroom tyrosinase, were used to prepare adhesive hydrogels bearing simultaneously self-healing properties.[99] An *in vivo* residue-specific incorporation strategy was also adopted to create engineered MAPs in *Escherichia coli* with high DOPA content (16.5 mol%), comparable to that of natural MAPs.[100] However in general, the expression of MAPs *via* recombinant DNA technology is complicated, troublesome, and expensive, which has fueled the booming development of polymer mimetics. Generally a two-step process is needed for creating mussel-inspired self-healing polymeric materials, in which various chemical reactions are used first to incorporate the mussel functional moieties into polymer backbones, followed by a cross-linking of these functionalized polymers to 3D networks through various interactions. During the past few years catechol groups have been successfully incorporated into various polymer backbones including polyethylene glycol,[40] polyallylamine,[39] chitosan,[101] polyurethane,[102] and peptide[103] to form self-healing polymeric networks through catechol–Fe^{3+} coordination, which could be potentially used in biomedical applications as drug delivery platforms or tissue adhesives.

Readily synthesized by nitration of dopamine, nitrodopamine kept all the advantages of catechol groups while obtaining an additional functionality from the *o*-nitrophenyl moiety: the photo-cleavability. Cross-linking of a nitrocatechol polymer prepared by end-functionalizing a four-arm star-poly(ethyleneglycol) with nitrodopamine created self-healing gels and coatings underwater that could be photodegraded upon exposure to light,[104] which could be widely used in multiple cell and medical applications.

Birkedal *et al.* devised a very simple and low-cost method to synthesize self-healing hydrogel through the interactions between natural polyphenol tannic acid (TA), iron ions, and polyallylamine (PAA), with TA and PAA acting as models for the catechol and lysine-rich Mfp-3 and -5.[105] Due to the very high catechol content in TA, the resulting material transformed into a sticky solid *via* formation of bis and tris complexes at pH above 3, which was much lower than the value (pH ~9) obtained from systems involving DOPA-functionalized polymers cross-linked by Fe^{3+}.[39,40] Flexible and lightweight fibers can also be spun from the concentrated Fe^{3+}–TA–PAA mixtures and could be potentially applied as water-sensitive mechanical locks.

5.3.1.2 *Catechol–B^{3+} Dynamic Covalent Coordination*

Catechol–B^{3+} dynamic covalent coordination is another catechol–metal coordination which has been commonly employed in preparing self-healing hydrogels. Generally speaking, the strength and reversibility of this coordination in aqueous media is governed by an equilibrium which depends heavily on the media pH and the pK_a of the boronic acid component. The formation of this coordination is favored at neutral and alkaline pH (above the pK_a of the boronic acid component) while a dissociation is favored by the equilibrium in an acidic environment.[106,107] The reversible formation and dissociation of the boronate ester, or the self-healing behavior of the resultant cross-linked network, is most effective when the pH is near the pK_a.[108]

He *et al.* constructed a self-healing polymeric network by reacting 1,3-benzenediboronic acid (BDBA) with catechol-functionalized 4-arm PEG under basic aqueous conditions (pH = 9), which was determined as being between the pK_a of catechol (9.3) and that of BDBA (8.7).[109] The dynamic tetrahedral borate ester gave rise to hydrogel networks which could exhibit covalent gel-like behavior and self-heal after mechanical disruption. Harada and co-workers devised pH and sugar-responsive gel assemblies under basic conditions (pH = 10), through the formation of cyclic boronate esters between poly(acrylamide) gels carrying phenylboronic acid moieties and catechol moieties respectively.[110] As the assembly and disassembly could be controlled by pH or competitive saccharide molecules, this system can potentially work for a biomolecule-responsive gel assembly in biomedical fields. Stadler and co-workers found differences in the behavior of dry and humid catechol–borate complexes.[111] By complexation of a catechol-functionalized polymer with $NaBH_4$ at pH ~9, covalent gels were prepared but the network could not restructure dynamically after mechanical disruption. However, humidity can be adsorbed by these gels to change the catechol–borate covalent bonds from irreversible to reversible. This finding may be important for several biochemical and pharmaceutical systems utilizing this kind of bond. A rapid self-healing and triple stimuli responsive hydrogel was reported recently, based on interactions between poly(dopamine methacrylate-*co*-*N*-isopropylacrylamide) and boric acid (H_3BO_3),[112] and it was found coordination of B^{3+}–catechol was strongly dependent on the pH and catechol concentration (Figure 5.8).

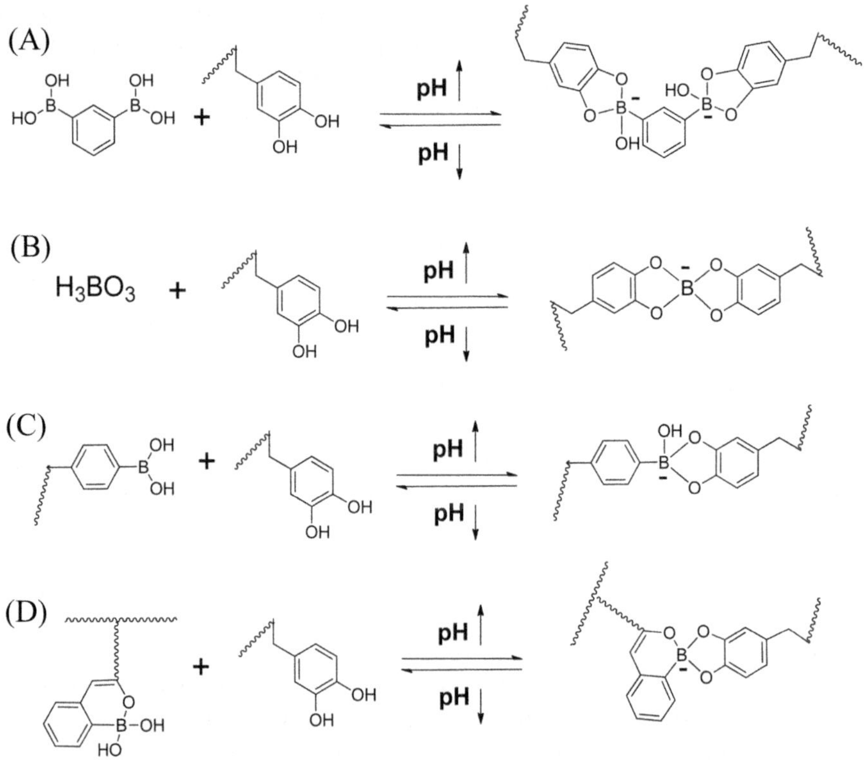

Figure 5.8 Cross-linking mechanism of catechol-functionalized polymers with different boronic acid derivatives. (A) 1,3-Benzenediboronic acid[109] (reproduced with permission from ref. 109, copyright 2011 Royal Society of Chemistry). (B) Boronic acid,[112] (C) phenylboronic acid[110] (reproduced with permission from ref. 110, copyright 2014 American Chemical Society). (D) 2-Acrylamidophenylboronic acid[113] (reproduced with permission from ref. 113, copyright 2015 American Chemical Society).

The fact that various dynamic catechol–B^{3+} conjugates formed at high pH would dissociate at low pH endows the above-mentioned materials great potential as drug delivery carriers, as many pathological changes including inflammation or tumor are accompanied by a pH decrease in the local microenvironment.[114] However, if healing can occur in a broad pH range or additional stability at relatively low pH values can be achieved, the possible applications of boronate ester hydrogels would be greatly expanded, especially in an acidic environment like the gastrointestinal tract.[115] One method to accomplish this is by using intramolecular coordination that can stabilize the formation of boronate esters at reduced pH.[116,117] To this end, Sumerlin and co-workers constructed boronate ester cross-linked hydrogels which can achieve healing over a wide pH range by reacting a catechol-functionalized copolymer with a 2-acrylamidophenylboronic acid (2APBA) copolymer, where the internal coordination facilitated the stabilization of cross-links formed at acidic and neutral pH.[113]

5.3.1.3 Other Catechol–Metal Coordination

Recently, Andersen and co-workers demonstrated how the mechanical properties of an aqueous metal-coordinating polymer network could be controlled by the metal ion identity, using simple catechol-modified PEG polymers and vanadium, iron, and aluminum salts as model systems.[118] VCl_3, $FeCl_3$, or $AlCl_3$ salts were all used to construct polymer networks with a catechol : metal ratio of 3 : 1. It was found that at fixed pH 8, V^{3+}–catechol polymer networks exhibited a 10-fold higher stability and significantly more solid-like properties than the Fe–catechol networks, mainly due to their differences in cross-link stoichiometry: V^{3+} was more inclined to induce tris coordination while Fe^{3+} induced primarily bis coordination at pH 8. Al^{3+}–catechol networks, on the other hand, displayed a 5-fold lower stability than that of Fe^{3+}–catechol networks, due to weaker non-coordinate Al^{3+}–catechol interactions attributed to the non-transition nature of aluminum. Birkedal *et al.* also reported that the mechanical properties and colors of mussel-inspired self-healing hydrogels based on DOPA–polyallylamine can be readily adjusted by systematically varying the coordinating metal from Al^{3+} to Ga^{3+} to In^{3+}.[119] These studies demonstrated the possibility of tuning viscoelastic properties and colors of hydrophilic polymer materials over several orders of magnitude purely by the choice of coordinating metal ions, according to different practical applications.

5.3.1.4 Catechol-Mediated Hydrogen Bonding and Aromatic Interactions

Despite the multiple metal coordination mechanisms mentioned above, biological self-healing in wet conditions, as occurs in self-assembled holdfast proteins in mussels, is generally thought to involve more than reversible metal chelates. Israelachvili, Waite, and their co-workers employed synthetic polyacrylate and polymethacrylate materials which were functionalized with mussel-inspired catechols blocked by silyl groups to study their self-healing behavior in metal-free water and it was found that the bisected polymer rods could only self-mend in acid solutions where the catechols were fully revealed due to the removal of silyl groups, and could not self-mend in neutral to basic environments where the catechols were blocked.[120] It was therefore concluded that the underwater self-healing in catechol-functionalized polyacrylates was achieved by hydrogen bonding between interfacial catechol moieties. Based on this mechanism, Stadler *et al.* manipulated copolymers of *N*-isopropylacrylamide (NIPAM) and dopamine methacrylate to establish a reversible, self-healing 3D network in aprotic solvents based on hydrogen bonding, which could be potentially used in the field of drug delivery.[121]

Recently, a novel injectable self-healing hydrogel with an anti-biofouling property based on self-assembly of an ABA tri-block copolymer (DNODN, Figure 5.9(A)) was reported by Li *et al.*, in which mussel-inspired catechol-functionalized PNIPAM and PEO were selected as the thermo-sensitive

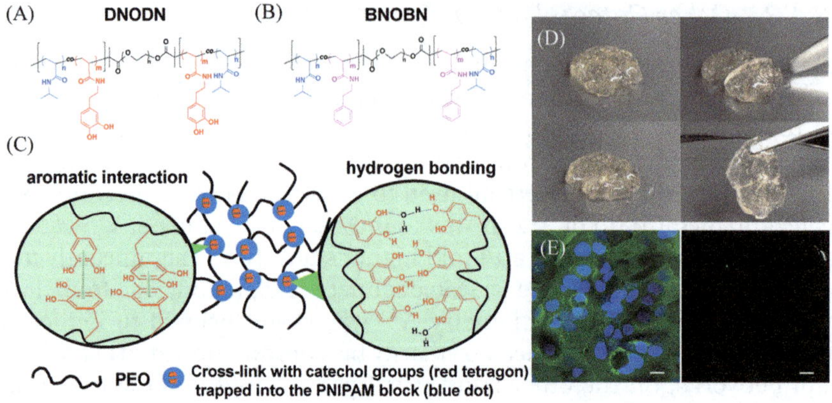

Figure 5.9 Chemical structure of the ABA tri-block copolymer DNODN (A) and BNOBN (B). (C) Schematic of a proposed structure of the DNODN hydrogel. (D) Optical evidence of self-healing. (E) Representative fluorescence microscopy images of uncoated (left) and DNODN hydrogel-coated (right) microwell dishes after exposure to Caco-2 cells for 48 h. The scale bar represents 20 µm.[122] Reproduced with permission from ref. 122, copyright 2015 WILEY-VCH.

A block and hydrophilic B block, respectively.[122] This hydrogel can withstand repeated deformation and recover its mechanical properties and structure within seconds, which was clearly related to catechol moieties because an absence or disfunction of catechol groups in this hydrogel would lead to a loss of self-healing property. To further elucidate whether hydrogen bonding between interfacial catechol moieties played the sole role in achieving self-healing, another ABA tri-block copolymer (BNOBN, Figure 5.9(B)) with A block from random copolymerization of phenylethylacrylamide and NIPAM and B block PEO was synthesized. The subdued while still tolerable self-healing performance of BNOBN hydrogel indicated that aromatic interactions, including quadrupolar interactions and π–π stacking interactions, could also contribute to the self-healing mechanism. Endowed with its thermo-sensitivity, injectability, inherent metal-free self-healing nature and anti-biofouling properties, the DNODN hydrogel exhibits great potential in various bioengineering applications (*e.g.* drug delivery) (Figure 5.9(C)–(E)).

5.3.2 Histidine–Metal Coordination

Histidine is an amino acid commonly found in the collagen fibrils of mussel byssus and its coordination with multiple divalent transition metal ions were believed to contribute to the mechanical properties as well as the self-healing behavior of the mussel byssus (Figure 5.10).[93,124,125] These coordination bonds, which are weaker than covalent bonds but stronger than non-covalent bonds, are proposed to serve as reversibly breakable sacrificial bonds in the threads that can rupture upon application of a load and reform when the applied load

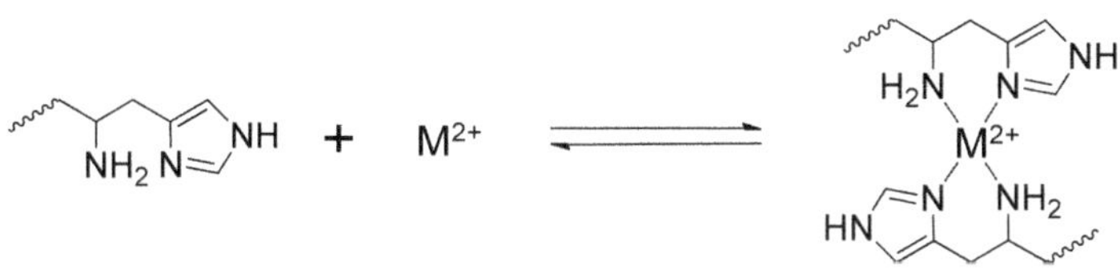

Figure 5.10 Cross-linking mechanism of histidine–metal coordination.[123] Reproduced with permission from ref. 123, copyright 2013 American Chemical Society.

is removed.[93] Transient network hydrogels cross-linked through histidine–metal coordination bonds using histidine-modified star poly(ethylene glycol) (PEG) polymers were reported.[123] The mechanical and relaxation properties of hydrogels cross-linked with different transition metal cations were studied by conventional oscillatory rheometry and it was found that the self-repair rate of the gels follow the order $Zn^{2+} > Cu^{2+} > Co^{2+} > Ni^{2+}$. Harrington *et al.* synthesized a histidine-rich peptide derived from mussel proteins (His_5-bys) and studied its reversible adhesive self-interaction in the presence and absence of metal ions using an atomic force microscope to investigate the behavior of histidine-rich protein at the molecular level.[126] Adhesion measurements were conducted between peptide-functionalized soft colloidal probes (SCPs) and peptide-coated glass slides and it was found that adhesion was only measured in the presence of metal ions and the stiffness of the modified SCPs got tripled, further confirming the contribution of histidine–metal coordination to the stiffness, high hysteresis, and self-healing properties of byssal threads.

5.4 Case Studies of Self-Healing Polymeric Materials for Environmental Applications

In this section, case studies of self-healing polymeric materials for environmental applications including pollutant adsorbents and water filtrations are introduced. Generally, the self-healing property was incorporated into intrinsic functional systems such as graphene or clay-based networks to extend their lifespan and enhance their performance and stability. Lu *et al.* constructed a self-healing pollutant-adsorbing hydrogel using polydopamine-modified clay as the main building block and Fe^{3+} ions as the physical cross-linkers (Figure 5.11(A)).[127] This hydrogel possesses self-healing capability *via* reformation of damaged catechol–Fe^{3+}complexes. Furthermore, this hydrogel can work as an adsorbent to effectively remove an organic pollutant Rhodamine 6G (Rh6G) from water with an adsorption capacity of 150 mg g^{-1}, and the main adsorption mechanism is attributed to the high specific surface area of clay and various interactions including hydrogen bonding and π–π stacking between the aromatic moieties of polydopamine and Rh6G. As demonstrated in Figure 5.11(B), dye-containing water was slowly poured through a D-clay/Fe^{3+} hydrogel-filled syringe and the clear water obtained at the end of the syringe indicated the complete removal of the Rh6G dye.

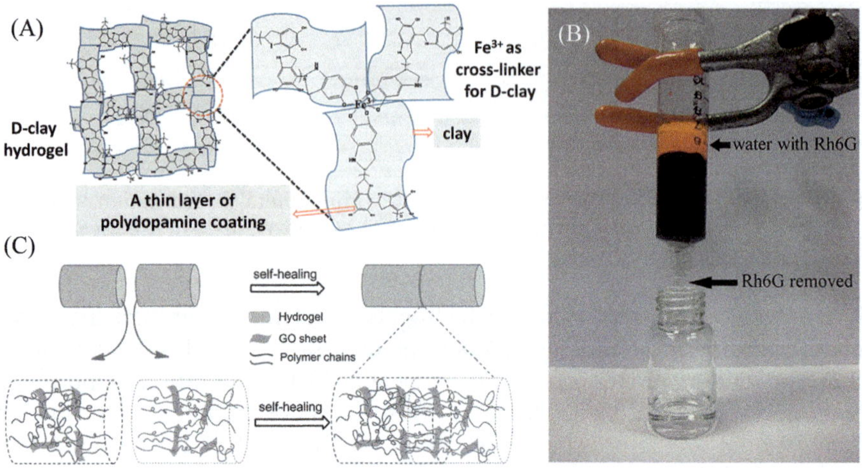

Figure 5.11 (A) Schematic description of the formation mechanism for D-clay hydrogels. (B) Photograph showing effective removal of Rh6G by the D-clay hydrogel. The D-clay content is 4 wt% and the $Fe(NO_3)_3{\cdot}9H_2O$/D-clay weight ratio is 4/96[127] (reproduced with permission from ref. 127, copyright 2013 American Chemical Society). (C) The proposed self-healing mechanism of the GO composite hydrogels[131] (reproduced with permission from ref. 131, copyright 2013 WILEY-VCH).

Graphene oxide (GO)-based materials have been commonly employed in water treatment due to their high specific surface area and abundant functional groups which could actively participate in the binding of various metal ions and organic compounds.[128,129] Cong *et al.* prepared a GO/poly(acryloyl-6-aminocaproic acid) (PAACA) composite hydrogel with enhanced mechanical properties and self-healing capability to pH stimulus.[130] GO nano sheets and calcium ions worked both as cross-linkers to trigger the formation of double networks including a Ca^{2+}-induced coordination network and a hydrogen bonding network, accounting for the enhanced mechanical properties. By simply adding some drops of acid solutions to the ruptured surfaces and bringing them into contact, the ruptured hydrogel pieces could heal into one integral piece rapidly, due to the reformation of reversible hydrogen bonding between the polar groups of the polymer side chains. A tough self-healing GO composite hydrogel fabricated by using graphene peroxide as polyfunctional initiating and cross-linking centers was reported.[131] This GO composite hydrogel showed excellent self-healing ability at ambient temperature with a tensile strength recovery degree up to 88% at prolonged healing time, and the final healed hydrogel could exhibit very high tensile strengths up to 0.35 MPa and extremely high elongations up to 4900%, which could be attributed to the mutual diffusion of long polymer chains at the interface of two damaged gel samples and their subsequent interactions with each other and with GO sheets though formation of hydrogen bonds (Figure 5.11(C)). These above-mentioned GO-based self-healing materials all show strong potential as filters, membranes, or coatings for environmental applications such as water treatment.

5.5 Conclusions and Outlook

The past decade has witnessed the booming development of self-healing materials. This chapter has presented examples of intrinsic self-healing polymeric materials constructed through reversible bond formation, including dynamic covalent bonding, hydrogen bonding, ionic interactions, π–π stacking, and metal–ligand coordination. With improved understanding of the basic mechanisms and principles of the self-repairing behaviors of mussel byssal threads, various mussel-inspired self-healing polymeric materials for different applications have been successfully developed, and are also reviewed in detail in this chapter. Until now the self-healing concept has referred primarily to the structural properties: the restoration of original structure and recovery of mechanical properties. Self-healing may also be employed in other material properties and performance matrices such as conductivity, optical transparency, and chemical stability. The self-mendable materials are expected to have great potential in engineering and environmental applications.[132] Another development direction for the intrinsic self-healing polymeric materials is to further emulate nature. A new generation of bio-inspired self-healing materials may evolve to incorporate both reversible bonds and a continuous circulatory system transporting the healing agents and building blocks to the damaged sites, to achieve long-term effective self-healing performance and practical engineering and environmental applications.

Acknowledgements

This work was supported by the Natural Sciences and Engineering Research Council of Canada (NSERC) (H. Zeng).

References

1. M. D. Hager, P. Greil, C. Leyens, S. van der Zwaag and U. S. Schubert, *Adv. Mater.*, 2010, **22**, 5424–5430.
2. B. Blaiszik, S. Kramer, S. Olugebefola, J. S. Moore, N. R. Sottos and S. R. White, *Annu. Rev. Mater. Res.*, 2010, **40**, 179–211.
3. S. R. White, N. Sottos, P. Geubelle, J. Moore, M. R. Kessler, S. Sriram, E. Brown and S. Viswanathan, *Nature*, 2001, **409**, 794–797.
4. N. K. Guimard, K. K. Oehlenschlaeger, J. Zhou, S. Hilf, F. G. Schmidt and C. Barner-Kowollik, *Macromol. Chem. Phys.*, 2012, **213**, 131–143.
5. F. Herbst, D. Döhler, P. Michael and W. H. Binder, *Macromol. Rapid Commun.*, 2013, **34**, 203–220.
6. J. A. Syrett, C. R. Becer and D. M. Haddleton, *Polym. Chem.*, 2010, **1**, 978–987.

7. W. Hayes and B. W. Greenland, *Healable Polymer Systems*, Royal Society of Chemistry, 2013.
8. R. J. Wojtecki, M. A. Meador and S. J. Rowan, *Nat. Mater.*, 2011, **10**, 14–27.
9. J. Sauer and M. Hara, in *Crazing in Polymers Vol. 2*, Springer, 1990, pp. 69–118.
10. Z. Wei, J. H. Yang, J. Zhou, F. Xu, M. Zrínyi, P. H. Dussault, Y. Osada and Y. M. Chen, *Chem. Soc. Rev.*, 2014, **43**, 8114–8131.
11. M. Q. Zhang and M. Z. Rong, *Polym. Chem.*, 2013, **4**, 4878–4884.
12. L. Engle and K. Wagener, *J. Macromol. Sci., Polym. Rev.*, 1993, **33**, 239–257.
13. A. Sanyal, *Macromol. Chem. Phys.*, 2010, **211**, 1417–1425.
14. C. Goussé, A. Gandini and P. Hodge, *Macromolecules*, 1998, **31**, 314–321.
15. X. Chen, M. A. Dam, K. Ono, A. Mal, H. Shen, S. R. Nutt, K. Sheran and F. Wudl, *Science*, 2002, **295**, 1698–1702.
16. X. Chen, F. Wudl, A. K. Mal, H. Shen and S. R. Nutt, *Macromolecules*, 2003, **36**, 1802–1807.
17. H. Laita, S. Boufi and A. Gandini, *Eur. Polym. J.*, 1997, **33**, 1203–1211.
18. Y. L. Liu and C. Y. Hsieh, *J. Polym. Sci., Part A: Polym. Chem.*, 2006, **44**, 905–913.
19. Y. L. Liu and Y. W. Chen, *Macromol. Chem. Phys.*, 2007, **208**, 224–232.
20. A. A. Kavitha and N. K. Singha, *ACS Appl. Mater. Interfaces*, 2009, **1**, 1427–1436.
21. A. M. Peterson, R. E. Jensen and G. R. Palmese, *Compos. Sci. Technol.*, 2011, **71**, 586–592.
22. A. M. Peterson, R. E. Jensen and G. R. Palmese, *ACS Appl. Mater. Interfaces*, 2013, **5**, 815–821.
23. U. Lafont, H. Van Zeijl and S. Van Der Zwaag, *ACS Appl. Mater. Interfaces*, 2012, **4**, 6280–6288.
24. J. Canadell, H. Goossens and B. Klumperman, *Macromolecules*, 2011, **44**, 2536–2541.
25. Y. Amamoto, H. Otsuka, A. Takahara and K. Matyjaszewski, *Adv. Mater.*, 2012, **24**, 3975–3980.
26. H. Otsuka, S. Nagano, Y. Kobashi, T. Maeda and A. Takahara, *Chem. Commun.*, 2010, **46**, 1150–1152.
27. N. V. Tsarevsky and K. Matyjaszewski, *Macromolecules*, 2002, **35**, 9009–9014.
28. P. Ghezzi, V. Bonetto and M. Fratelli, *Antioxid. Redox Signaling*, 2005, **7**, 964–972.
29. J. A. Yoon, J. Kamada, K. Koynov, J. Mohin, R. Nicolaÿ, Y. Zhang, A. C. Balazs, T. Kowalewski and K. Matyjaszewski, *Macromolecules*, 2011, **45**, 142–149.
30. Y. Yang and M. W. Urban, *Chem. Soc. Rev.*, 2013, **42**, 7446–7467.
31. C. Yuan, M. Z. Rong, M. Q. Zhang, Z. P. Zhang and Y. C. Yuan, *Chem. Mater.*, 2011, **23**, 5076–5081.
32. I. Willner and S. Rubin, *Angew. Chem., Int. Ed. Engl.*, 1996, **35**, 367–385.
33. C.-M. Chung, Y.-S. Roh, S.-Y. Cho and J.-G. Kim, *Chem. Mater.*, 2004, **16**, 3982–3984.

34. J. Ling, M. Z. Rong and M. Q. Zhang, *Polymer*, 2012, **53**, 2691–2698.
35. P. Cordier, F. Tournilhac, C. Soulie-Ziakovic and L. Leibler, *Nature*, 2008, **451**, 977–980.
36. S. J. Kalista Jr, T. C. Ward and Z. Oyetunji, *Mech. Adv. Mater. Struct.*, 2007, **14**, 391–397.
37. S. J. Kalista and T. C. Ward, *J. R. Soc., Interface*, 2007, **4**, 405–411.
38. M. Burnworth, L. Tang, J. R. Kumpfer, A. J. Duncan, F. L. Beyer, G. L. Fiore, S. J. Rowan and C. Weder, *Nature*, 2011, **472**, 334–337.
39. M. Krogsgaard, M. A. Behrens, J. S. Pedersen and H. Birkedal, *Biomacromolecules*, 2013, **14**, 297–301.
40. N. Holten-Andersen, M. J. Harrington, H. Birkedal, B. P. Lee, P. B. Messersmith, K. Y. C. Lee and J. H. Waite, *Proc. Natl. Acad. Sci.*, 2011, **108**, 2651–2655.
41. S. Burattini, B. W. Greenland, D. H. Merino, W. Weng, J. Seppala, H. M. Colquhoun, W. Hayes, M. E. Mackay, I. W. Hamley and S. J. Rowan, *J. Am. Chem. Soc.*, 2010, **132**, 12051–12058.
42. S. Burattini, H. M. Colquhoun, B. W. Greenland and W. Hayes, *Faraday Discuss.*, 2009, **143**, 251–264.
43. L. Brunsveld, B. Folmer, E. Meijer and R. Sijbesma, *Chem. Rev.*, 2001, **101**, 4071–4098.
44. M. J. Krische and J.-M. Lehn, *Molecular Self-Assembly Organic Versus Inorganic Approaches*, Springer, 2000, pp. 3–29.
45. D. C. Sherrington and K. A. Taskinen, *Chem. Soc. Rev.*, 2001, **30**, 83–93.
46. W. H. Binder, S. Bernstorff, C. Kluger, L. Petraru and M. J. Kunz, *Adv. Mater.*, 2005, **17**, 2824–2828.
47. S. Djurdjevic, D. A. Leigh, H. McNab, S. Parsons, G. Teobaldi and F. Zerbetto, *J. Am. Chem. Soc.*, 2007, **129**, 476–477.
48. F. H. Beijer, H. Kooijman, A. L. Spek, R. P. Sijbesma and E. Meijer, *Angew. Chem., Int. Ed.*, 1998, **37**, 75–78.
49. T. Park, S. C. Zimmerman and S. Nakashima, *J. Am. Chem. Soc.*, 2005, **127**, 6520–6521.
50. F. Maes, D. Montarnal, S. Cantournet, F. Tournilhac, L. Corté and L. Leibler, *Soft Matter*, 2012, **8**, 1681–1687.
51. R. Zhang, T. Yan, B.-D. Lechner, K. Schröter, Y. Liang, B. Li, F. Furtado, P. Sun and K. Saalwächter, *Macromolecules*, 2013, **46**, 1841–1850.
52. R. P. Sijbesma, F. H. Beijer, L. Brunsveld, B. J. Folmer, J. K. Hirschberg, R. F. Lange, J. K. Lowe and E. Meijer, *Science*, 1997, **278**, 1601–1604.
53. A. Faghihnejad, K. E. Feldman, J. Yu, M. V. Tirrell, J. N. Israelachvili, C. J. Hawker, E. J. Kramer and H. Zeng, *Adv. Funct. Mater.*, 2014, **24**, 2322–2333.
54. P. Y. Dankers, E. N. van Leeuwen, G. M. van Gemert, A. Spiering, M. C. Harmsen, L. A. Brouwer, H. M. Janssen, A. W. Bosman, M. J. van Luyn and E. Meijer, *Biomaterials*, 2006, **27**, 5490–5501.
55. H. Colquhoun and B. Klumperman, *Polym. Chem.*, 2013, **4**, 4832–4833.
56. A. W. Bosman, R. P. Sijbesma and E. Meijer, *Mater. Today*, 2004, 7, 34–39.
57. G. M. van Gemert, J. W. Peeters, S. H. Söntjens, H. M. Janssen and A. W. Bosman, *Macromol. Chem. Phys.*, 2012, **213**, 234–242.

58. K. E. Feldman, M. J. Kade, T. F. de Greef, E. Meijer, E. J. Kramer and C. J. Hawker, *Macromolecules*, 2008, **41**, 4694–4700.
59. Y. Chen, A. M. Kushner, G. A. Williams and Z. Guan, *Nat. Chem.*, 2012, **4**, 467–472.
60. J. Hentschel, A. M. Kushner, J. Ziller and Z. Guan, *Angew. Chem.*, 2012, **124**, 10713–10717.
61. A. Phadke, C. Zhang, B. Arman, C.-C. Hsu, R. A. Mashelkar, A. K. Lele, M. J. Tauber, G. Arya and S. Varghese, *Proc. Natl. Acad. Sci.*, 2012, **109**, 4383–4388.
62. J. Cui and A. del Campo, *Chem. Commun.*, 2012, **48**, 9302–9304.
63. K. Mequanint, A. Patel and D. Bezuidenhout, *Biomacromolecules*, 2006, **7**, 883–891.
64. Y. Chen, N. Ballard and S. A. Bon, *Polym. Chem.*, 2013, **4**, 387–392.
65. M. Guo, L. M. Pitet, H. M. Wyss, M. Vos, P. Y. W. Dankers and E. W. Meijer, *J. Am. Chem. Soc.*, 2014, **136**, 6969–6977.
66. B. Yang, Y. Zhang, X. Zhang, L. Tao, S. Li and Y. Wei, *Polym. Chem.*, 2012, **3**, 3235–3238.
67. A. B. Brochu, S. L. Craig and W. M. Reichert, *J. Biomed. Mater. Res., Part A*, 2011, **96**, 492–506.
68. B. C. Tee, C. Wang, R. Allen and Z. Bao, *Nat. Nanotechnol.*, 2012, **7**, 825–832.
69. A. Eisenberg and M. Rinaudo, *Polym. Bull.*, 1990, **24**, 671.
70. J.-S. Kim, R. J. Jackman and A. Eisenberg, *Macromolecules*, 1994, **27**, 2789–2803.
71. M. R. Tant, K. A. Mauritz and G. L. Wilkes, *Ionomers: synthesis, structure, properties and applications*, Springer Science & Business Media, 1997.
72. S. J. Kalista, J. R. Pflug and R. J. Varley, *Polym. Chem.*, 2013, **4**, 4910–4926.
73. S. D. Bergman and F. Wudl, *J. Mater. Chem.*, 2008, **18**, 41–62.
74. F. R. Kersey, D. M. Loveless and S. L. Craig, *J. R. Soc., Interface*, 2007, **4**, 373–380.
75. J. B. Beck, J. M. Ineman and S. J. Rowan, *Macromolecules*, 2005, **38**, 5060–5068.
76. M. Chiper, D. Fournier, R. Hoogenboom and U. S. Schubert, *Macromol. Rapid Commun.*, 2008, **29**, 1640–1647.
77. J. R. Kumpfer, J. J. Wie, J. P. Swanson, F. L. Beyer, M. E. Mackay and S. J. Rowan, *Macromolecules*, 2011, **45**, 473–480.
78. Z. Wang and M. W. Urban, *Polym. Chem.*, 2013, **4**, 4897–4901.
79. E. A. Appel, J. del Barrio, X. J. Loh and O. A. Scherman, *Chem. Soc. Rev.*, 2012, **41**, 6195–6214.
80. Z. Wei, J. He, T. Liang, H. Oh, J. Athas, Z. Tong, C. Wang and Z. Nie, *Polym. Chem.*, 2013, **4**, 4601–4605.
81. S. Basak, J. Nanda and A. Banerjee, *Chem. Commun.*, 2014, **50**, 2356–2359.
82. L. Fang, M. A. Olson, D. Benítez, E. Tkatchouk, W. A. Goddard III and J. F. Stoddart, *Chem. Soc. Rev.*, 2010, **39**, 17–29.
83. Z. Zhu, C. J. Cardin, Y. Gan, C. A. Murray, A. J. White, D. J. Williams and H. M. Colquhoun, *J. Am. Chem. Soc.*, 2011, **133**, 19442–19447.

84. H. M. Colquhoun, Z. Zhu, C. J. Cardin, Y. Gan and M. G. Drew, *J. Am. Chem. Soc.*, 2007, **129**, 16163–16174.
85. S. Burattini, H. M. Colquhoun, J. D. Fox, D. Friedmann, B. W. Greenland, P. J. Harris, W. Hayes, M. E. Mackay and S. J. Rowan, *Chem. Commun.*, 2009, 6717–6719.
86. Z. Xu, J. Peng, N. Yan, H. Yu, S. Zhang, K. Liu and Y. Fang, *Soft Matter*, 2013, **9**, 1091–1099.
87. R. Vaiyapuri, B. W. Greenland, S. J. Rowan, H. M. Colquhoun, J. M. Elliott and W. Hayes, *Macromolecules*, 2012, **45**, 5567–5574.
88. J. H. Kavouras and J. S. Maki, *Invertebr. Biol.*, 2003, **122**, 138–151.
89. J. Kavouras and J. Maki, *J. Appl. Microbiol.*, 2004, **97**, 1236–1246.
90. J. H. Waite, *Int. J. Adhes. Adhes.*, 1987, 7, 9–14.
91. J. H. Waite, *Structure, cellular synthesis and assembly of biopolymers*, Springer, 1992, pp. 27–54.
92. B. P. Lee, P. B. Messersmith, J. N. Israelachvili and J. H. Waite, *Annu. Rev. Mater. Res.*, 2011, **41**, 99.
93. E. Vaccaro and J. H. Waite, *Biomacromolecules*, 2001, **2**, 906–911.
94. M. Yu, J. Hwang and T. J. Deming, *J. Am. Chem. Soc.*, 1999, **121**, 5825–5826.
95. J. Waite and M. Tanzer, *Biochem. Biophys. Res. Commun.*, 1980, **96**, 1554–1561.
96. H. Zeng, D. S. Hwang, J. N. Israelachvili and J. H. Waite, *Proc. Natl. Acad. Sci.*, 2010, **107**, 12850–12853.
97. Q. Lu, D. X. Oh, Y. Lee, Y. Jho, D. S. Hwang and H. Zeng, *Angew. Chem.*, 2013, **125**, 4036–4040.
98. Q. Lu, D. S. Hwang, Y. Liu and H. Zeng, *Biomaterials*, 2012, **33**, 1903–1911.
99. B. J. Kim, D. X. Oh, S. Kim, J. H. Seo, D. S. Hwang, A. Masic, D. K. Han and H. J. Cha, *Biomacromolecules*, 2014, **15**, 1579–1585.
100. B. Yang, N. Ayyadurai, H. Yun, Y. S. Choi, B. H. Hwang, J. Huang, Q. Lu, H. Zeng and H. J. Cha, *Angew. Chem., Int. Ed.*, 2014, **53**, 13360–13364.
101. P. S. Yavvari and A. Srivastava, *J. Mater. Chem. B*, 2015, **3**, 899–910.
102. P. Sun, J. Wang, X. Yao, Y. Peng, X. Tu, P. Du, Z. Zheng and X. Wang, *ACS Appl. Mater. Interfaces*, 2014, **6**, 12495–12504.
103. H. Ceylan, M. Urel, T. S. Erkal, A. B. Tekinay, A. Dana and M. O. Guler, *Adv. Funct. Mater.*, 2013, **23**, 2081–2090.
104. Z. Shafiq, J. Cui, L. Pastor-Pérez, V. San Miguel, R. A. Gropeanu, C. Serrano and A. del Campo, *Angew. Chem.*, 2012, **124**, 4408–4411.
105. M. Krogsgaard, A. Andersen and H. Birkedal, *Chem. Commun.*, 2014, **50**, 13278–13281.
106. D. Roy, J. N. Cambre and B. S. Sumerlin, *Chem. Commun.*, 2009, 2106–2108.
107. J. N. Cambre, D. Roy and B. S. Sumerlin, *J. Polym. Sci., Part A: Polym. Chem.*, 2012, **50**, 3373–3382.
108. J. Yan, G. Springsteen, S. Deeter and B. Wang, *Tetrahedron*, 2004, **60**, 11205–11209.
109. L. He, D. E. Fullenkamp, J. G. Rivera and P. B. Messersmith, *Chem. Commun.*, 2011, **47**, 7497–7499.

110. M. Nakahata, S. Mori, Y. Takashima, A. Hashidzume, H. Yamaguchi and A. Harada, *ACS Macro Lett.*, 2014, **3**, 337–340.
111. M. Vatankhah-Varnoosfaderani, A. GhavamiNejad, S. Hashmi and F. J. Stadler, *Chem. Commun.*, 2013, **49**, 4685–4687.
112. M. Vatankhah-Varnoosfaderani, S. Hashmi, A. GhavamiNejad and F. J. Stadler, *Polym. Chem.*, 2014, **5**, 512–523.
113. C. C. Deng, W. L. Brooks, K. A. Abboud and B. S. Sumerlin, *ACS Macro Lett.*, 2015, **4**, 220–224.
114. J. Su, F. Chen, V. L. Cryns and P. B. Messersmith, *J. Am. Chem. Soc.*, 2011, **133**, 11850–11853.
115. D. Tarus, E. Hachet, L. Messager, B. Catargi, V. Ravaine and R. Auzély-Velty, *Macromol. Rapid Commun.*, 2014, **35**, 2089–2095.
116. D. Roy and B. S. Sumerlin, *ACS Macro Lett.*, 2012, **1**, 529–532.
117. K. T. Kim, J. J. L. M. Cornelissen, R. J. M. Nolte and J. C. M. van Hest, *J. Am. Chem. Soc.*, 2009, **131**, 13908–13909.
118. N. Holten-Andersen, A. Jaishankar, M. J. Harrington, D. E. Fullenkamp, G. DiMarco, L. He, G. H. McKinley, P. B. Messersmith and K. Y. C. Lee, *J. Mater. Chem. B*, 2014, **2**, 2467–2472.
119. M. Krogsgaard, M. R. Hansen and H. Birkedal, *J. Mater. Chem. B*, 2014, **2**, 8292–8297.
120. B. K. Ahn, D. W. Lee, J. N. Israelachvili and J. H. Waite, *Nat. Mater.*, 2014, **13**, 867–872.
121. M. Vatankhah-Varnoosfaderani, A. GhavamiNejad, S. Hashmi and F. J. Stadler, *Macromol. Rapid Commun.*, 2015, **36**, 447–452.
122. L. Li, B. Yan, J. Yang, L. Chen and H. Zeng, *Adv. Mater.*, 2015, **27**, 1294–1299.
123. D. E. Fullenkamp, L. He, D. G. Barrett, W. R. Burghardt and P. B. Messersmith, *Macromolecules*, 2013, **46**, 1167–1174.
124. M. J. Harrington and J. H. Waite, *J. Exp. Biol.*, 2007, **210**, 4307–4318.
125. M. J. Harrington, H. S. Gupta, P. Fratzl and J. H. Waite, *J. Struct. Biol.*, 2009, **167**, 47–54.
126. S. Schmidt, A. Reinecke, F. Wojcik, D. Pussak, L. Hartmann and M. J. Harrington, *Biomacromolecules*, 2014, **15**, 1644–1652.
127. S. Huang, L. Yang, M. Liu, S. L. Phua, W. A. Yee, W. Liu, R. Zhou and X. Lu, *Langmuir*, 2013, **29**, 1238–1244.
128. Y. Fu, J. Wang, Q. Liu and H. Zeng, *Carbon*, 2014, **77**, 710–721.
129. J. Zhang, M. S. Azam, C. Shi, J. Huang, B. Yan, Q. Liu and H. Zeng, *RSC Adv.*, 2015, **5**, 32272–32282.
130. H.-P. Cong, P. Wang and S.-H. Yu, *Chem. Mater.*, 2013, **25**, 3357–3362.
131. J. Liu, G. Song, C. He and H. Wang, *Macromol. Rapid Commun.*, 2013, **34**, 1002–1007.
132. K. A. Williams, A. J. Boydston and C. W. Bielawski, *J. R. Soc., Interface*, 2007, **4**, 359–362.

CHAPTER 6

Biomimetic Materials for Efficient Atmospheric Water Collection

LIANBIN ZHANG[a] AND PENG WANG*[a]

[a]Water Desalination and Reuse Center (WDRC), Division of Biological and Environmental Sciences and Engineering, King Abdullah University of Science and Technology (KAUST), Thuwal, Saudi Arabia
*E-mail: Peng.Wang@kaust.edu.sa

6.1 Introduction

Water is the foundation of life. Although 75% of the Earth's surface is covered by water, 97% of the water is salty water, leaving only 3% as fresh water. However, two-thirds of the fresh water on Earth is frozen in snow and ice in glaciers and ice caps, which leaves only about 1% of all the Earth's water as liquid and fresh, making it a very scarce natural resource. Nowadays, water scarcity is a severely growing problem in semi-arid desert regions, land-scarce countries, and in countries with high levels of economic activity. As more people put ever increasing demands on limited supplies, the cost and effort to build or even maintain access to water is increasing. So far to meet the increasing freshwater requirements, many countries depend on desalination technologies to obtain fresh water should economic conditions allow.[1–5] Particularly in the Middle East, seawater desalination is

RSC Smart Materials No. 20
Smart Materials for Advanced Environmental Applications
Edited by Peng Wang

Published by the Royal Society of Chemistry, www.rsc.org

a vital freshwater resource in countries such as Saudi Arabia, United Arab Emirates, and Kuwait.[5] However, due to the high energy consumption in these desalination processes, rising energy costs, and diminishing fossil fuel stock, more sustainable desalination methods that utilize alternative energy resources are still highly desired.[6–10] On the other hand, in remote and rural areas where access to centralized drinking water supply and even electricity is unavailable, the desalination method is not suited.[11–15]

To face the conventional water shortage and rising water prices, the search and use of alternative sources of water has been attempted over the last decades. Among all the alternative water sources, fog has been considered as a promising one.[16–23] Recently, the UN water crisis report by the Food and Agriculture Organization has pointed out that, "fog collection technology appears to be an extremely promising and low-cost water harvesting system for drinking water, crop irrigation, livestock beverage, and forest restoration in dryland mountains."[24] The collection of fog for the purpose of the production of fresh water has attracted increasing attention over the past few decades. As one part of the big solution to the water crisis, fog harvesting has already been successfully demonstrated in the field in more than 17 countries over recent decades.[25]

Technically, the collection of fog is not complicated, and it can be generally realized by exposing collecting mesh materials to the atmosphere, and the fog is pushed through the mesh by the wind.[23,26] A fraction of the fog droplets are then captured on the mesh material by impaction. When more fog droplets are captured, they merge to form larger droplets, run down the mesh material into gutters and eventually into a storage tank due to gravity. Although the mesh-based fog collection is effective, further improvement of the fog collection efficiency is still highly desired.

To adapt to the living environment, natural creatures have evolved various elaborate biomaterials with superior performances.[27] In nature, some native animals and plants in many of the arid regions readily cope with insufficient access to fresh water or lack of precipitation through fog or dew harvest as well as water vapor absorption.[28–31] For example, in the dry Namib Desert, desert *Stenocara* beetles can survive by collecting water from foggy air based on the micrometer-sized patterns of hydrophobic and hydrophilic regions on their backs.[28] In the same region, dune grasses can also intercept tiny water drops contained in fog and transport them to their roots by employing an anisotropic microstructure on its long and thin leaves.[29] With special structures consisting of periodic spindle-knots, spider silks are also endowed with water collecting properties due to the directional movement of water drops from joints to spindle-knots.[30] The cactus *Opuntia microdasys* has an integrated fog collection system arising from the evenly distributed clusters of spines and trichomes on the cactus stem.[31] Learning from nature is a shortcut for designing new functional materials. In this chapter, the fabrication of the biomimetic fog harvesting materials is presented.

6.2 Desert Beetle-Inspired Surface with Patterned Wettability for Fog Collection

6.2.1 Introduction

The *Stenocara* beetle, a native of Namib Desert in southwest Africa, one of the most arid habitats on earth, lives in a location that receives a mere 40 mm of rain a year yet it can harvest drinking water from early morning fogs.[28,32,33] Figure 6.1 shows a desert beetle tilting its body forward to capture small water droplets in fog (Figure 6.1d).[33] This unique behavior is termed fog-basking. Closer observation reveals that the *Stenocara* beetle's elytra has a unique array of raised hydrophilic non-wax-coated bumps approximately 0.5–1.5 mm separation and 0.5 mm in diameter (Figure 6.1b and c), on a superhydrophobic wax-coated background with microstructures, which allows for efficient condensation of water from fog-laden wind. Water condenses on the hydrophilic bumps when the surface is cooler than the surrounding air, forming droplets that grow in size yet are pinned to their location due to the superhydrophobic exoskeleton background. These water droplets grow until they detach and roll off the surface and into the mouth of the *Stenocara* beetle, providing a reliable supply of drinking water. Furthermore, the size and density of the raised hydrophilic bumps provide a good balance between water droplet detachment from the surface and inhibition of its re-evaporation to the atmosphere. The efficient water collection capability induced by the hydrophilic/superhydrophobic patterned surface of the *Stenocara* beetle has inspired researchers to mimic its surface pattern for the purpose of efficient water collection. Besides, such a patterned surface can also be of use for collecting drinking water from fog which is in abundant presence in coastal regions.

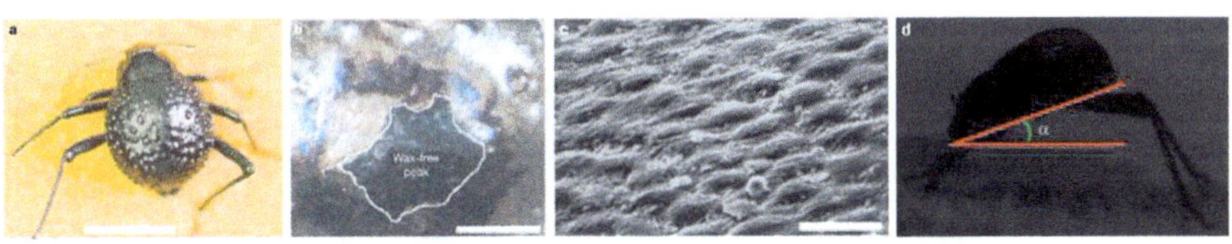

Figure 6.1 The water-capturing surface of the fused elytra of the desert *Stenocara* beetle. (a) Adult female, dorsal view; peaks and troughs are evident on the surface of the elytra. (b) A hydrophilic wax-free "bump" on the elytra, (c) scanning electron micrograph of the textured surface of the depressed areas. Scale bars, (a) 10 mm; (b) 0.2 mm; (c) 10 μm. (d) photograph of a fog-basking beetle inside a fog chamber exhibiting a characteristic fog-basking head stand. This posture allows fog water collected on the beetle's dorsal surface to trickle down to its mouth. ((a–c) Reprinted by permission from Macmillan Publishers Ltd: *Nature*, 2001, **414**, 33–34,[28] copyright 2001. (d) Reprinted with permission from BioMed Central Ltd: *Front. Zool.*, 2010, 7, 23.[33] Copyright 2010.)

6.2.2 Traditional Lithographic Methods for the Fabrication of Biomimetic Patterned Surfaces for Fog Collection

The most commonly used strategy for the fabrication of a surface with patterned wettability is the mask-based photolithographic method. Photolithographic methods all share the same working principle: exposure of an appropriate material to electromagnetic radiation (UV, DUV, EUV, or X-ray) introduces a pattern into the material as a result of a set of chemical changes in its molecular structure.[34–36] When masks are used, the lithographic process yields a replica of the pattern on the mask. Figure 6.2 shows a typical photolithography-based fabrication of the patterned wetting surface by Zahner *et al.*[37] First, a superhydrophobic, microporous poly(butyl methacrylate*co*-ethylene dimethacrylate) (BMA-EDMA) film was prepared by UV-initiated radical polymerization of a prepolymer mixture containing monomers, cross-linkers, porogens, and a UV initiator onto a glass plate. Then, the superhydrophobic BMA-EDMA film was wetted with a mixture composed of a hydrophilic methacrylate monomer (*e.g.*, [2-(methacryloyloxyl)ethyl]

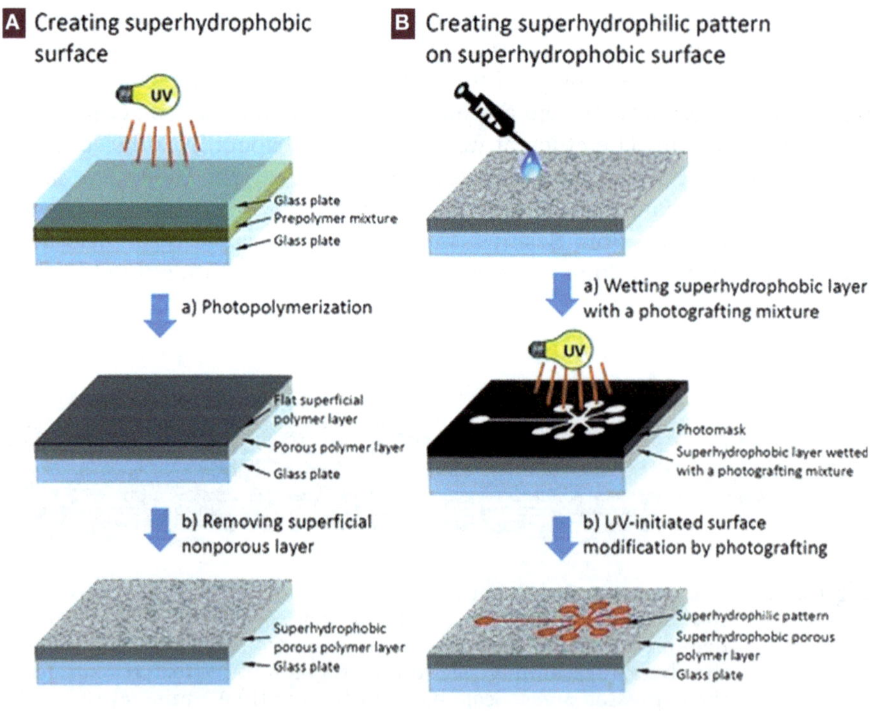

Figure 6.2 Schematic illustration of typical preparation of patterned wetting surfaces by a photolithographic method. (A) Making superhydrophobic porous polymer films on a glass support and (B) creating superhydrophilic micropatterns by UV-initiated photografting. (Adapted with permission from D. Zahner, J. Abagat, F. Svec, J. M. J. Fréchet and P. A. Levkin, A Facile Approach to Superhydrophilic–Superhydrophobic Patterns in Porous Polymer Films, *Adv. Mater.*, 2011, 23, 3030–3034. Copyright © [2011] John Wiley and Sons.)

trimethylammonium hydrochloride (META) or 2-acrylamido-2-methyl-1-propanesulfonic acid), benzophenone as the initiator, and a mixture of *tert*-butanol/water, and then was irradiated with UV light through a photomask. The photografting of the hydrophilic polymer only occurred at the area exposed to the UV light, and after mask removal and washing with solvents, hydrophilic patterns were obtained on the superhydrophobic surface. This method can be used to create superhydrophilic patterns on the superhydrophobic background with feature sizes as small as a few micrometers.

Based on this photolithographic method, Garrod *et al.* first fabricated patterned a superhydrophobic–superhydrophilic surface to mimic the desert beetle's back for fog collection.[38] As shown in Figure 6.3a, they created patterned hydrophilic polymers on a superhydrophobic background using plasma-chemical deposition through patterned grids as mask. They explored the effect of various hydrophobic–hydrophilic surface functionalities and pattern sizes on water-collection efficiency and found that hydrophilic spots with a size of about 500 μm exhibited optimal water collection performance in the study, whereas spots exceeding 1200 μm showed little to no condensation (Figure 6.3b). For the larger spots (exceeding 1200 μm), it was observed that a critical size for droplet detachment could not be reached: the rate of water loss through processes such as evaporation exceeds the rate of fog droplet condensation. This means that the hydrophilic domains are so large that they act in a way similar to that of purely hydrophilic surfaces. In addition, the authors also revealed there is a minimum hydrophilic spot diameter required to facilitate micro-condensation: below 400 μm, droplets were seen to form and quickly reach critical size. However, they lacked sufficient mass to overcome the surface tension, and thus remained attached to the surface. These results are consistent with the *Stenocara* beetle's back, which has bumps approximately 600 μm in diameter and 500 μm to 1.5 mm apart.

By using a combination of microcontact printing lithography techniques and surface modification, Varanasi and coworkers fabricated two different hybrid surfaces: flat surfaces with repeating hydrophobic and hydrophilic

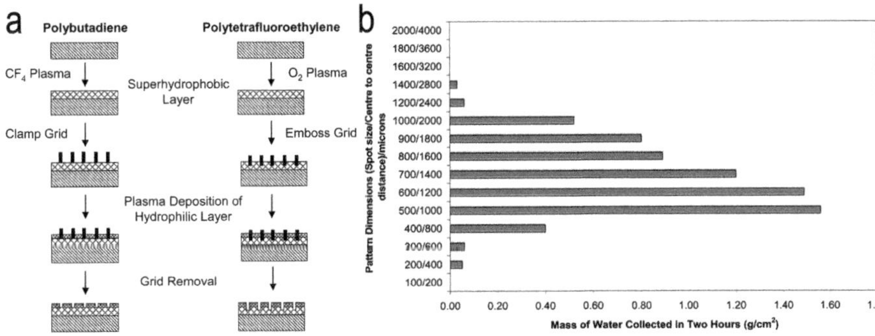

Figure 6.3 (a) Schematic illustration of micropatterning a hydrophilic plasma polymer onto a superhydrophobic background. (b) Water condensation study for different sized arrays. (Reprinted with permission from *Langmuir*, 2007, **23**, 689–693. Copyright 2007 American Chemical Society.)

regions, and textured surfaces consisting of pillars that had hydrophobic sidewalls and hydrophilic top surfaces.[39] Their results also demonstrated that textured hydrophobic surfaces with hydrophilic tops promote nucleation and growth of Cassie-type droplets and therefore exhibited superior droplet shedding properties under condensation.

Also by using the photolithography-based method, Chen and coworkers prepared hierarchical nanograss micropyramidal architectures with hybrid wettability.[40] Globally, the structured surfaces were superhydrophobic, favoring the departure of the condensing droplets (Figure 6.4). Locally, they had wettable patches that facilitated droplet growth and preferential coalescence. With such 3D structured hybrid surfaces, they showed that the number density of droplets was enhanced by approximately 65% and the spontaneous removal volume of droplets was dramatically increased by approximately 450%.

Recently, Bai and coworkers reported the fabrication a novel kind of surface with star-shaped wettability patterns.[41] In their study, a superhydrophilic surface was first fabricated by depositing TiO_2 slurry onto a bare glass slide *via* a spin-coating method. Then the film is treated with a

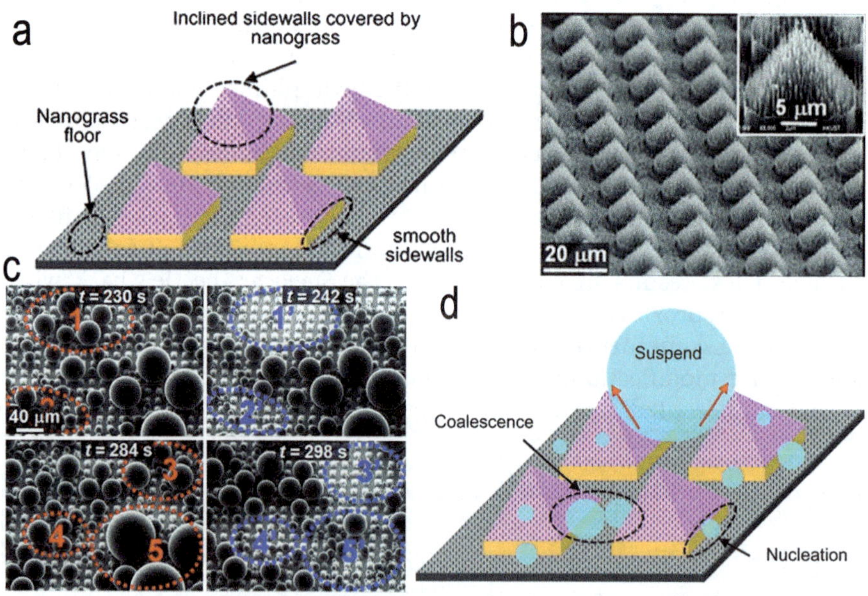

Figure 6.4 (a) Schematic drawing showing different surface roughness on the smooth sidewalls of the pyramids, nanograssed sidewalls, and nanograssed floor. (b) SEM image showing the as-fabricated nanograssed micropyramid arrays on a Si wafer. (c) Drop departure dynamics on surfaces with hierarchical nanograssed micropyramid arrays. (d) Schematic drawing of drops sitting on the nanograssed micropyramid arrays. The synergy between the micro- and nanoscale roughness features resulted in a stable Cassie state and upwards surface tension force, which assisted in drop departure. Adapted with permission from X. Chen, J. Wu, R. Ma, M. Hua, N. Koratkar, S. Yao and Z. Wang, Nanograssed Micropyramidal Architectures for Continuous Dropwise Condensation, *Adv. Funct. Mater.*, 2011, **21**, 4617–4623. Copyright © [2011], John Wiley and Sons.

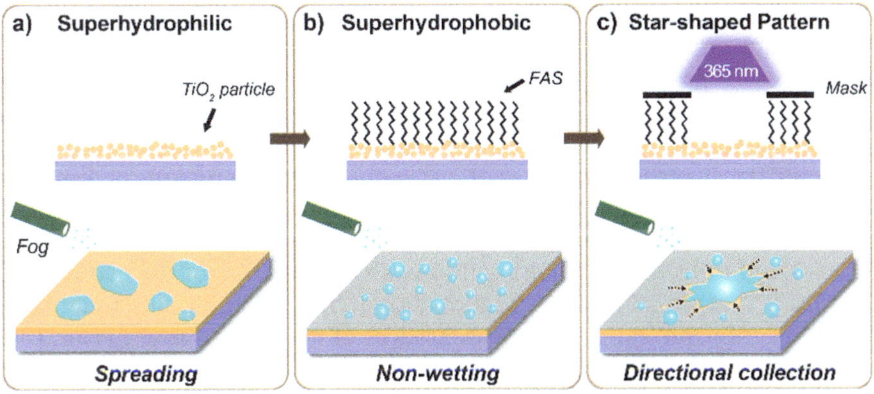

Figure 6.5 Schematic illustration of the fabrication process of bioinspired surfaces with star-shaped wettability patterns. a) Superhydrophilic surface composed of TiO_2 nanoparticles, where fog droplets spread (bottom). b) Superhydrophobic surface modified with FAS showing non-wetting property to fog droplets (bottom). c) Bioinspired gradient surface with a star-shaped wettability pattern. It is realized by illuminating the FAS-modified film under UV light with a photomask. The fog droplets would be collected directionally toward the star-shape region, which is more wettable (bottom). Reprinted with permission from H. Bai, L. Wang, J. Ju, R. Sun, Y. Zheng and L. Jiang, Efficient Water Collection on Integrative Bioinspired Surfaces with Star-Shaped Wettability Patterns, *Adv. Mater.*, 2014, **26**, 5025–5030. Copyright © [2014] John Wiley and Sons.

heptadecafluorodecyl-trimethoxysilane (FAS) to change the wettability from superhydrophilic (Figure 6.5a) to superhydrophobic (Figure 6.5b), where fog droplets hardly wet the surface and remain in a spherical shape (Figure 6.5b). In the next step, photomasks with a circle-shaped pattern or 4-, 5-, 6-, and 8-pointed star-shaped patterns were used to obtain features of wettability patterns *via* selective illumination of UV light, which decomposed the FAS monolayer (Figure 6.5c) by the underneath photocatalytic TiO_2. On the surfaces with star-shaped wettability patterns, fog droplets could be directionally collected toward more wettable regions (Figure 6.5c). They found that such surfaces with star-shaped patterns realize more efficient water collection compared to other surfaces that are uniformly superhydrophilic, uniformly superhydrophobic, or even circle patterned. The improved water collection by the star-shaped pattern is because the tips of the star generate a Laplace pressure gradient from the shape gradient, which further enhances this directional movement of water droplets toward the center of the stars.

6.2.3 Direct Methods for Creating Patterned Wettability for Fog Harvesting

Although these photolithography-based methods for creating such patterned surfaces for fog collection are effective, these methods mainly rely on a mask-based strategy, where a pre-designed mask preparation is involved followed by a subsequent pattern transfer, which is multistep, indirect, and expensive. To produce stable hydrophilic micro-sized patterns on superhydrophobic substrates for practical applications, a direct method (*i.e.*, one-step and thus mask-free) of depositing hydrophilic species in well-defined micropatterns is still highly desired. The difficulty of direct deposition of hydrophilic species onto the superhydrophobic surface to obtain the micropatterned surface lies in the following: (1) the ultralow surface energy of the superhydrophobic substrate surface greatly decreases the adhesion between the hydrophilic species and the surface, resulting in unstable deposition of the hydrophilic species;[42–44] (2) in addition, a superhydrophobic surface entails rough surface structure and trapped in the interstices of the rough surface is air. Due to the presence of those discrete air pockets, a water droplet sitting on such a solid-air composite surface interacts only with a small fraction of the solid surface, which is widely known as Cassie's wetting state.[45,46] Therefore, in this case, both limited interaction area and adhesion between the hydrophilic species in the aqueous droplets and the superhydrophobic substrate make it impossible to directly pattern superhydrophobic surfaces. Thus, a direct and general method to this end has to enhance the interaction strength (*i.e.*, adhesion) between the hydrophilic species in the aqueous droplets and the superhydrophobic surface and to increase the contacting area of solid surface and liquid droplets.

Zhai and coworkers first prepared hydrophilic patterns on superhydrophobic surfaces using a layer-by-layer assembly of polyelectrolytes in a mixed water/2-propanol solvent.[47] This approach can be used to create patterned surfaces that mimic the water harvesting structure of the Stenocara beetle's back. However, this method is not suited for the large-scale fabrication of the patterned surface, and is difficult to control the pattern size.

Recently, inspired by high adhesive mussel protein,[48] Zhang *et al.* have reported a facile and, most importantly, direct method for the preparation of a micropatterned surface for fog harvesting by an inkjet printing strategy (Figure 6.6).[49] By directly inkjet printing a mussel-inspired ink of dopamine solution with delicately optimized solution composition, stable Wenzel's microdroplets of dopamine solution with well-defined micropatterns are obtained onto the superhydrophobic surfaces. Upon the formation of polydopamine *via in situ* polymerization, superhydrophilic micropatterns with well-controlled pattern dimension can be readily achieved on the superhydrophobic surfaces (Figure 6.6a and b). This method is very convenient for the design and fabrication of patterns with different sizes and configurations, and is thus suitable for the optimization and selection of the pattern for fog collection purposes. Their results have revealed that the micropatterned superhydrophobic surface with a similar pattern dimension to that of the desert beetle's back (*i.e.*, ~500 μm pattern size and ~1000 μm separation distance), exhibited the highest water collection efficiency among the uniformly superhydrophobic surface, superhydrophilic surface, and other patterned surfaces with different pattern sizes. With this convenient strategy, the design and large-scale fabrication of patterned superhydrophobic surfaces

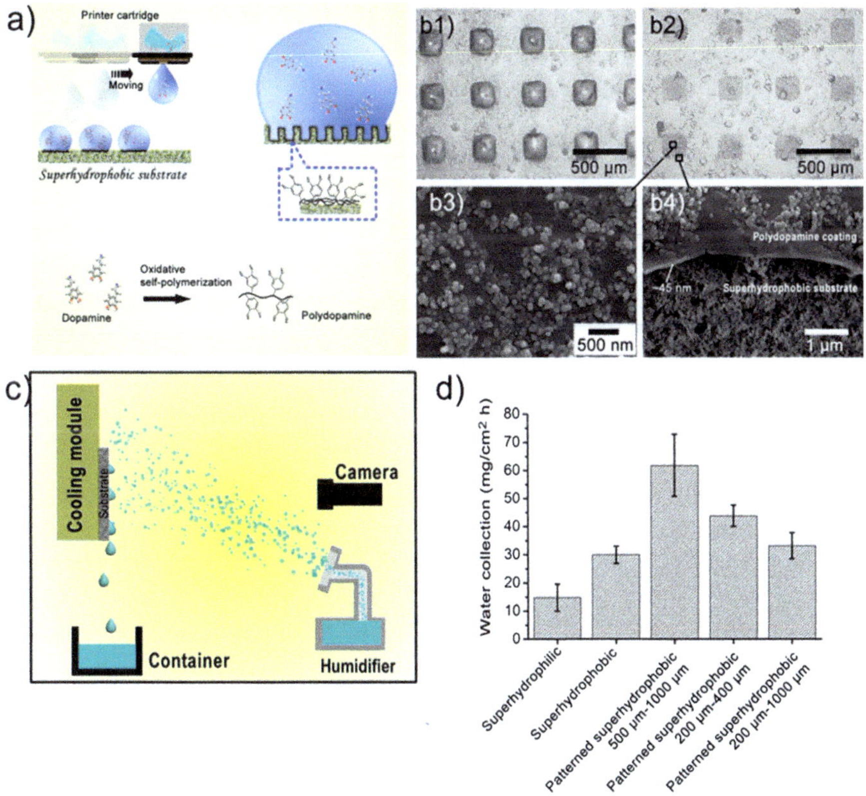

Figure 6.6 (a) Scheme of the inkjet printing method for the micropatterning of superhydrophilicity on superhydrophobic surfaces. Optical microscopic photograph of the as-printed dopamine droplet (b1) and (b2) on the superhydrophobic surface. (b3) and (b4), SEM images of the polydopamine patterns. (c) Experimental setup of water collection system from fog. (d) Water collection efficiency by five different surfaces. L. Zhang, J. Wu, M. N. Hedhili, X. Yang and P. Wang, *J. Mater. Chem. A*, 2015, **3**, 2844–2852. Reproduced by permission of The Royal Society of Chemistry.

would be greatly facilitated, which opens up a new avenue for the patterned superhydrophobic surface in practical applications, such as fog harvesting and water condensation/collection in thermal desalination processes.

To further improve the practical applicability, Thickett *et al.* prepared thin polymer coatings with hybrid hydrophobic and hydrophilic regions by a dewetting method.[50] In their study, polymer bilayers were prepared on clean, smooth silicon substrates consisting of a polystyrene (PS) under layer and a poly(4-vinylpyridine) (P4VP) top layer by sequential spin coating. Annealing was performed above 160 °C (above the glass transition temperature for both polymers), and a resultant surface morphology consisting of a series of isolated droplets and interconnected cylinders of P4VP on a PS background

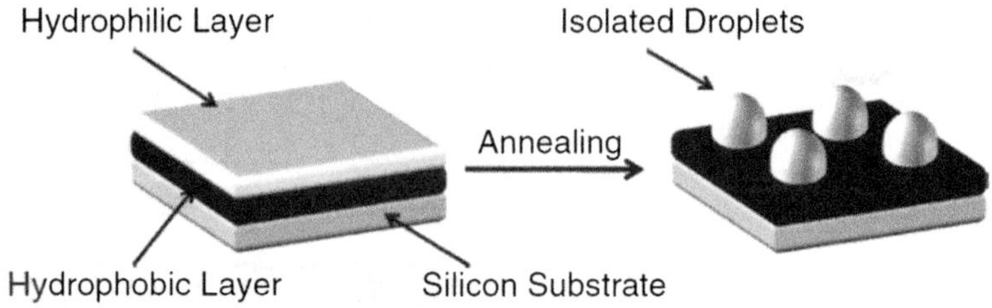

Figure 6.7 Proposed formation of micropatterned surfaces *via* the dewetting of polymer bilayer films. The top layer is an unstable P4VP film, which spontaneously dewets on the bottom PS layer. Reprinted with permission from Biomimetic Surface Coatings for S. C. Thickett, C. Neto and A. T. Harris, Atmospheric Water Capture Prepared by Dewetting of Polymer Films, *Adv. Mater.*, 2011, **23**, 3718–3722. Copyright © [2011] John Wiley and Sons.

was obtained directly (Figure 6.7). They demonstrated that such hybrid films could achieve an enhanced rate of surface water condensation. Such coatings can be prepared on substrates of any shape using conventional techniques such as spin coating, dip coating, or spraying and the procedure is readily scalable, allowing for fabrication on the meter scale.

6.3 Spider Silk-Inspired Fibers for Atmospheric Water Collection

Another important example of nature for water droplet collection is spider silk (Figure 6.8a). Zheng and coworkers' findings revealed that the water-collecting ability of the silk of the cribellate spider *Uloborus walckenaerius* is the result of a unique fiber structure that forms after wetting, with the 'wet-rebuilt' fibers characterized by periodic spindle-knots made of random nanofibrils and separated by joints made of aligned nanofibrils (Figure 6.8b).[30] The spindle-knots are composed of highly random nanofibrils, thus having a rough surface structure. The joints are composed of aligned nanofibrils, with less roughness than the knots. Therefore, the spindle-knot is more hydrophilic and has a higher apparent surface energy than the joint (Figure 6.8c). The force generated by a surface energy gradient that arises from a difference in surface roughness is given by eqn (6.1).

$$F = \int_{L_j}^{L_k} \gamma\left(\cos\theta_A - \cos\theta_R\right)\mathrm{d}l \tag{6.1}$$

where γ is the surface tension of water, θ_A and θ_R are the advancing and receding angles of water drop on spider silk, respectively, and dl is the integrating variable along the length from the joint (L_j) to the spindle-knot (L_k). The surface energy gradient arising from differences in roughness will thus drive water drops to move from the less hydrophilic region (joint with relative lower surface energy) to the more hydrophilic region(spindle-knot with high surface energy) (Figure 6.8b).

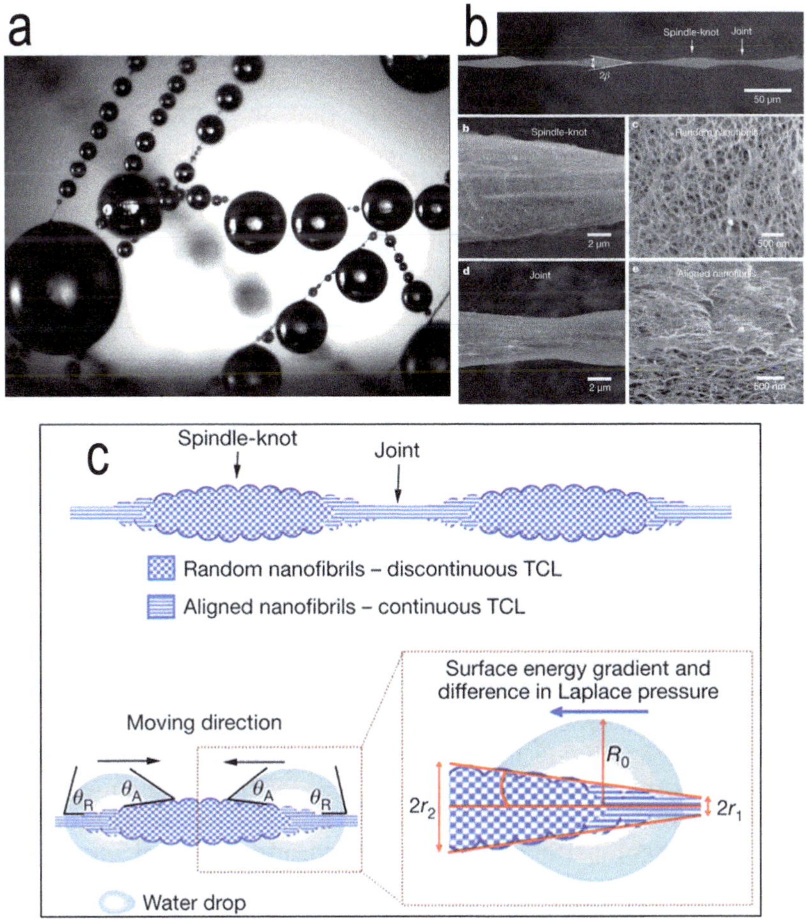

Figure 6.8 Directional water movement inspired by water-collecting spider silk. (a) Water drops chain on a spider's web in mist. (b) SEM image showing periodic spindle-knots and joints in *Uloborus walckenaerius* spider silk. (c) Schematic illustration of the directional water-collection mechanism on spider silk. Adapted with permission from Macmillan Publishers Ltd., Nature Publishing Group: *Nature*, 2010, **463**, 640–643, copyright (2010).

Furthermore, the authors have also analyzed another possible driving force for directional water drop movement arising from the spindle-shaped geometry of the knots, which will generate a difference in Laplace pressure. As illustrated in Figure 6.8c, a spindle-knot can be thought of as two oppositely curved and joined conical objects. Such a conical shape with a curvature gradient will give rise to a difference in Laplace pressure (ΔP) acting on a water drop, as described by eqn (6.2).

$$\Delta P = -\int_{r_1}^{r_2} \frac{2\gamma}{(r + R_0)^2} \sin\beta \, \mathrm{d}z \tag{6.2}$$

where r is the local radius, R_0 is the drop radius, β is the half apex-angle of the spindle-knot, and z is the integrating variable along the diameter of the spindle-knot. The Laplace pressure on the high curvature site (the joint with local radius r_1) is larger than that on the low curvature site (the spindle-knot with local radius r_2) because r_1 is smaller than r_2, and the resultant non-equilibrium Laplace pressure difference within a water drop will propel the drop to move from the joint to the spindle-knot. The overall result is that the surface energy gradient arising from the anisotropic surface structures and the difference in Laplace pressure act cooperatively to drive condensing and growing water drops from the joint to the spindle-knot.

After the mechanism of the unique water collection capability of spider silk was revealed, a series of artificial fibers that mimic the structural features of wet-rebuilt spider silk have been reported.[30,51–55] Jiang and coworkers first designed and fabricated artificial polymeric fibers with similar periodic microstructures.[30] In follow-up work, they further demonstrated that tiny water droplets could be driven with a controllable direction by tuning the design of the surface curvature and the chemical and roughness gradients on the surfaces of fibers. In one of their studies, they further prepared a series of artificial spider silks with spindle-knots in which the chemical compositions and surface nanostructures were subtly designed.[53] They used a nylon fiber and different polymers including poly(vinylacetate) (PVAc), poly(methyl methacrylate) (PMMA), polystyrene (PS), and poly(vinylidene fluoride) (PVDF), to fabricate artificial spider silks. On the other hand, the surface roughness (porous nanostructures) of the spindle-knots was also designed through phase separation. Their investigations demonstrated that tiny water drops (tens of picoliters) could be driven with controllable direction ("toward" or "away from" the knot) by optimizing the cooperation of curvature, chemical, and roughness gradients on the fiber surfaces (Figure 6.9). This study provided a good way for designing smart materials and devices to drive tiny water drops in a controllable manner.

6.4 Desert Plants-Inspired Water Collection

In order to survive in an extremely hot living environment, plants also have evolved complex structures that give them special capabilities to collect and transport water.[29,31,56] Recently, it was discovered by Ju and coworkers that cactus species living in deserts have an integrated multifunctional system that facilitates efficient fog collection, which can be attributed to the integration of its multi-level surface structures.[31] As shown by Figure 6.10, the cactus stem shows a well-distributed array of clusters of needle-like spines and trichomes on the surface. SEM imaging of the spines reveals that a spine is composed of three parts with different structural features: the tip contains oriented barbs, the middle contains gradient grooves, and the base contains

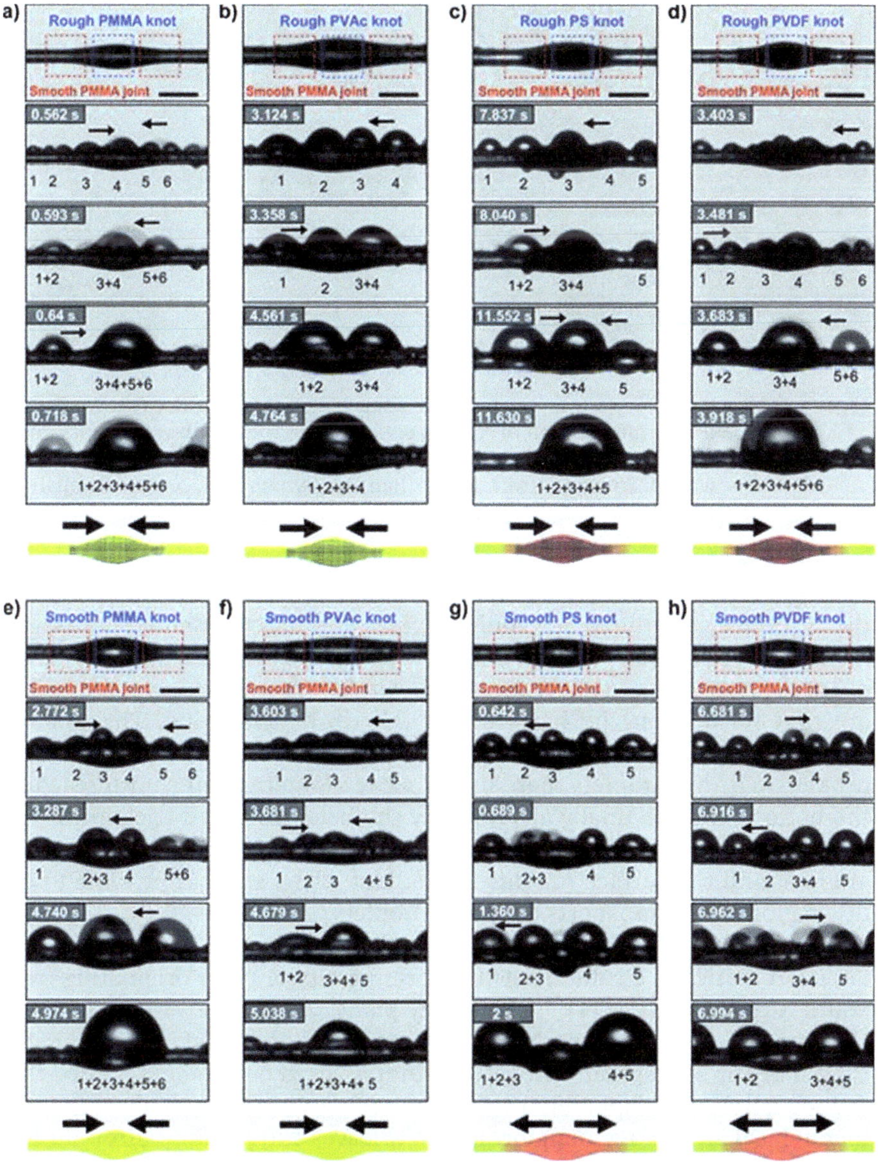

Figure 6.9 *In situ* optical observation of driving tiny water drops with controllable direction on artificial spider silks. Tiny water drops were driven toward the knots on the fibers with rough (a) PMMA, (b) PVAc, (c) PS, and (d) PVDF spindle-knots. After the fibers were smoothened, drops were driven in different directions: toward the knots on the fibers with smooth (e) PMMA and (f) PVAc spindle-knots or away from the knots on the fibers with smooth (g) PS and (h) PVDF spindle-knots. Reprinted with permission from H. Bai, X. Tian, Y. Zheng, J. Ju, Y. Zhao and L. Jiang, Direction Controlled Driving of Tiny Water Drops on Bioinspired Artificial Spider Silks, *Adv. Mater.*, 2010, **22**, 5521–5525. Copyright © [2010] John Wiley and Sons.

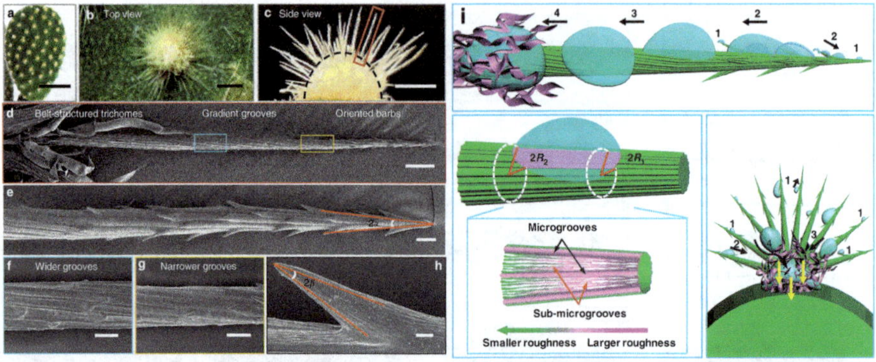

Figure 6.10 (a)–(h) Appearance and SEM images structures of the cactus. (i) Proposed mechanism of the fog collection of the cactus. Scale bars, 5 cm (a), 500 mm (b and c), 100 mm (d), 20 mm (e–g), and 2 mm (h). Adapted with permission from Macmillan Publishers Ltd., Nature Publishing Group: *Nat. Commun.*, 2012, **3**, 1247. Copyright (2012).

belt-structured trichomes. The authors have shown that the integration of these surface structures (*i.e.*, conical spines and barbs, oriented barbs, gradient grooves and belt-structured trichomes) is critical for the superior ability of these structures to collect fog. As shown by Figure 6.10i, small droplets are first formed on the barb and at the tip of the spine. These small droplets move directionally along the grooved spine and coalesce into big droplets. The big droplets then further move along the spine until reaching the bottom and are finally absorbed by the belt-structured trichomes. The authors have shown that, similar to spider silk, both the Laplace pressure gradient and the surface roughness gradient of the spine contribute to the driving force that transports the water droplets.

Another example of a desert plant with fog collection capability is *Cotula fallax*, which is a tufted, cushion-forming alpine plant originating from South Africa (Figure 6.11).[56] This hardy plant can survive various kinds of climates and can be grown throughout the world, thus making it a suitable candidate for widespread cultivation. Andrews and coworkers revealed that the fog-harvesting capability of the *Cotula fallax* plant is attributed to the hierarchical 3D arrangement formed by its leaves and the tiny hairs on them, which assist the collection and retention of water droplets on the foliage for extended periods of time, avoiding the dependency upon the direction of fog-bearing winds.[57] Water is collected throughout the entirety of the plant and held within the foliage itself. The key feature of the fine hairs is to maximize the amount of water captured and prevent the water from falling off.

Motivated by the unique water collection capability of these desert plants, artificial water collection structures have been reported.[58–62] Heng *et al.* fabricated a branch-structured ZnO structure which consisted of a large ZnO wire and an array of small ZnO wires on the large wires to mimic the cactus structure.[58] These ZnO wires are conically shaped, with the diameters

Figure 6.11 Images of *Cotula fallax*: (a) dry plant, (b) water collected by a dry plant, (c) water collected by a plant later in the day (scale bar = 20 mm), and (d) schematic representation of water collection by *Cotula fallax* over the course of a day, showing an interior mat-like structure formed by overlapping hairs. Reprinted with permission from H. G. Andrews, E. A. Eccles, W. C. E. Schofield and J. P. S. Badyal, *Langmuir*, 2011, **27**, 3798–3802. Copyright (2011) American Chemical Society.

gradually increasing from the tip to the root of a wire, as shown in Figure 6.12. They found that water drops condensed on the tips of wires are driven to the root by a capillary force induced by this diameter gradient. These artificial ZnO structures exhibited higher water collection efficiency than that of the cactus.

Also inspired by the unique structure of the cactus, Cao *et al.* reported the fabrication of an artificial fog collector through integrating cactus spine-like hydrophobic conical micro-tip arrays with a hydrophilic cotton matrix. In their work, the authors first employed a magnetic particle-assisted molding method to fabricate cactus-inspired conical micro-tips on a large scale. A magnetic suspension containing polydimethylsiloxane (PDMS) and cobalt magnetic particles (MPs) was used to prepare the conical tips. Under the external magnetic field, the MPs in the blend tend to arrange along the direction of magnetic field, resulting in the ordered conical micro-tip arrays (Figure 6.13). Later on, the authors, through integrating hydrophobic conical micro-tip arrays with the hydrophilic cotton matrix, constructed a cactus-like spherical fog collector on a large scale, as illustrated in Figure 6.13. This cactus-like fog collector could spontaneously and continuously achieve the collection, transportation, and preservation of fog water. They found that the cactus-inspired fog collector can harvest about 3 mL water in 10 min under a normal fog velocity of 45–50 cm s^{-1}. This indicated that the basic water requirement (2.5 liters per person per day under moderate climatic conditions) for maintaining human survival could be solved by using 100 cactus-like fog collectors to collect drinking water from fog-laden wind for only

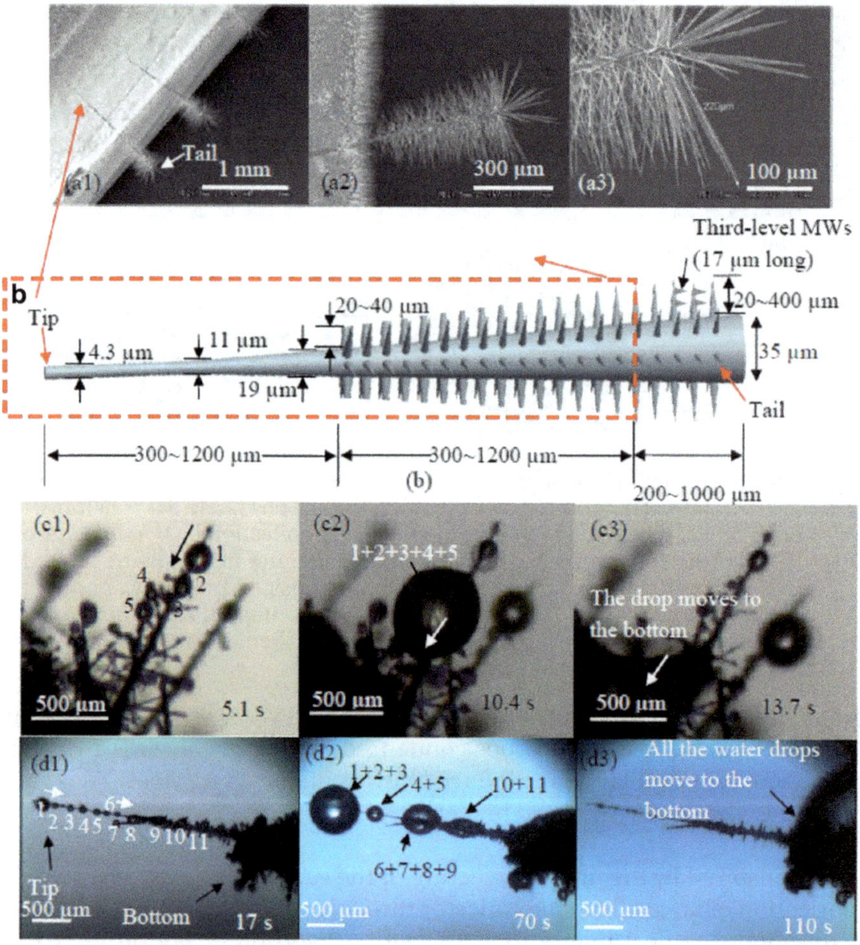

Figure 6.12 Synthesized branched ZnO wire structures (a1–a3, b) and the transportation (c1–c3) and merging (d1–d3) of water drops on a synthesized branched wire structure. Reprinted with permission from X. Heng, M. Xiang, Z. Lu and C. Luo, *ACS Appl. Mater. Interfaces*, 2014, **6**, 8032–8041. Copyright (2014) American Chemical Society.

1.5 hours per day. This cactus-inspired fog collector demonstrates promising applications in the regions with drinking water scarcity.

6.5 Summary and Outlook

In this chapter we have presented some examples of the biomimetic fabrication of various kinds of materials and devices for effective fog harvesting. Microstructure and wettability are very important factors to be considered in designing efficient biomimetic fog-collecting systems. Through millions of years of evolution, biological systems have developed unique and highly

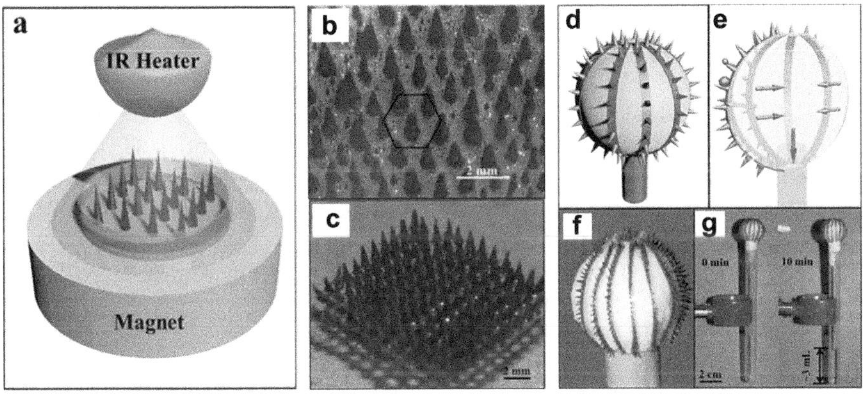

Figure 6.13 Scheme (a) and optical photographs (b,c) of the prepared micro-tip array. The illustration of the cactus-inspired device (d–g). Adapted with permission from M. Cao, J. Ju, K. Li, S. Dou, K. Liu and L. Jiang, Facile and Large-Scale Fabrication of a Cactus-Inspired Continuous Fog Collector, *Adv. Funct. Mater.*, 2014, **24**, 3235–3240. Copyright © [2014] John Wiley and Sons.

efficient material systems with delicate microstructures and composition for the purpose of fog collection to survive in harsh conditions, and these bioinspirations have stimulated the development of the artificial functional materials. We believe in the future that the biomimetic strategy is still one of the most promising sources for the design and fabrication of functional materials and devices for the solution of the global water crisis, and there are still great challenges and opportunities for the practical applications. On the other hand, although we specifically highlighted the bioinspired structures and properties, practical applications still further require the complete material systems to have good mechanical and chemical stability. Large-scale fabrication is another important aspect for practical applications. Nature always offers a wide range of vivid inspirations for many different material properties. Therefore, in the long term, a systematic understanding of the synergistic functions of biological systems and simultaneously learning from different biological systems would be highly desirable to combine their unique features to enable integrated high-performance applications.

References

1. B. E. Logan and M. Elimelech, *Nature*, 2012, **488**, 313–319.
2. A. D. Khawaji, I. K. Kutubkhanah and J.-M. Wie, *Desalination*, 2008, **221**, 47–69.
3. T. Matsuura, *Desalination*, 2001, **134**, 47–54.
4. M. Elimelech and W. A. Phillip, *Science*, 2011, **333**, 712–717.
5. N. Ghaffour, T. M. Missimer and G. L. Amy, *Desalination*, 2013, **309**, 197–207.

6. E. Mathioulakis, V. Belessiotis and E. Delyannis, *Desalination*, 2007, **203**, 346–365.
7. E. Tzen and R. Morris, *Solar Energy*, 2003, **75**, 375–379.
8. C. Charcosset, *Desalination*, 2009, **245**, 214–231.
9. A. Subramani, M. Badruzzaman, J. Oppenheimer and J. G. Jacangelo, *Water Res.*, 2011, **45**, 1907–1920.
10. V. G. Gude, N. Nirmalakhandan and S. Deng, *Renewable Sustainable Energy Rev.*, 2010, **14**, 2641–2654.
11. H. Y. Yang, Z. J. Han, S. F. Yu, K. L. Pey, K. Ostrikov and R. Karnik, *Nat. Commun.*, 2013, **4**, 2220.
12. H. Kunze, *Desalination*, 2001, **139**, 35–41.
13. E. Butler, A. Silva, K. Horton, Z. Rom, M. Chwatko, A. Havasov and J. R. McCutcheon, *Desalination*, 2013, **312**, 23–30.
14. K. N. Knust, D. Hlushkou, R. K. Anand, U. Tallarek and R. M. Crooks, *Angew. Chem., Int. Ed.*, 2013, **52**, 8107–8110.
15. S. J. Kim, S. H. Ko, K. H. Kang and J. Han, *Nat. Nanotechnol.*, 2010, **5**, 297–301.
16. I. Lekouch, K. Lekouch, M. Muselli, A. Mongruel, B. Kabbachi and D. Beysens, *J Hydrol.*, 2012, **448–449**, 60–72.
17. I. Lekouch, M. Muselli, B. Kabbachi, J. Ouazzani, I. Melnytchouk-Milimouk and D. Beysensf, *Energy*, 2011, **36**, 2257–2265.
18. P. Gandhidasan and H. I. Abualhamayel, *Water Environ. J.*, 2007, **21**, 19–25.
19. R. S. Schemenauer and P. I. Joe, *Atmos. Res.*, 1989, **24**, 53–69.
20. J. D. D. Rivera, *Atmos. Res.*, 2011, **102**, 335–342.
21. O. Klemm, R. S. Schemenauer, A. Lummerich, P. Cereceda, V. Marzol, D. Corell, J. van Heerden, D. Reinhard, T. Gherezghiher, J. Olivier, P. Osses, J. Sarsour, E. Frost, M. J. Estrela, J. A. Valiente and G. M. Fessehaye, *AMBIO*, 2012, **41**, 221–234.
22. N. Agam and P. R. Berliner, *J. Arid Environ.*, 2006, **65**, 572–590.
23. K.-C. Park, S. S. Chhatre, S. Srinivasan, R. E. Cohen and G. H. McKinley, *Langmuir*, 2013, **29**, 13269–13277.
24. Food and Agricultural Organization for the United Nations, United Nations Convention to Combat Desertification, Mountain Partnership Secretariat, Swiss Agency for Development and Cooperation, and Centre for Development and Environment. Highlands and Drylands – Mountains, a Source of Resilience in Arid Regions, 2011, Rome.
25. United Nations Educational Scientific and Cultural Organization; The United Nations World Water Development Report 4: Managing Water under Uncertainty and Risk, 2012, Paris.
26. *FogQuest: Sustainable Water Solutions*, http://www.fogquest.org, (accessed August 2010).
27. A. K. Goel, D. A. McAdams and R. B. Stone, *Biologically Inspired Design*, Springer, London, 2013.
28. A. R. Parker and C. R. Lawrence, *Nature*, 2001, **414**, 33–34.
29. A. Roth-Nebelsick, M. Ebner, T. Miranda, V. Gottschalk, D. Voigt, S. Gorb, T. Stegmaier, J. Sarsour, M. Linke and W. Konrad, *J. R. Soc., Interface*, 2012, **9**, 1965–1974.

30. Y. Zheng, H. Bai, Z. Huang, X. Tian, F.-Q. Nie, Y. Zhao, J. Zhai and L. Jiang, *Nature*, 2010, **463**, 640–643.
31. J. Ju, H. Bai, Y. Zheng, T. Zhao, R. Fang and L. Jiang, *Nat. Commun.*, 2012, **3**, 1247.
32. W. J. Hamilton and M. K. Seely, *Nature*, 1976, **262**, 284–285.
33. T. Nørgaard and M. Dacke, *Front. Zool.*, 2010, **7**, 23.
34. W. M. Moreau, *Semiconductor Lithography: Principles and Materials*, Plenum, New York, 1988.
35. D. Brambley, B. Martin and P. D. Prewett, *Adv. Mater. Opt. Electron.*, 1994, **4**, 55–74.
36. Y. Xia, J. A. Rogers, K. E. Paul and G. M. Whitesides, *Chem. Rev.*, 1999, **99**, 1823–1848.
37. D. Zahner, J. Abagat, F. Svec, J. M. J. Fréchet and P. A. Levkin, *Adv. Mater.*, 2011, **23**, 3030–3034.
38. R. P. Garrod, L. G. Harris, W. C. E. Schofield, J. McGettrick, L. J. Ward, D. O. H. Teare and J. P. S. Badyal, *Langmuir*, 2007, **23**, 689–693.
39. K. K. Varanasi, M. Hsu, N. Bhate, W. Yang and T. Deng, *Appl. Phys. Lett.*, 2009, **95**, 094101.
40. X. Chen, J. Wu, R. Ma, M. Hua, N. Koratkar, S. Yao and Z. Wang, *Adv. Funct. Mater.*, 2011, **21**, 4617–4623.
41. H. Bai, L. Wang, J. Ju, R. Sun, Y. Zheng and L. Jiang, *Adv. Mater.*, 2014, **26**, 5025–5030.
42. T. L. Sun, L. Feng, X. F. Gao and L. Jiang, *Acc. Chem. Res.*, 2005, **38**, 644–652.
43. X. Zhang, F. Shi, J. Niu, Y. G. Jiang and Z. Q. Wang, *J. Mater. Chem.*, 2008, **18**, 621–633.
44. X. M. Li, D. Reinhoudt and M. Crego-Calama, *Chem. Soc. Rev.*, 2007, **36**, 1350–1368.
45. A. B. D. Cassie and S. Baxter, *Trans. Faraday Soc.*, 1944, **40**, 546–551.
46. S. Wang and L. Jiang, *Adv. Mater.*, 2007, **19**, 3423–3424.
47. L. Zhai, M. C. Berg, F. C. Cebeci, Y. Kim, J. M. Milwid, M. F. Rubner and R. E. Cohen, *Nano Lett.*, 2006, **6**, 1213–1217.
48. H. Lee, S. M. Dellatore, W. M. Miller and P. B. Messersmith, *Science*, 2007, **318**, 426–430.
49. L. Zhang, J. Wu, M. N. Hedhili, X. Yang and P. Wang, *J. Mater. Chem. A*, 2015, **3**, 2844–2852.
50. S. C. Thickett, C. Neto and A. T. Harris, *Adv. Mater.*, 2011, **23**, 3718–3722.
51. H. Bai, X. Tian, Y. Zheng, J. Ju, Y. Zhao and L. Jiang, *Adv. Mater.*, 2010, **22**, 5521–5525.
52. H. Bai, J. Ju, R. Sun, Y. Chen, Y. Zheng and L. Jiang, *Adv. Mater.*, 2011, **23**, 3708–3711.
53. H. Bai, J. Ju, Y. Zheng and L. Jiang, *Adv. Mater.*, 2012, **24**, 2786–2791.
54. Y. Chen, L. Wang, Y. Xue, Y. Zheng and L. Jiang, *Soft Matter*, 2012, **8**, 11450–11454.
55. Y. Hou, Y. Chen, Y. Xue, Y. Zheng and L. Jiang, *Langmuir*, 2012, **28**, 4737–4743.
56. N. Hind, *Curtis's Bot. Mag.*, 2006, **23**, 307–313.

57. H. G. Andrews, E. A. Eccles, W. C. E. Schofield and J. P. S. Badyal, *Langmuir*, 2011, **27**, 3798–3802.
58. X. Heng, M. Xiang, Z. Lu and C. Luo, *ACS Appl. Mater. Interfaces*, 2014, **6**, 8032–8041.
59. M. Cao, J. Ju, K. Li, S. Dou, K. Liu and L. Jiang, *Adv. Funct. Mater.*, 2014, **24**, 3235–3240.
60. J. Ju, X. Yao, S. Yang, L. Wang, R. Sun, Y. He and L. Jiang, *Adv. Funct. Mater.*, 2014, **24**, 6933–6938.
61. J. Ju, K. Xiao, X. Yao, H. Bai and L. Jiang, *Adv. Mater.*, 2013, **25**, 5937–5942.
62. F. Bai, J. Wu, G. Gong and L. Guo, *Adv. Sci.*, 2015, **1**, 1500047.

CHAPTER 7

"Slippery" Liquid-Infused Surfaces Inspired by Nature

NICOLE S. ZACHARIA*[a]

[a]Department of Polymer Engineering, University of Akron, Akron, OH 44325, USA
*E-mail: nzacharia@uakron.edu

7.1 Introduction and Background

7.1.1 Introduction

Creation of smart materials and surfaces with control of wetting behavior is an important problem that has occupied scientists for decades. Surface wetting behavior controls numerous phenomena that we experience on a daily basis such as the application of creams and cosmetics, rain hitting buildings or car windshields, the staining of textiles and fabrics, or condensation on heating or cooling equipment. Ideally one would like to be able to apply coatings that can resist water or other liquids, or a "slippery" coating, to any arbitrary substrate be it metal, polymer, ceramic, or fabric. This coating should also have mechanical robustness and be able to perform over a large temperature range, to be washed or folded, depending on the substrate, or to withstand mechanical stress and strain. Lastly, such a coating should be made without materials of questionable safety, such as long-chain perfluorinated acids that are environmentally persistent and some of which are bioaccumulative as well.[1]

RSC Smart Materials No. 20
Smart Materials for Advanced Environmental Applications
Edited by Peng Wang

Published by the Royal Society of Chemistry, www.rsc.org

Nature provides us with any number of examples of surfaces from which it is possible to take inspiration, with special types of surface wetting such as the well-known lotus leaf,[2] rose petals,[3] butterfly wings,[4] the namib beetle which is able to collect water from the atmosphere,[5] shark skin,[6] and so on. Insects such as water striders use special types of wettability on the surfaces of their legs to walk over the surface of bodies of water.[7] Although the fabrication of such materials and surfaces has been long studied, challenges remain, and recently a new class of surfaces has gained a great deal of attention. These surfaces are made by infusing a liquid or a lubricant into a porous or textured surface, imbuing the surface with liquid-like properties.[8] This exciting new class of materials has been shown to have self-cleaning properties, anti-ice properties, to be resistant to biofouling, and even to have other types of properties such as drag resistance as well as mechanical robustness. They can be made with a number of different textured or porous surfaces and materials, can be infused with different types of liquids, and can therefore be applied to a large number of different types of substrates or objects.

7.1.2 Background and Biomimetic Inspiration

What is being referred to here as "special wettability" can be classified into a few categories. These include superhydrophobic[9,10] surfaces (static water contact angle >150°), superhydrophillic[11] surfaces (static water contact angle <10°), self-cleaning, which means that drops are easily shed from the surface, or water-pinning surfaces, whereby drops are held to the surface even if that surface is completely inverted. Water-pinning surfaces, just like self-cleaning or water-shedding surfaces, can have very high static contact angles. The above-mentioned examples from the natural world have properties that are created primarily with texture. The chemistry that nature uses in plants and animals is somewhat limited, for example excluding the possibility of the creation of low surface energy surfaces with fluorine. Roughening of a surface is known to enhance that surface's inherent wetting properties; that is, roughening a smooth hydrophilic surface makes it more hydrophilic and conversely roughening a smooth hydrophobic surface makes it more hydrophobic. Amongst rough surfaces with a high contact angle, one can imagine a case when the drop sits on top of the surface feature not coming into contact with a large proportion of the surface (Figure 7.1a) or the case in which the drop permeates into the surface features (Figure 7.1b). The former case is known as the Cassie–Baxter[10] state and results in a drop that is easily shed from a surface as it has relatively low contact area with the surface. The latter case is the Wenzel[9] state and results in a drop that is pinned to the surface.

Nature provides examples of both of these wetting states. The rose, shown in Figure 7.2b,[12] and other flower petals are known to be water pinning while the leaves of the lotus plant, shown in Figure 7.2a,[13] which is one of numerous botanical exemplars, is a superhydrophobic, water-shedding surface. Water drops on a lotus leaf surface exist in the Cassie–Baxter state with a layer of air trapped between the water drop and the lower points on the lotus

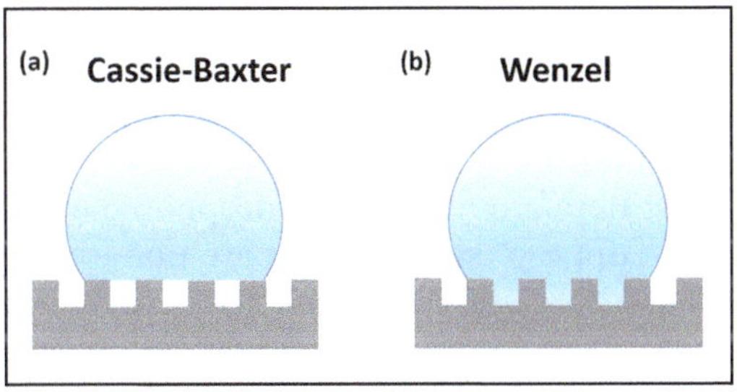

Figure 7.1 (a) A drop sitting on top of a textured surface that is then able to be easily shed from that surface and (b) a drop penetrating into the surface features of a textured surface, which results in a pinned drop.

Figure 7.2 (a) Left column shows the superhydrophobic lotus leaf[13] and an SEM image of a replica of the leaf's surface. (b) Right column shows a rose (*Rosa landora*)[12] and an SEM image of a replica of a rose petal surface. SEM images adapted with permission from R. Fürstner and W. Barthlott, *Langmuir*, 2005, **21**, 956. Copyright American Chemical Society 2005.

leaf's surface and water drops on the rose petal surface exist in the Wenzel state. Figure 7.2 also shows SEM images of exact replicas of these surfaces made by casting and curing polydimethylsiloxane on the plant surfaces.[14] It can be seen that both the rose petal and the lotus leaf surface have hierarchical textures with larger micro-scaled features covered in smaller nano-scaled features. Differences in water pinning *versus* water shedding are largely due to differences in the density and geometry of the nano-scaled features, although water-pinning surfaces often have chemically hydrophilic surface defects to promote adhesion.

Of specific interest here are not superhydrophobic or water-pinning surfaces, but so-called "slippery" surfaces that are non-wetting or resist adhesion of or fouling by a range of different materials, including but not limited to water. Such surfaces have great potential application in a wide diversity of areas. This includes consumer products such as clothes that resist staining, coatings for car windshields, or ketchup bottles that can easily be emptied. It also includes industrial applications where dropwise condensation and roll off might be desirable, such as for turbine engine blades, creation of low drag surfaces under flow, or the ability to resist biofouling on a medical implant or the inner surface of pipes. A great deal of research has been done to create such surfaces, most commonly by mimicking the famous lotus leaf structure.[15–17] These lotus leaf surfaces are not necessarily oleophobic, and a great deal of work has also gone into designing surfaces that are oleophobic,[18] or both hydrophobic and oleophobic (omniphobic).[19]

For a surface to be self-cleaning, it is desirable for the liquid not only to not wet it (*i.e.* to bead up on the surface) but also to be easily able to leave or slip off of the surface. In order to characterize this "slipperiness", a parameter more relevant than the static contact angle is the roll-off angle, defined as the tilt required for the droplet to leave the surface. This generally can be understood by some balance of the forces pinning the droplet to the surface and gravitational forces moving the droplet away. Related to this is the contact angle hysteresis, which is defined as the difference between the advancing and receding contact angles as the drop is moving. These two parameters are shown in Figure 7.3. The origin of this hysteresis is generally attributed to surface roughness and chemical heterogeneities, although these contributions are not quantifiable. Hysteresis is the result of pinning at the

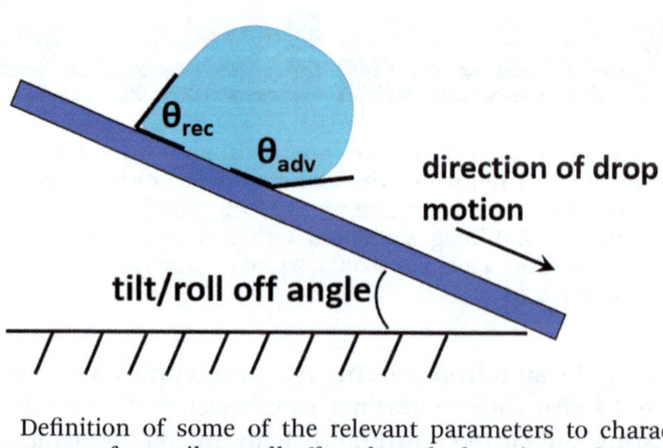

Figure 7.3 Definition of some of the relevant parameters to characterize slippery surfaces; tilt or roll off angle and advancing and receding contact angles. The contact angle hysteresis is the difference between advancing and receding angles.

three-phase contact line (a liquid drop in air on a solid surface). Observation of the contact line allows one to observe "stick-slip" behavior wherein the contact line "sticks" and then periodically is caused to "slip" by the forces of wetting and then sticks again.[20] A "slippery" surface should have both low roll-off angles as well as contact angle hysteresis, not exhibiting much of the "sticking" behavior. A lotus leaf has a water roll-off angle of 10° or less, and very low contact angle hysteresis. For the self-cleaning that occurs with lotus leaf-type surfaces, a water drop rolls instead of sliding off the surface, and picks up dirt onto the drop's surface as it rolls. However, lotus leaves or surfaces replicating the lotus leaf are not necessarily able to achieve the same for oily liquids, or even for hot water (as water's surface energy changes with temperature[21] and the wax on a lotus leaf may melt at higher temperatures). Other types of hydrophobic or oleophobic surfaces do not always possess this low roll-off angle (*e.g.* the "rose petal effect" which alludes to surfaces with a very high water contact angle as well as pinning forces). A slippery surface is much more able to repel liquids or even the adhesion of ice than a surface merely possessing a high static contact angle. Even in the case of a textured surface such as the lotus leaf with air pockets creating the Cassie state there are a number of possible failure modes, for example under conditions of immersion into a fluid, high humidity, or high pressure. At cold temperatures small drops of water may condense in the air pockets of a lotus leaf structured surface, ruining the water-shedding properties of the surface. Under these conditions the air pockets may collapse or be wetted with water and the structure loses some of its special wetting properties.

7.1.3 Introduction of the SLIPS Concept

One proposed solution to the problem of creating a robust slippery surface is to use an incompressible liquid to replace the compressible gas layer that lies between water drops and the lotus leaf surface.[8,22] Clearly for this to be effective, the liquid infused into the surface must be immiscible with the liquid of interest to be repelled. Such surfaces have been called slippery liquid-infused porous surfaces (SLIPS). This solution takes inspiration from the natural world. For example, the pitcher plant (Figure 7.4)[23] when wet has such a slippery surface with which it traps its prey. Another example is a fish's scales, which are slippery only when wet. Similarly, once these liquid-infused surfaces have a suitable liquid introduced into them their slipperiness increases. Figure 7.4 shows a pitcher plant (Figure 7.4a) and SEM images of its surface's microstructure (Figure 7.4c);[24] the plant has a hierarchical structure similar to other natural examples of superwetting surfaces. Figure 7.4b shows the basic concept for fabrication of a SLIPS surface;[8] a surface that is textured or porous can be infused with a liquid to create a SLIPS surface. This can be understood as such; if the lubricant wets the textured surface (*i.e.* with a contact angle $\theta = 0°$) even in the presence of water (or the other droplet to be repelled) it creates an atomically smooth layer and essentially removes pinning forces.[25] This is because there are no contact lines between

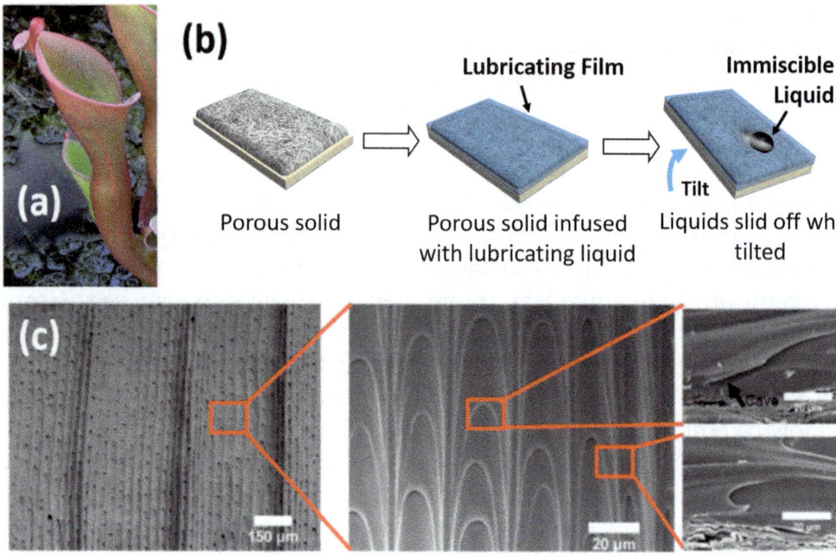

Figure 7.4 SLIPS surfaces take inspiration from (a) the pitcher plant that is slippery when wet.[23] (b) A textured or porous surface is imbibed with a lubricant to create synthetic versions of SLIPS materials. Adapted with permission from Macmillan Publishers Ltd., Nature Publishing Group: (T. S. Wong, S. H. Kang, S. K. Y. Tang, E. J. Smythe, B. D. Hatton, A. Crinthal and J. Aizenberg, *Nature*, 2011, **477**, 443), copyright (2011). (c) SEM micrographs of peristome surface and longitudinal section. Top view (left and middle) and longitudinal view (right) along the radial direction. With kind permission from Springer Science and Business Media.[24]

the droplet and the surface, and it is the contact lines largely that govern droplet pinning.[26] The role of the surface texture is to lock the liquid or lubricant in place with capillary forces.

The idea of making SLIPS systems was first described by the groups of Quéré and Aizenberg.[8,22] A surface with micro or nano structures on its surface such as a textured superhydrophobic surface can be considered to be a suitable porous or textured material. If a liquid has sufficiently large affinity for the substrate (as defined by the static contact angle) it will wick into the features of the surface. The texture of the solid keeps the liquid in place, even against gravity if the surface is tilted. Depending on the degree of affinity for the surface, the textured features might be totally encapsulated by the liquid, or they might partially protrude from the liquid, uncoated.[25] A number of suitable porous substrates for liquid/lubricant-infused surfaces have been reported, including lithographically defined materials, self-assembled materials, as well as random structures varying from ceramics to metals to polymers. A number of infusing liquids have also been demonstrated, including ionic liquids, various oils, and even water.[8] The main design rules for this type of system are that the liquid to be infused into the porous medium must be able to be infiltrated into the surface and that the target liquid to be repelled must be immiscible with the infused liquid. Otherwise the surface will not have the desired function, and there sometimes needs to be some chemical tuning of the surface to meet these criteria.[8,27–29]

For a well-defined surface texture (such as lithographically defined posts) the critical contact angle that determines whether the liquid will wet into the textured surface can be defined by parameters such as post width, height, and density on the surface. When a second liquid droplet comes in contact with the wetted surface, it may sink into the lubricant layer and become pinned on the surface, or it may float on top of the oil layer.[22,25] The speed of water drops sliding across the surface can be tuned by changing the viscosity of the oil, implying that the oil layer flows with the water drop. As the surfaces are reported to be self-cleaning in some circumstances, it can be assumed that the drop's motion takes place at least partially by rolling, which is the case for lotus-leaf type surfaces. As the drop comes into contact with a piece of dust or other debris, the piece of debris becomes adhered to the drop. If the adhesion between the drop and the piece of dust is stronger than the adhesion between the dust and the surface itself, the dust will continue to travel will the rolling drop, depicted in Figure 7.5b. In most self-cleaning surfaces the dust is in contact largely with air pockets and therefore the adhesion between the dust and the surface is not large. If the drop were only sliding on the surface, the dust would just be pushed along and not necessarily removed. It is not unusual for SLIPS surfaces to be reported to shed drops with contact angles much lower than 150°, especially in the case of organic liquids. This on the other hand suggests sliding of drops exclusively. It is most likely that both sliding and rolling cases are possible, as well as mixed slipping and rolling, depending on the system, as depicted in Figure 7.5c. The properties of the infused liquid likely will be a determining factor in this type of behavior. By observing water drops impacting on liquid-infused surfaces, it has also been seen that higher viscosity oils are more stable and more difficult to displace by water drops.

7.2 Self-Cleaning SLIPS Surfaces

In the first reports of liquid-infused surfaces, Quéré and coworkers[22] reported a self-cleaning material made from a solid that was made from an acrylate with a pattern of regular triangular spikes wet by a silicone oil. Figure 7.5a shows a drop sliding on a freestanding SLIPS surface.[38] Aizenberg and coworkers in their first report demonstrated use of a substrate with a random texture (porous Teflon membrane) infused with Krytox,[8] a Dupont lubricant which is a poly(perfluoroether). This SLIPS surface showed ability to shed both water and organic solvents. This was the case even though in some circumstances (for example, hexane) the static contact angle of the drop on the SLIPS surface was rather low. In this same report repellency to multiple organic solvents and ice, and even crude oil and blood, was also demonstrated. It was shown that even if one of these fluids was pinned by the textured Teflon membrane without the lubricant, it was shed by the final

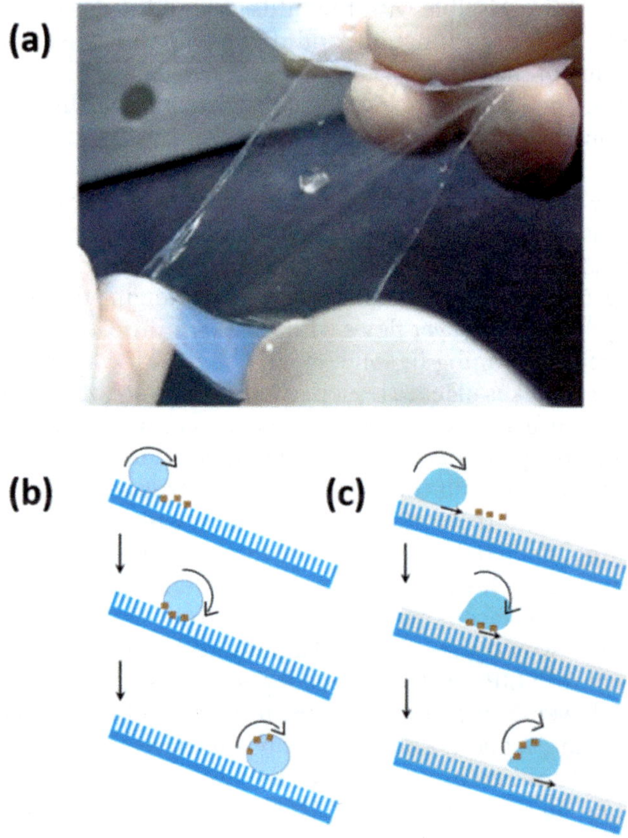

Figure 7.5 (a) Drop sliding on a freestanding, transparent SLIPS surface. Adapted with permission from I. Okada and S. Shiratori, *ACS Appl. Mater. Interfaces*, 2014, **6**, 1502. Copyright 2013 American Chemical Society. (b) Self-cleaning on a superhydrophobic surface is enabled by the rolling of drops along the surface collecting dust or other debris. (c) The liquid-infused surface may have different modes of self-cleaning, including rolling, sliding, or a combination or slip-stick motion.

SLIPS structure. By using a fluorinated solid substrate and fluorinated lubricant, it is possible to fabricate an omniphobic surface without use of a fluorinated silane, the use of which in many consumer products has recently come into question. Furthermore, an epoxy resin-based substrate was also used to create an optically clear SLIPS surface. Contact angle hysteresis and roll-off angles are both very low for these SLIPS surfaces, regardless of the magnitude of the static contact angle on the surface. This again confirms that likelihood that there is not purely drop rolling as they are shed, but rather sliding or some type of stick-slip behavior.

Beyond Teflon membranes and polymer replicas of lithographically defined surfaces, SLIPS materials have been fabricated directly from lithographically defined textured silicon surfaces,[25,27,28] sol–gel metal oxide

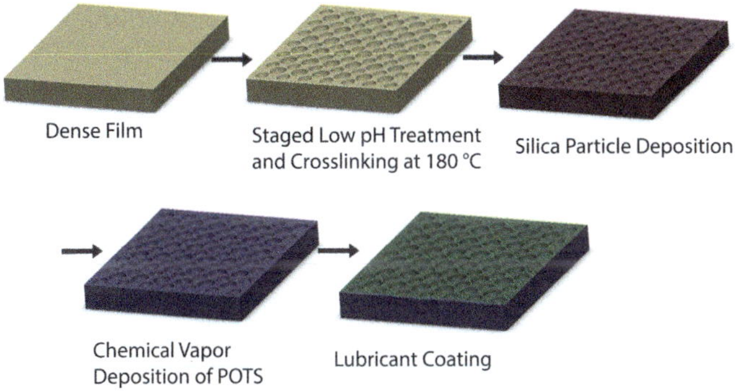

Figure 7.6 Strategy for creating LbL SLIPS surface starting from a dense polyelectrolyte multilayer, creating a hierarchical texture, fluorination, and wicking the lubricant into the surface. Adapted with permission from X. Y. Huang, J. D. Chrisman and N. S. Zacharia, *ACS Macro Lett.*, 2013, **2**, 826. Copyright 2013 American Chemical Society.

surfaces,[29,30] electrochemically deposited polypyrrole surfaces,[31] inverse colloidal crystals,[32] metal surfaces roughened by acid etching or electrochemical means,[33–35] and several types of these coatings have been deposited onto fabrics as well.[36,37] One particularly elegant method was demonstrated by Shiratori and coworkers using polymer phase separation to create a porous poly(vinylidene fluoride-*co*-hexafluoropropylene) structure that can then be filled with a fluorinated lubricant, resulting in an optically transparent, free-standing SLIPS surface fabricated at room temperature in a matter of a few minutes.[38] They have been infused with various silicone oils, fluorinated liquids, ionic liquids, water, and organic solvents.[8,25] It has been demonstrated that these surfaces are still effective at high temperature.[39]

Amongst the various techniques for fabricating a SLIPS surface, the layer-by-layer (LbL)[40] method of assembling polyelectrolytes and charged colloids into films is gaining in popularity. In this method of making coatings, a substrate is exposed sequentially to oppositely charged solutions of polymers or colloidal particles, and at each of these steps a layer of material is deposited onto the surface. Zacharia and coworkers first reported the use of the LbL method to fabricate a SLIPS-type surface.[41] A hierarchically textured surface is first made by an acid treatment of a polyelectrolyte multilayer followed by deposition of silica nanoparticles. The film is then fluorinated with a perfluoroalkyl silane before Krytox is imbibed into film. The complete strategy depicted schematically in Figure 7.6. This system has low contact angle hysteresis (~2–3°) as well as roll of angle for both water and decane (also ~2–3°), but due to the micro-scaled features of the substrate that scatter light the surface is not transparent. The transparency issue was later solved by the various groups. Aizenberg and coworkers presented a system made using the LbL assembly of poly(diallyldimethyl ammonium chloride) and negatively charged silica nanoparticles with an average size of 20 nm.[42] After removal of the organic component by calcination, a nanoporous film remains. This film is fluorinated and then lubricant is imbibed into it. Shiratori *et al.* used chitin nanofibers and silica nanoparticles to create with LbL a SLIPS surface that is antireflective with high transmittance (~97%) and a low refractive index (~1.20) because of nanoporosity in the material.[43] Lastly, Lynn and coworkers have used a so-called reactive LbL that covalently bonds the two components together to produce a transparent SLIPS surface as well.[44] One advantage of the LbL technique is that in addition to coating hard, planar surfaces such as glass or silica, curved surfaces such as an NMR tube can be coated, or soft substrates such as paper or fabric can also be coated without large changes in processing conditions.

7.3 More Than Omniphobicity: Extra Functionality

SLIPS surfaces are omniphobic, but their functionality goes beyond resisting drops of water or organic solvents, making them truly "smart" surfaces. There are a number of reports focusing on the extra functionalities possible in composite liquid–solid surfaces, ranging from increased heat transfer to improvements in electrowetting to applications as low bio-fouling surfaces.

7.3.1 SLIPS for Anti-Icing Surfaces

As previously mentioned, SLIPS surfaces have been reported as effective coatings to protect from ice formation. There are two basic parts to the problem of anti-icing; first the nucleation and formation of the ice and second the adhesion of the ice, once formed, to the surface. A good anti-icing surface should delay ice nucleation for as long as possible, and when that ice does form it should be easily shed from the surface. Ice formation on surfaces can be quite a serious problem; for example on power lines or airplanes. As a strategy for creating functional surfaces that delay or even altogether prevent the nucleation of ice, superhydrophobic surfaces have been closely examined.[45] It had been thought that lotus leaf-type surfaces would be favorable for limiting ice formation as these surfaces energetically favor a limited contact area with water. However, under conditions of high atmospheric humidity frost is usually able to form on superhydrophobic surfaces, and can decrease the hydrophobicity of the surface or make the surface into a water-pinning surface with a large contact angle hysteresis.[45] This is due to the fact under atmospheric conditions below freezing but with high humidity small droplets of water are able to condense in the air pockets of the surface's texture. Once water completely wets the hierarchical texture the surface loses its water-shedding properties. On the other hand, with a SLIPS surface, there is already a liquid in that texture so water condensation is not able to reduce the slipperiness. Figure 7.7 shows the behavior of a lotus leaf-type superhydrophobic structure and the corresponding SLIPS surface made by infusing Krytox into it. The surface is a polyelectrolyte multilayer fabricated by LbL as described in ref. 40 that has been made superhydrophobic with a fluorinated silane. Figure 7.7a shows that at room temperature a water drop bounces off

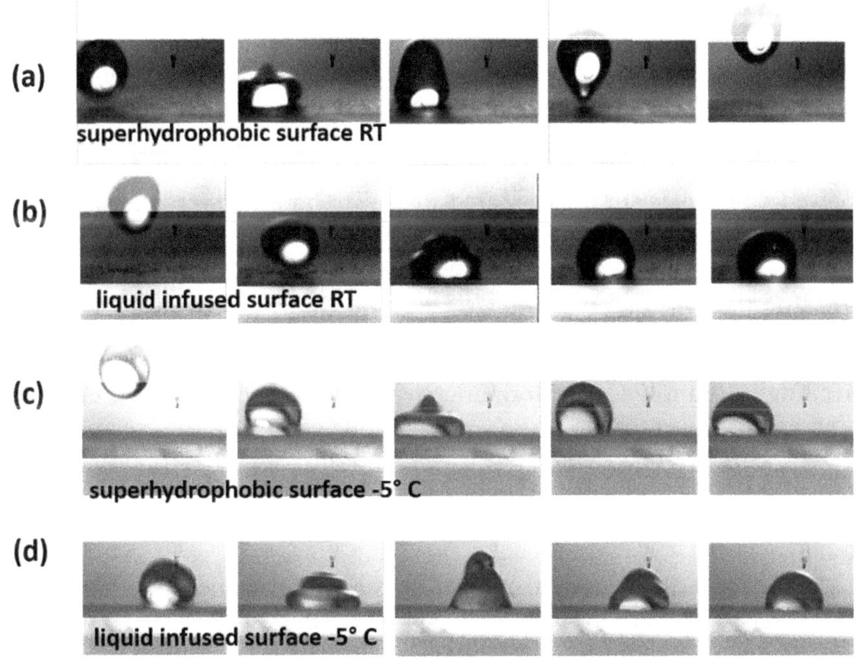

Figure 7.7 (a) A water drop bouncing off of a superhydrophobic surface at room temperature. (b) A water drop impacting a LIS made from infusing Krytox into the same superhydrophobic surface at room temperature. (c) A water drop impacting the superhydrophobic surface at −5 °C. (d) A water drop impacting the LIS at −5 °C. Images taken at a frame rate of 1000 frames per second.

the surface; the surface repels water. However, at low temperature, −5 °C, the drop sticks to the superhydrophobic surface. These images are taken at 40% relative humidity. The SLIPS surface is fabricated by taking the same superhydrophobic surface and infusing it with Krytox. The freezing point of Krytox is far below the observation temperature, and therefore the behavior of the SLIPS surface is the same at either temperature. Note that the drop impact behavior is completely different than the superhydrophobic case; even at room temperature the drop does not bounce off the surface. Upon impact it is immobilized. Drop shedding is based on the drop being able to flow along with the lubricant across the surface. Aizenberg and coworkers have coated DSC pans with SLIPS surfaces and demonstrated that there is a meaningful increase in the ability to supercool water compared to other types of surfaces, showing that SLIPS surfaces should effectively delay ice formation.[46]

Varanasi *et al.*[27] measured the strength of ice adhesion to SLIPS surfaces with different underlying textures using lithographically defined posts in silicon infused with liquids with different surface tensions. They report that, perhaps unsurprisingly, the existence of an excess of the oil or lubricant will significantly reduce the adhesion of ice to the surface. However, when there is a thermodynamically stable amount of lubricant present the surface tension of the infusing liquid does not make a significant different in the adhesion of ice to the surface, and that by increasing the density of features on the underlying textured substrate it is possible to reduce the adhesion of ice to the surface. This is explained as the edges of the surface features acting as stress concentrators, weakening the adhesion of the ice sheet. They also showed their lubricant-impregnated surface to have a lower adhesion to ice compared to smooth, hydrophobic surfaces.

One obvious limitation to SLIPS type surfaces is that there is only a small amount of liquid at the surface. Once freezing has taken place at that surface, that liquid can no longer provide an anti-icing benefit. Furthermore, it has been shown that ice formation can cause the draining of the infused liquid, meaning that anti-icing performance would be limited to one time. In an approach by Rykaczewski *et al.*,[47] a SLIPS type of surface that secretes antifreeze was developed. This surface has a porous, superhydrophobic "skin" with a reservoir of antifreeze below it. This surface secretes antifreeze in response to the presence of ice, overcoming these limitations of other SLIPS systems.

Electrosprayed polymeric anti-ice SLIPS surfaces have also been reported,[48] as well as anti-icing surfaces based on a textured surface infused with a hygroscopic polymer.[49] Although this last strategy is counter to the general SLIPS design rules to have the infused liquid/lubricant being completely immiscible with the liquid to be repelled, with this strategy an extremely low adhesion between ice and the surface has been achieved. It is said to be "self-lubricating" as the hygroscopic polymer swells as temperatures are lowered, bulging out of the porous structured surface, creating a hydrogel layer on top of the surface that is wet and helps reduce ice adhesion significantly compared to both superhydrophilic and superhydrophobic flat surfaces. The useful temperature range can be tuned by the activity of water in the hygroscopic polymer. This surface also shows good self-healing and resistance to wear.

7.3.2 SLIPS for Anti-Fouling Surfaces

Beyond repellency to organic solvents and water, SLIPS surfaces can repel complex fluids; crude oil, jam, blood were all initially demonstrated.[8] One of the layer-by-layer SLIPS made using biocompatible materials (including almond oil as the lubricant) has been reported not only to repel blood but to be antithrombogenic[50] and to greatly increase the time for blood to clot on this surface as compared to a glass slide or the hydrophobilized underlying LbL film. Therefore, SLIPS are an obvious choice to investigate resistance to adhesion with respect to a wider range of materials. For example, a startup company, LiquiGlide,[51] promises to provide the "only permanently wet slippery surface." They advertise such applications as SLIPS applied to the inside of glue bottles or ketchup bottles, as a way to allow viscous fluids to more easily be poured or squeezed out of bottles.

With respect to fouling from biofilms, it is known that bacterial adhesion to surfaces is a multistep process generally proceeded by excretion and deposition of various macromolecules by the bacteria onto the surface.[52] This extracellular material is generally some combination of extracellular DNA, proteins, and polysaccharides. A surface that can resist the adherence of these materials should be able to successfully resist bacterial growth on its surface. Literature reports exist of liquid-infused surfaces for resistance to bacterial adhesion in aqueous environments. Aizenberg *et al.* show reduced adhesion for their SLIPS surface made from Krytox-infused Teflon membranes for *Pseudomonas aeruginosa* biofilm attachment over a 7 day period, as well as for *Staphylococcus aureus* and *Escherichia coli*, as compared to either the Teflon membrane alone or a dry, superhydrophobic surface.[53] They also excluded cytotoxicity of the fluorinated lubricant as the reason for reduced biofilm attachment.[53] In other reports, a surface based on a macroporous poly(butyl methacrylate-*co*-ethylene dimethacrylate) infused with a perfluoropolyether (fluorinert) was shown to be effective at resisting biofilm formation of several types of bacteria under low nutrient conditions and to be able to resist *P. aeruginosa* biofilm formation for 7 days under high nutrition conditions as well, although somewhat less successfully.[54] More specifically one particular strain of *P. aeruginosa* was only able to cover ~1% of the SLIPS surface after 7 days while *P. aeruginosa* isolated from wastewater covered a little more than 10% of the SLIPS surface under high nutrient conditions. The same group also demonstrated that this surface was effective against adhesion of algal *Ulva linza* zoospores and cypris larvae of the barnacle *Balanus amphitrite*.[55] In another example, a SLIPS surface was formed on a roughened aluminum substrate, also with a perfluoropolyether as the infused lubricant, in order to protect from sulfate-reducing bacteria.[33] These bacteria are one of the chief culprits for corrosion caused by organisms, which is an important mechanism in corrosion for engineering structures in marine environments, accounting for perhaps 20% of total mass losses due to corrosion. These reports show that SLIPS surfaces may be useful in marine and wastewater environments where other types of antibacterial surfaces (such as those that release an antibacterial agent) might lose efficacy as that antibacterial agent diffuses out into the water.

In addition to adhesion of biofilms, there have been results showing that SLIPS surfaces are suitable for resisting deposition of mineral scale such as calcium carbonate, which can be a serious problem in a number of industrial applications such as desalination or oil production.[56,57] Current methods for dealing with scale buildup include chemical dissolution or mechanical removal which can be inefficient, expensive, and/or environmentally damaging. In one example, porous polypyrrole surfaces, formed by electropolymerization onto steel, were fluorinated *via* reactive ion etching and infused with low surface energy liquids such as fluorinated oils or ionic liquids and exposed to aqueous environments that promote scale buildup. It was observed that the SLIPS surfaces infused with either low surface liquid had an order of magnitude less scale deposited onto them than plain stainless steel surfaces. A similar study using a patterned silicon wafer impregnated with silicone oil attributes the scale-reducing properties to reduced salt crystal nucleation on the liquid-infused surface. Further design criteria for successfully reducing crystal nucleation are given. If the lubricant has a positive spreading coefficient on the substrate it will coat all of the texture's features and reduce the presence of nucleation sites. Conversely, if the lubricant has a negative spreading coefficient some features of the surface's texture will not be covered by the lubricant and will be able to serve as crystal nucleation sites.

7.3.3 Beyond Slippery Surfaces

The applications described thus far for different SLIPS surfaces rely on the "slippery" properties of these surfaces; that is, their ability to resist adhesion of one material or another, be that water in any of its phases, oil, biomacromolecules, or salts. There are, however, other potential applications for these surfaces. These surfaces represent a new class of materials, wherein both solid and liquid phases are present. One could therefore imagine coatings that are solid but that have the properties of liquids. This might be advantageous for heat transfer, for example. One report shows that liquid infusion may reduce the voltages necessary for creating electrowetting phenomena due to the low contact angle hysteresis at liquid–liquid interfaces.[58] The types of fluids that have thus far been infused into these surfaces has been somewhat limited and a great deal of potential lies in expanding the materials sets that are used to create SLIPS or liquid-infused surfaces more generally.

Improved condensation on liquid-infused surfaces compared to smooth or superhydrophobic surfaces has been demonstrated.[28,59] This improved property is due to the low adhesion of drops. A more recent report, however, has demonstrated further enhanced properties with a variation on the SLIPS concept.[60] This work uses a heterogeneous surface of hydrophilic defects in a background of hydrophobic oil. Water drops are then nucleated within the oil, creating a situation with an overall greatly increased amount of drop nucleation which is favorable to heat transfer. This strategy was demonstrated on CuO surfaces on copper pipes to create an order of magnitude more drop nucleation than on the same surface not infused with oil, and the overall heat transfer is 100% greater than that achieved by other state of the art dropwise condensing surfaces.

Several examples show the ability of SLIPS surfaces to control the flow of fluids or the motion of drops. For example, Aizenberg *et al.* have shown that infusing surfaces with liquids of varying viscosity can change the speed at which drops are able to move along the SLIPS surface.[39] This result is in line with the idea that the drops are flowing along with a thin layer of the lubricant, and the more viscous that lubricant is the more slowly it will flow. Figure 7.8 shows a series of SLIPS surfaces fabricated by infusing fluorinated liquids into strips of Teflon membranes. These liquids range in viscosity, and it can be seen that water drops (colored differently on each SLIPS surface) roll more quickly

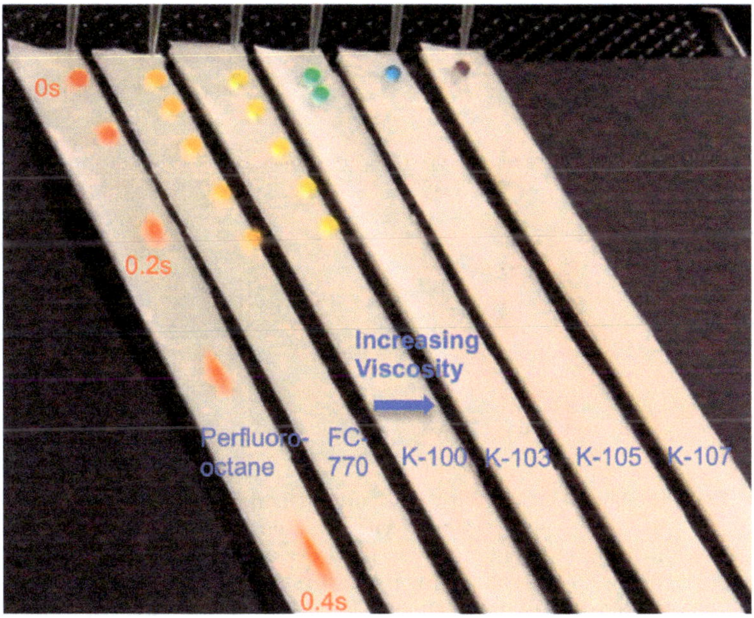

Figure 7.8 Mobility of drops on SLIPS surfaces infused with lubricants of increasing viscosity. It can be seen that with an increase of viscosity there is a decrease in drop speed. The drops are colored water and the surfaces are Teflon membranes infused from left to right with perfluoro-octane, fluorinert, and then a series of increasingly viscous Krytox oils. Reprinted with permission from D. Daniel, M. N. Mankin, R. A. Belisle, T. S. Wong and J. Aizenberg, *Appl. Phys. Lett.*, 2013, **102**, 231603. Copyright 2013, AIP Publishing LLC.

on less viscous surfaces and more slowly on more viscous surfaces. One could envision patterned SLIPS surfaces designed to guide droplet mobility.

In a similar vein, SLIPS surfaces can be thought of as being used to guide the flow not only of drops but of an entire flow. Varanasi *et al.* have proposed that SLIPS surfaces might be able to reduce surface drag in laminar flows,[61] which is more difficult than to reduce drag in turbulent flows. These surfaces should also be useful for drag reduction in turbulent flows, so one could envision energy-saving coatings on airplanes or ships that reduce drag.

In a totally different application, liquid-infused membranes have been reported as useful for separations in multiphase systems or as a tunable gating system.[62] By filling the pores of a membrane with a liquid (instead of a gas as is most often present) one can create a tunable way to control transport of these different phases through the pores of the membrane. Figure 7.9 shows the mechanism for this gated transport. In the case of a membrane with no infused liquid, gas can freely flow through the pores without any applied pressure. However, liquid will only flow through the pores with some applied pressure. On the other hand, with a liquid-filled pore, there is a positive pressure required to overcome the capillary force with which the liquid is

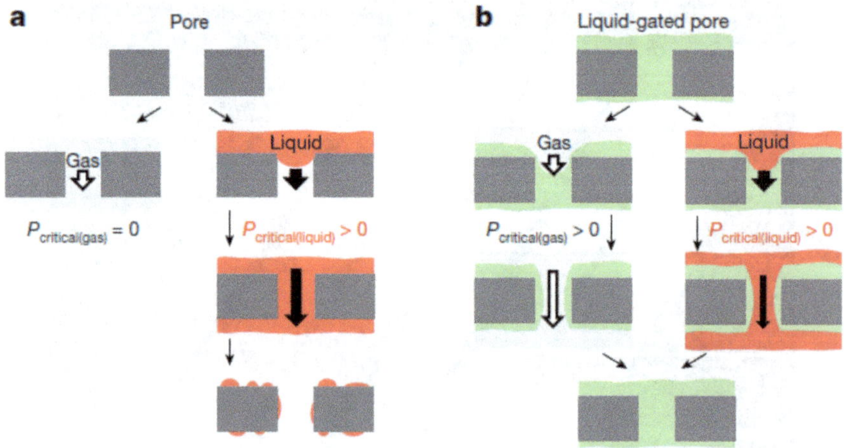

Figure 7.9 Mechanism of multiphase "gating" using a liquid-infused membrane. (a) In a porous membrane with air-filled pores, gas can travel across the membrane freely. However, if (b) the membrane is filled with a liquid there requires some critical pressure for the gas to flow through, allowing for control over transport across the membrane. Reprinted by permission from Macmillan Publishers Ltd: (X. Hou, Y. H. Hu, A. Grinthal, M. Khan and J. Aizenberg, *Nature*, 2015, **519**, 70), copyright 2015.

held in the pore and allow both gas and liquid to flow across the membrane. The liquid is dynamic in nature and can be pushed aside for either gas or other liquid to flow past and afterwards fills the pores again. Once a positive pressure is required for even the gas to flow across the membrane, which material flows past the membrane can be finely controlled. This membrane also benefits from the more conventional properties of SLIPS materials and has good non-fouling properties. This example shows the potential of liquid-infused surfaces to perform tasks beyond repellency of liquids.

7.4 Thermodynamics and Stability

Some of the most important questions and challenges regarding SLIPS surfaces if they are to be more widely adopted are their stability, both thermodynamic and mechanical. Questions about drop dynamics on these surfaces have only begun to be examined. The innovation of SLIPS surfaces lies in the presence of a liquid phase that has properties superior to solid phases in certain aspects. However, liquids cannot hold their shape and are intrinsically mechanically weak. Therefore strategies to increase mechanically robustness are required.

7.4.1 Thermodynamic Description of SLIPS Surfaces

Although wetting at surfaces is a well-described and investigated phenomenon, SLIPS surfaces present some additional considerations. First of all, there are four phases present in SLIPS surfaces; the textured or porous solid, the infused liquid, the liquid that is being repelled, and air. A paper by Varanasi *et al.* describes the thermodynamics of liquid-infused surfaces, using lithographically defined silicon posts infused with several different liquids for experimental verification, and they demonstrate as many as 12 different thermodynamic states possible for a drop on a liquid-infused surface.

In many initial reports SLIPS surfaces are described as having a smooth, perfect liquid surface. This is not always the case; although physically placing an excess of lubricant at the surface is possible, it will drain away by gravity to some thermodynamically determined amount. This may leave a texture that is completely submerged under the liquid, or one that with surface features that are not completely coated with lubricant, emerging from the liquid layer. The mode of applying the liquid to the surface may determine whether or not an excess of liquid is infused into the surface. For example, dip coating may result in an excess of lubricant or infused liquid while the forces encountered during spin coating may remove any excess. In addition to loss of lubricant by gravitational forces, lubricant may also be lost as water or other liquids flow across SLIPS surfaces, if the lubricant can spread on the water. This is clearly important in terms of long-term use of a SLIPS surface. Varanasi *et al.* also report that for anti-icing, SLIPS with a thermodynamic excess of lubricant perform better than SLIPS with only the thermodynamically stable amount of lubricant.[25] Examples like this show why understanding the morphology of SLIPS surfaces is key, but also because the morphology determines the contact line, which then sets the pinning forces and mobility of drops on the surface.

In this work it was demonstrated that if the lubricant completely wets the underlying substrate with a contact angle of zero, even in the presence of the liquid to be repelled, this can also imply the lubricant will spread onto that droplet.

The spreading coefficient can be defined as

$$S_{\mathrm{lw(s)}} \equiv \gamma_{\mathrm{ws}} - \gamma_{\mathrm{wl}} - \gamma_{\mathrm{ls}}$$

if we consider the three phases to be the solid substrate (s), the infused liquid (l), and the water (w) that sits on top of the surface.

We can say that the infused liquid will spread onto the water drop if $S_{\mathrm{lw(s)}} > 0$, and the opposite will happen for a negative spreading coefficient. Young's equation is not applicable in this case, as a spreading surface is not in equilibrium and no equilibrium contact line is in existence. Lubricating liquids with different surface tensions may be chosen in order to satisfy this criteria. The can be demonstrated for silicone oil (positive $S_{\mathrm{lw(s)}}$) impregnated silicon surfaces that are modified with a hydrophobic silane; the silicone oil wets the substrate very well and also cloaks or envelops the water droplet. On the other hand, an imidazolium-based ionic liquid (negative $S_{\mathrm{wl(s)}}$) was also studied. It was not as able to wet the underlying surface, and as a consequence does not spread onto the water drop. However, the roll off angle is higher for the ionic liquid than the silicone oil as it is not able to completely wet the surface, showing that there will be a trade off in lubricant stability as well as "slipperiness" of the surface.

To describe all of the possible configurations of a drop on a SLIPS surface, one must consider what the configuration is at the interface. There are two different interfaces; on the surface away from the drop and then underneath the drop. Underneath the drop the surface can be dry (lubricant not spreading on it), the surface can be submerged completely by oil, and the surface can be partially submerged with some emergent features of the texture not being covered in oil. Which of these three states will exist is based on which configuration results in the lowest total interfacial energy. Under the drop itself the surface can be submerged, partially submerged with some emergent features, or the drop can be impaled on the surface, and which of these states will exist will also be the result of minimizing interfacial energy. This explains how SLIPS surfaces can be water pinning, even when infused with low surface energy lubricants.

From considerations of spreading coefficients and surface tensions values, specific criteria for 12 different cases can be laid out. Figure 7.10 shows these 12 states.

First, the lubricant will either cloak the drop or not, and then for each of these two states there are 6 different possibilities.

Understanding what the state of the morphology is at the air/surface/drop interface is important to being able to design SLIPS surfaces with good properties. If the surface features are emergent, then they will influence the

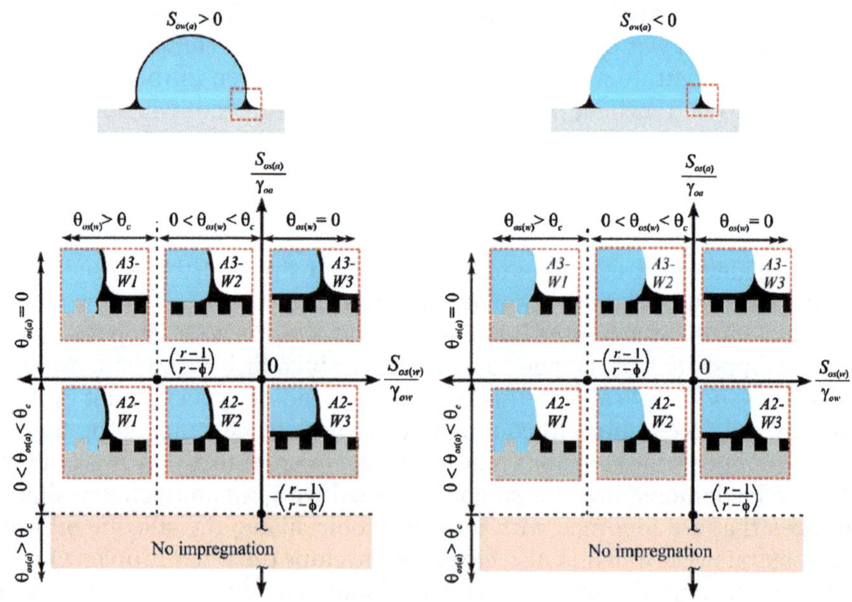

Figure 7.10 The 12 different states possible for a drop on a SLIPS surface. The lubricant will either cloak the drop (left) or not (right), and then for each of these states there are 6 distinct possibilities. Reproduced from ref. 25 with permission from the Royal Chemistry Society.

mobility of the drops, otherwise not. This might explain different results reported in the literature as to whether or not the specifics of the underlying texture are important. For example, Aizenberg and coworkers reported that a hierarchical texture is not important to achieve good slipperiness,[30] but Zacharia and coworkers found for their LbL system that an underlying hierarchical texture gives the best results.[41]

7.4.2 Stability of SLIPS Surfaces

SLIPS surfaces have been shown to be operational at low temperatures, which hierarchically structured superhydrophobic surfaces are not well able to do. They can repel hot water and liquids, which some superhydrophobic surfaces also cannot do. It has been demonstrated that applying a large amount of hydrostatic pressure does not collapse a SLIPS surface[8] and that they retain their properties of undergoing shear forces, not under conditions of flow (*e.g.* samples being spun on a spin coater repeatedly[41]). In fact, such treatment of the surface will remove the excess lubricant and leave only that which is thermodynamically stable, resulting in a surface with more constant properties as a function of use. The hydrostatic pressure specifically is a failure mode for hierarchically textured superhydrophobic surfaces, so a SLIPS surface may be better suited for underwater use. The fact that the air layer has been replaced with an incompressible liquid layer allows for all of these properties. A number of SLIPS surfaces have also been reported to be self-healing. If the film is scratched or damaged, the lubricant can flow into the damaged area to recreate the smooth liquid surface. Presumably there is some limit as to the total volume of damaged area that can be healed, based on the amount of liquid captured within the SLIPS film, but this is one of the advantages of having a surface with a liquid component. This in particular gives SLIPS surfaces a significant advantage over superhydrophobic surfaces. Because of the requirement for a specific type of hierarchical texture, a modest amount of wear or damage can ruin the properties of a superhydrophobic or lotus leaf-type surface.

The loss of lubricant, however, from the SLIPS surface due to gravity or during use of the surface is clearly a key challenge to long-term use and stability. Using different viscosities of lubricant has been investigated by imaging drop impact on these surfaces from overhead. One can see ripples in the lubricant as drops impact the surface, and that for higher viscosity lubricants it is more difficult to displace the lubricant.[63] However, higher viscosity lubricants also leads to lower droplet mobility. Similarly, as discussed above, choosing a lubricant that does not cloak the drops to be shed can reduce lubricant loss but may also reduce slipperiness of the surface. Studying dynamics of drops interacting with SLIPS surfaces also may lead to insights regarding engulfment or "cloaking" of the water drop by the surface. It has been shown that at short times a water drop is deformed by the oil layer, similar to what can be seen in Figure 7.7, which spreads at the oil–water interface and which then results in longer times in a stable thin oil film under the drop or a rupture of that oil film.[64] It was further hypothesized that drainage of the oil/lubricant is the result of long-range intermolecular forces destabilizing the oil–water interface.

For applications using SLIPS in a fluid flow (as opposed to using SLIPS to repel drops) this has also been investigated. A recent report looking at SLIPS surfaces made from a porous sol–gel material on the inside surface of tubing showed that in this closed system (fluid flowing through the pipe) very little lubricant was lost.[65] This was true with multiple types of infused lubricant and at physiologically relevant conditions with flows of 10, 50, or 90 mL min^{-1} which are representative of flow rates used in processes such as neonatal hemodialysis. At the same time, in another work by Stone *et al.*, the failure method of SLIPS surfaces under a shear flow is discussed, namely lubricant drainage from the surface in response to shear.[66] In this work it is shown that there will be some critical shear stress from an external flow that will cause drainage, but also that design of the surface can raise this critical shear stress. By studying the flow of oil in lithographically defined microscale grooves, it was determined that by placing periodic barriers in those channels it is possible to increase the retention of that oil.

These results imply that SLIPS surfaces might well be suitable for the interior of pipes or microfluidic devices where the entire surface would be coated by the flow. In another recent work, it has been suggested that using a polymer as the substrate that can be swelled by the lubricant is one way to slow or reduce the loss of lubricant. It has been shown that by modifying fabrics with a PDMS layer and then infusing with an oil led to a greatly reduced amount of lubricant loss and loss of omniphobic properties compared to just infusing the fabric directly or by modifying the fabrics with nanoparticles before infusing with oil.[37] The PDMS is swollen by the lubricants used in that work, acting as a delayed release layer and to prevent loss by the high affinity between the PDMS and the lubricant. The stone work likens the length of lubricant that can be retained on a surface under flow to a capillary raise, and it is yet to be fully explored on other types of substrates whether the feature size of density can improve fluid retention by changing the capillary forces. Other strategies to help "lock" the lubricant in place or to provide a reservoir of lubricant will surely be required to use these materials in commercial applications. One example of utilizing a reservoir strategy is a "vascularized" anti-fouling surface.[67] As lubricant can be lost in aqueous environments, a continuous self-replenishing supply of lubricant was provided to the surface by embedding a 3D vascular network into PDMS and filling it with silicone oil. As the oil is lost from the surface it is replenished by the vascular network. This type of SLIPS material was shown to resist biofilm formation by a number of bacteria and performed better compared to a non-vascularized SLIPS surface after being exposed for 48 hours to conditions that promote lubricant evaporation.

7.5 Conclusions and Outlook

Slippery SLIPS surfaces, based on the concept of infusing a liquid into a porous or textured surface, have rapidly proven to represent a potentially transformative new class of materials. Their fabrication follows simple design rules. These are:

- the solid surface should have sufficient texture to lock the lubricant into place
- the lubricant should have an affinity for that underlying solid
- the lubricant should be immiscible with whatever liquid is to be repelled.

SLIPS surfaces have been fabricated with textiles, polymers, ceramics, and materials using a range of oils and other liquids as the infused phase. Ordered substrates, such as those lithographically fabricated, are useful for fundamental studies to determine the role of texture for the overall properties of SLIPS surfaces. However, simpler fabrication methods will be more applicable to most commercial uses of these materials. A number of surface/lubricant materials combinations have been shown effective, including those without long-chain perfluorinated acids that raise environmental and health questions.

SLIPS surfaces combine the properties of both solids and liquids. These composite surfaces exhibit the best of both worlds with properties such as improved heat transfer or self-healing due to the presence of the liquid phase, but more mechanical robustness than a coating of oil due to the solid substrate. In fact, in addition to being self-healing these surfaces are somewhat more tolerant of wear or damage than superhydrophobic surfaces are. Increasing the range of materials used in these surfaces can only increase the possible applications of SLIPS surfaces.

Although the concept is simple, a complete thermodynamic description of SLIPS materials is not entirely straightforward. Certainly a good understanding of these thermodynamic properties will be required to make forward progress in the rational design of SLIPS. For example, understanding the thermodynamics of a particular materials system will enable one to predict whether it results in a water-pinning or a water-shedding surface. Another important parameter that will determine surface properties is whether or not the surface is completely coated by the lubricant or if some of the surface features are emergent. Controlling how much lubricant is thermodynamically stable on these surfaces will be important, as will understanding the failure mechanisms whereby lubricant can be lost from these surfaces, including drainage and loss to evaporation. Studying the dynamics of water drops and fluid flows interacting with these surfaces may help in understanding this last point, and such studies have only begun to be reported. A better fundamental understanding of these surfaces can only lead to more complete design rules for their implementation.

References

1. J. M. Conder, R. A. Hoke, W. de Wolf, M. H. Russell and R. C. Buck, *Environ. Sci. Technol.*, 2008, **42**, 995.
2. W. Barthlott and C. Neinhuis, *Planta*, 1997, **202**, 1.
3. M. J. Liu, Y. M. Zheng, J. Zhai and L. Jiang, *Acc. Chem. Res.*, 2010, **43**, 368.
4. Y. Zheng, X. Gao and L. Jiang, *Soft Matter*, 2007, **3**, 178–182.
5. A. R. Parker and C. R. Lawrence, *Nature*, 2001, **414**, 33–34.
6. D. W. Bechert, M. Bruse, W. Hage and R. Meyer, *Naturwissenschaften*, 2000, **87**, 157.
7. X. Gao and L. Jiang, *Nature*, 2004, **432**, 26.
8. T. S. Wong, S. H. Kang, S. K. Y. Tang, E. J. Smythe, B. D. Hatton, A. Crinthal and J. Aizenberg, *Nature*, 2011, **477**, 443.
9. R. N. Wenzel, *Ind. Eng. Chem.*, 1936, **28**(8), 988.
10. A. B. D. Cassie and S. Baxter, *Trans. Faraday Soc.*, 1944, **40**, 546.
11. K. E. Tettey, M. I. Dafinone and D. Lee, *Mater. Express*, 2011, **1**, 89.
12. http://commons.wikimedia.org/wiki/Rosa_%27Landora%27.
13. http://commons.wikimedia.org/wiki/Lotus_effect.
14. R. Fürstner and W. Barthlott, *Langmuir*, 2005, **21**, 956.
15. S. Nishimoto and B. Bhushan, *RSC Adv.*, 2013, **3**, 671.
16. P. Taylor, *Curr. Opin. Colloid Interface Sci.*, 2011, **16**, 326.
17. K. S. Liu, X. Yao and L. Liang, *Chem. Soc. Rev.*, 2010, **39**, 3240.
18. K. S. Liu, Y. Tian and L. Jiang, *Prog. Mater. Sci.*, 2013, **58**, 503.
19. A. Tuteja, W. Choi, J. M. Mabry, G. H. McKinley and R. E. Cohen, *Proc. Natl. Acad. Sci. U. S. A.*, 2008, **105**, 18200.
20. M. E. R. Shanahan, *Langmuir*, 1995, **11**, 1041.
21. Y. Liu, X. Chen and J. H. Xin, *J. Mater. Chem.*, 2009, **19**, 5602.
22. A. Lafuma and D. Quéré, *Europhys. Lett.*, 2011, **96**, 56001.
23. http://commons.wikimedia.org/wiki/Pitcher_(plant)#/media/File:Kew.gardens.pitcher.plant.heliamphora.arp.jpg.
24. C. P. Hsu, Y. M. Lin and P. Y. Chen, *JOM*, 2015, **67**, 744.
25. J. D. Smith, R. Dhiman, S. Anand, E. Reza-Garduno, R. E. Cohen, G. H. McKinley and K. K. Varanasi, *Soft Matter*, 2013, **9**, 1772.
26. J. F. Joanny and P. G. de Gennes, *J. Chem. Phys.*, 1984, **81**, 552.
27. S. B. Subramanyam, K. Rycazewski and K. K. Varanasi, *Langmuir*, 2013, **29**, 13414.
28. S. Anand, A. T. Paxson, R. Dhiman, J. D. Smith and K. K. Varanasi, *ACS Nano*, 2012, **6**, 10122.
29. W. Ma, Y. Higaki, H. Otsuka and A. Takahara, *Chem. Commun.*, 2013, **49**, 597.
30. P. Kim, M. J. Kreder, J. Alvarenga and J. Aizenberg, *Nano Lett.*, 2013, **13**, 1793.
31. P. Kim, T. S. Wong, J. Alvarenga, M. J. Kreder, W. E. Adorno-Martinez and J. Aizenberg, *ACS Nano*, 2012, **6**, 6569.
32. N. Vogel, R. A. Belisle, B. Hatton, T. S. Wong and J. Aizenberg, *Nat. Commun.*, 2013, **4**, 2176.

33. J. Yang, H. J. Song, H. Y. Ji and B. B. Chen, *J. Adhes. Sci. Technol.*, 2014, **28**, 1949.
34. P. Wang, Z. Lu and D. Zhang, Slippery liquid-infused porous surfaces fabricated on aluminum as a barrier to corrosion induced by sulfate reducing bacteria, *Corros. Sci.*, 2015, **93**, 159.
35. S. S. Yang, R. Qiu, H. Q. Song, P. Wang, Z. Q. Shi and Y. F. Wang, *Appl. Surf. Sci.*, 2015, **328**, 491.
36. C. Shillingford, N. MacCallum, T. S. Wong, P. Kim and J. Aizenberg, *Nanotechnology*, 2014, **25**, 014019.
37. V. G. Damle, A. Tummala, S. Chandrashekar, C. Kido, A. Roopesh, X. D. Sun, K. Doudrick, J. Chinn, J. R. Lee, T. P. Burgin and K. Rykaczewski, *ACS Appl. Mater. Interfaces*, 2015, **7**, 4224.
38. I. Okada and S. Shiratori, *ACS Appl. Mater. Interfaces*, 2014, **6**, 1502.
39. D. Daniel, M. N. Mankin, R. A. Belisle, T. S. Wong and J. Aizenberg, *Appl. Phys. Lett.*, 2013, **102**, 231603.
40. G. Decher, *Science*, 1997, **277**, 1232.
41. X. Y. Huang, J. D. Chrisman and N. S. Zacharia, *ACS Macro Lett.*, 2013, **2**, 826.
42. S. Sunny, N. Vogel, C. Howell, T. L. Vu and J. Aizenberg, *J. Adv. Funct. Mater.*, 2014, **24**, 6658.
43. K. Manabe, S. Nishizawa, K. H. Kyung and S. Shiratori, *ACS Appl. Mater. Interfaces*, 2014, **6**, 13985.
44. U. Manna and D. M. Lynn, *Adv. Mater.*, 2015, **27**, 3007.
45. H. A. Stone, *ACS Nano*, 2012, **6**, 6536.
46. P. W. Wilson, W. Z. Lu, H. J. Xu, P. Kim, M. J. Kreder, J. Alvarenga and J. Aizenberg, *Phys. Chem. Chem. Phys.*, 2013, **2**, 581.
47. X. D. Sun, V. G. Damle, S. L. Z. Liu and K. Rykaczewski, *Adv. Mater. Interfaces*, 2015, **2**, DOI: 10.1002/admi.201400479.
48. Q. Liu, Y. Yang, M. Huang, Y. X. Zhou, Y. Y. Liu and X. D. Liang, *Appl. Surf. Sci.*, 2015, **346**, 68.
49. J. Chen, R. M. Dou, D. P. Cui, Q. L. Zhang, Y. F. Zhang, F. J. Xu, X. Zhou, J. J. Wang, Y. L. Song and L. Jiang, *ACS Appl. Mater. Interfaces*, 2013, **5**, 4026–4030.
50. K. Manabe, K. H. Kyung and S. Shiratori, *ACS Appl. Mater. Interfaces*, 2015, **7**, 4763–4771.
51. www.liquiglide.com.
52. S. A. Blenkinsopp and J. W. Costerton, *Trends Biotechnol.*, 1991, **9**, 138.
53. A. K. Epstein, T. S. Wong, R. A. Belisle, E. M. Boggs and J. Aizenberg, *Proc. Natl. Acad. Sci. U. S. A.*, 2012, **109**, 13182.
54. J. S. Li, T. Kleintschek, A. Rieder, Y. Cheng, T. Baumbach, U. Obst, T. Schwartz and P. A. Levkin, *ACS Appl. Mater. Interfaces*, 2013, **5**, 6704.
55. L. L. Xiao, J. S. Li, S. Mieszkin, A. Di Fino, A. S. Clare, M. E. Callow, J. A. Callow, M. Grunze, A. Rosenhahn and P. A. Levkin, *ACS Appl. Mater. Interfaces*, 2013, **5**, 10074.
56. T. V. J. Charpentier, A. Neville, S. Baudin, M. J. Smith, M. Euvrard, A. Bell, C. Wang and R. Barker, *J. Colloid Interface Sci.*, 2015, **444**, 81.

57. S. B. Subramanyam, G. Azimi and K. K. Varanasi, *Adv. Mater. Interfaces*, 2014, **1**, 1300068.
58. E. Bormashenko, R. Pogreb, Y. Bormashenko, H. Aharoni, E. Shulzinger, R. Grinev, D. Rozenman and R. Rozenman, *RSC Adv.*, 2015, **5**, 32491–32496.
59. K. Rykaczewski, A. T. Paxson, M. Staymates, M. L. Walker, X. Sun, S. Anand, S. Srinivasan, G. H. McKinley, J. Chinn, J. H. J. Scott and K. K. Varanasi, *Sci. Rep.*, 2014, **4**, 4158.
60. R. Xiao, N. Miljkovic, R. Enright and E. N. Wang, *Sci. Rep.*, 2013, **3**, 1988.
61. B. R. Solomon, K. S. Khalil and K. K. Varanasi, *Langmuir*, 2014, **30**, 10970.
62. X. Hou, Y. H. Hu, A. Grinthal, M. Khan and J. Aizenberg, *Nature*, 2015, **519**, 70.
63. C. Lee, H. Kim and Y. Nam, *Langmuir*, 2014, **30**, 8400.
64. A. Carlson, P. Kim, G. Amberg and H. A. Stone, *Europhys. Lett.*, 2013, **104**, 34008.
65. C. Howell, T. L. Vu, C. P. Johnson, X. Hou, O. Ahanotu, J. Alvarenga, D. C. Leslie, O. Uzun, A. Waterhouse, P. Kim, M. Super, M. Aizenberg, D. E. Ingber and J. Aizenberg, *Chem. Mater.*, 2015, **27**, 1792.
66. J. S. Wexler, I. Jacobi and H. A. Stone, *Phys. Rev. Lett.*, 2015, **114**, 168301.
67. C. Howell, T. L. Vu, J. J. Lin, S. Kolle, N. Juthani, E. Watson, J. C. Weaver, J. Alvarenga and J. Aizenberg, *ACS Appl. Mater. Interfaces*, 2014, **6**, 13299.

CHAPTER 8

Challenges and Opportunities of Superhydrophobic/ Superamphiphobic Coatings in Real Applications

MAXIME PAVEN[a], LENA MAMMEN[a], AND DORIS VOLLMER*[a]

[a]Max-Planck Institute, Mainz, Germany
*E-mail: vollmerd@mpip-mainz.mpg.de

8.1 Wetting

Understanding the wetting of solids by liquids is essential in many biological, medical and industrial processes. Examples range from artificial tissue engineering to practical applications like biofouling, non-fogging coatings, fog harvesting or self-cleaning textiles. Surfaces with high liquid repellency, often inspired by nature, have promising applications.

The wetting behaviour of an ideal, flat and chemically homogeneous surface can be described by the Young equation. This equation can be derived by balancing the interfacial forces at a three-phase contact line:[1]

$$\cos\theta = \frac{\gamma_S - \gamma_{SL}}{\gamma_L} \tag{8.1}$$

Here γ_L, γ_S and γ_{SL} are the liquid–gas, solid–gas and solid–liquid interfacial tensions. In reality, the surface is never completely flat and chemically

RSC Smart Materials No. 20
Smart Materials for Advanced Environmental Applications
Edited by Peng Wang

Published by the Royal Society of Chemistry, www.rsc.org

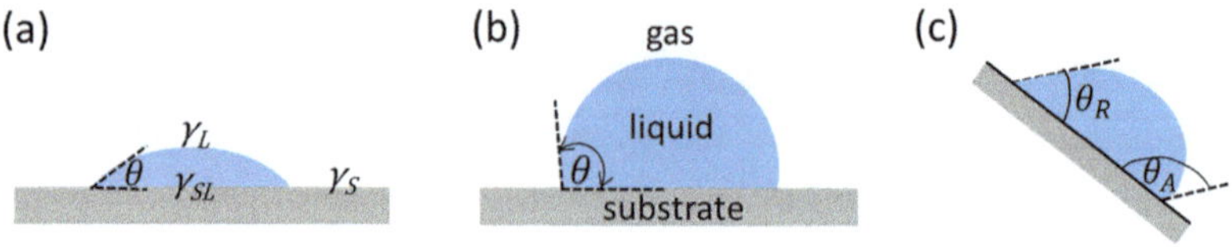

Figure 8.1 Sketch of a sessile drop on a surface. The material contact angle θ can be determined at the three-phase contact line of liquid, surface and vapour. (a) Hydrophilic substrate, (b) hydrophobic substrate, (c) the difference between the advancing θ_A and receding θ_R material contact angle determines the contact angle hysteresis.

homogeneous. To account for this non-ideality, we call the contact angle θ that is measured on a flat surface the material contact angle, instead of the Young contact angle. θ can be estimated by eye or quantified by optical microscopy (Figure 8.1). Depending on the value of θ one distinguishes between hydrophilic ($\theta < 90°$, $\gamma_S > \gamma_{SL}$) and hydrophobic ($\theta > 90°$, $\gamma_S < \gamma_{SL}$) surfaces (Figure 8.1a and b). For $\gamma_S = \gamma_{SL} + \gamma_L$ the liquid completely wets the surface ($\theta = 0°$), whereas for $\gamma_S < \gamma_{SL} + \gamma_L$ a finite contact angle is formed. A hydrophilic surface is also called a high energy surface. The surface energy can be lowered by coating a surface, for example with fluorinated molecules. In that case, contact angles of about 120° have been reported on a smooth homogeneous surface.[2,3]

The material's contact angle describes wetting at the macroscopic scale, meaning 0.1–1 µm away from the real wetting line. However, very close to the wetting line, *i.e.* in a range of 1–100 nm, surface forces, like van der Waals or electrostatic double-layer forces, come into play, leading to a deviating microscopic contact angle.[4]

In reality, a completely smooth and chemically homogeneous surface does not exist. The drop will start to move as soon as the surface is tilted above a certain angle, the so-termed roll-off angle α (Figure 8.1c). Tilting causes the contact angles at the front and rear side of the drop to differ. θ_A denotes the angle at the front side when the drops starts to advance. Correspondingly, θ_R denotes the angle at the rear when the rear side of the drop starts to advance. The reason that drop motion requires a certain tilting angle is caused by pinning of the drop on chemical inhomogeneities or surface protrusions. On a smooth surface, high adhesion often prevents the drop from rolling off. Alternatively, the advancing material contact angle θ_A can be determined by increasing the volume of a liquid drop until the three-phase contact line starts to advance. Correspondingly, to find the receding material contact angle θ_A the volume is reduced until the three-phase contact line recedes. The difference, $\theta_A - \theta_R$, is called contact angle hysteresis.

8.1.1 Rough Surface: Wenzel's and Cassie's Models

Wenzel suggested that the wetting behaviour of a rough and smooth surface are related: *"It is only necessary to apply the fact that, within a measured unit area on a rough surface, there is actually more surface, and in that sense therefore*

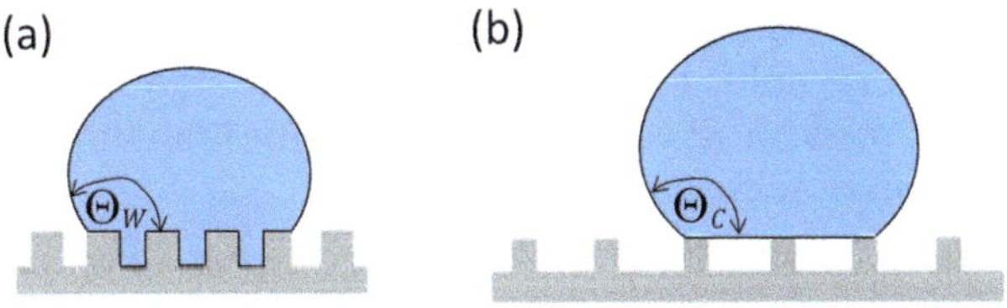

Figure 8.2 A sessile liquid drop resting on a rough surface in the fully wetted Wenzel state (a) and in the composite Cassie–Baxter state (b) respectively.

a greater intensity of surface energy, than in the same measured unit area on a smooth surface".[4] Based on the assumption of thermodynamic equilibrium and complete wetting of the surface's protrusions, Wenzel generalized the Young equation towards rough surfaces.[5]

$$\cos\Theta_W = r\cos\theta, \quad r \geq 1 \text{ and } |\cos\Theta_W \leq 1| \tag{8.2}$$

Θ_W is the apparent contact angle on a rough surface, r is the total surface area divided by the projected surface area, and θ is the material contact angle on a smooth surface with the same chemical composition (Figure 8.2a). In the following, we use the term "apparent contact angle" for drops on rough surfaces to distinguish those from the angle on its smooth counterpart. According to the Wenzel model, surface roughness increases the surface's hydrophobicity or hydrophilicity. The apparent contact angle decreases on a hydrophilic surface and increases on a hydrophobic surface.

However, for hydrophobic surfaces that are very rough (large r) the Wenzel formula predicts a total drying of the surface ($\Theta_W = 180°$) which is not physical due to the lack of contact between the drop and the substrate.[6] In this case it is energetically favourable that the liquid drop partially rests on air pockets and partially on top of the protrusions (Figure 8.2b). This composite state is referred to as the Cassie or Cassie–Baxter state. In thermodynamic equilibrium the apparent contact angle Θ_C on a composite surface consisting of two materials can be calculated with the Cassie–Baxter equation:[7]

$$\cos\Theta_C = f_1\cos\theta_1 + f_2\cos\theta_2 \tag{8.3}$$

Here f_1 and f_2 are the fractional surface areas of the two components that are in contact with the liquid drop and θ_1 and θ_2 are the corresponding material contact angles. The Cassie equation (eqn (8.3)) is derived from a minimization of the total free energy considering all interfacial and surface energies. If the second component is air ($f_2 = 1 - f_1 = f_{air}$; $\theta_2 = 180°$; $\theta_1 = \theta$), the Cassie equation can be rewritten as:

$$\cos\Theta_C = f(1 + \cos\theta) - 1; \text{ and } f_1 \equiv f \tag{8.4}$$

The adhesion of a drop on a surface differs significantly for the Wenzel or Cassie state, despite the fact that in both cases the apparent contact angle can exceed 150°. In the Wenzel state, adhesion is large due to the extended contact line. In general, a drop cannot roll-off but remains pinned on the surface even if $\alpha = 90°$. On the contrary, if the contact line is broken, the drop can move easily. This holds for a drop in the Cassie state. Adhesion is low if the drop partially rests on air cushions ($f_2 = f_{air}$). This is an important property for the preparation of self-cleaning surfaces.

Both the Wenzel equation and the Cassie equation predict exactly one apparent contact angle for a certain surface topography and surface material. Contrarily, the apparent contact angle can vary greatly from experiment to experiment, as its value depends on the way the drop is deposited and on the exact position on the surface. This dependence of the apparent contact angle on details of the experimental procedure and surface properties is caused by roughness and surface inhomogeneities, as these cause pinning of the contact line at certain positions.[4,8] The apparent contact angle can take every value between the apparent receding and apparent advancing contact angle, $\Theta_r^{app} \leq \Theta^{app} \leq \Theta_a^{app}$.

In the following we will not discuss the advancing and receding contact angles on a flat surface. Therefore, we will delete the indices "app", *i.e.* Θ_a and Θ_r refer to the apparent advancing and, respectively, receding contact angle on a rough surface. Often it is difficult to decide whether a drop is in the Wenzel or in the Cassie state as the contact angle can be large in both cases. In general a drop rolls off much easier in the Cassie state than in the Wenzel state. However, the roll-off angle can also exceed 20° for drops in the Cassie state. Furthermore, a drop can be partially in the Wenzel and partially in the Cassie state.

8.1.2 Laser Scanning Confocal Microscopy (LSCM)

Detailed information on the shape of the drop–air and drop–substrate interface can be obtained by laser scanning confocal microscopy, in the following referred to as LSCM (Figure 8.3). LSCM renders it possible to image the interface between a drop and a substrate with a resolution of 250 nm in the horizontal direction and 700 nm in the vertical direction. In this case the substrate needs to be transparent. To determine the drop contour and the contact angles, data analysis is easier if both the water drop and the substrate are fluorescently labelled. It is important that the dyes are not interfacially active. If more than one dye is used, care needs to be taken that the emission wavelengths of the dyes do not overlap. If the LSCM has more than one detector, the reflection of light from the interfaces with sufficient difference of the refractive indices can be measured simultaneously. No reflection is visible if the difference of the reflective indices, Δn, is too low, here for $\Delta n < 0.1$. To quantify the thickness of the air cushion or the height of the pillars, the indices of refraction of the objective and the medium should be identical. Otherwise, the heights need to be corrected by the refractive indices of the components. A dry objective allows determining the coordinates of the water–air interface, to measure the thickness of the air layer and to determine the apparent contact angle with high accuracy.

Micropillar arrays, *e.g.* made from poly-dimethylsilane (PDMS) or SU8, a commonly used epoxy-based negative photoresist, are well-suited model systems to obtain fundamental understanding about the wetting

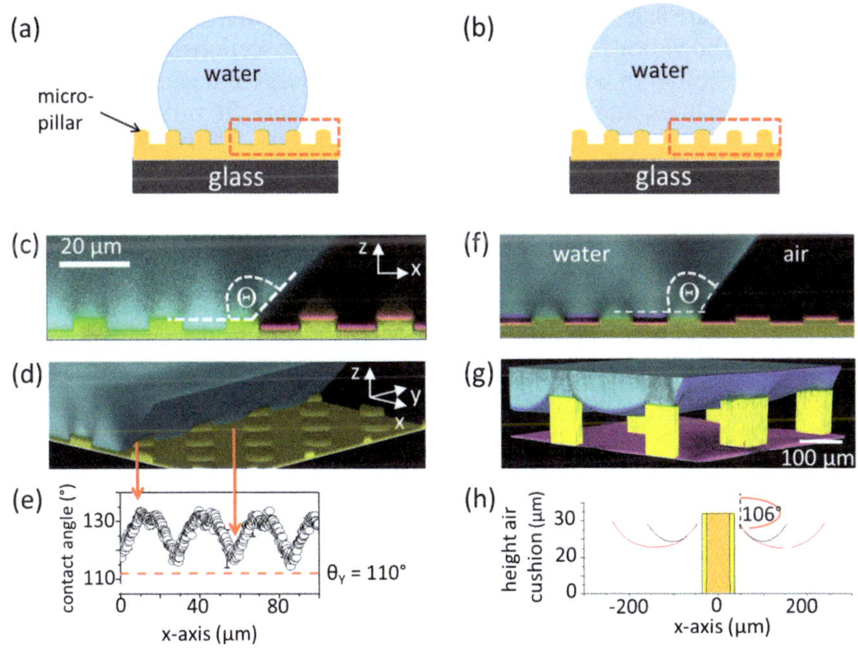

Figure 8.3 (a) Sketch of a drop in the Wenzel state and (b) in the Cassie state. To image the water–substrate–air interface by laser scanning confocal microscopy (LSCM), the pillars and the water droplet were fluorescently labelled. The hydrophobic perylene monoimide (PMI, emission maximum at λ_{PMI} = 540 nm) dye was added to PDMS or SU8 during soft-moulding. For water, perylenediimide tetrapyridoxy, WS-PDI, was used as it hardly changed the water–air interfacial tension, with γ > 71 mN m^{-1} at a concentration of 0.1 mg mL^{-1} ≈ 0.06 mM. Its emission maximum, $\lambda_{WS\text{-}PDI}$ = 600 nm, is well above that of PMI, so that the fluorescence from PDMS and water hardly overlap. (c) LSCM image of a vertical section of a water drop in the Wenzel state in contact with PDMS pillars and air. (d) A three-dimensional LSCM image of a drop in the Wenzel state. (e) The apparent contact angle Θ varies along the three-phase contact line. (f and g) LSCM image of a two dimensional section and three-dimensional view of a water droplet in the Cassie state in contact with (f) PDMS and (g) SU8 pillars and air. The dry objective allows one to accurately measure the thickness of the air layer, but the much higher index of refraction n of SU8, n_{SU8} = 1.6 causes the SU8 pillars to appear shorter by a factor of about 1.6. Therefore, artificial pillars with the actual height are inserted. (h) Variation of the thickness of the air cushion along the main axis (black) and diagonal (red). Water (blue), PDMS or SU8 (yellow) and light reflected at the PDMS–air interface (magenta). (c, d and f) The round PDMS pillars had the following dimensions: a = 10 µm, h = 5 µm, and d = 20 µm or d = 40 µm. (g) The square SU8 pillars had the following dimensions: a = 50 µm, h = 25 µm, and d = 200 µm. Adapted with permission from P. Papadopoulos, *et al.*, Wetting on the microscale: shape of a liquid drop on a microstructured surface at different length scales, *Langmuir*, 2012, **28**, 8392–8398. Copyright 2012 American Chemical Society. Adapted with permission from P. Papadopoulos, *et al.*, How superhydrophobicity breaks down, *Proc. Natl. Acad. Sci.*, 2013, **110**(9), 3254–3258.

dynamics. Figure 8.3a displays a sketch of a water drop in the Wenzel state and, Figure 8.3b shows one in the Cassie state. A 2D cross section along the x–z axis shows water (blue) in contact with the PDMS pillars (yellow) and air (black). The magenta lines mark the reflection of light from the pillar–air, and air–water interface. No reflection is visible from the water–PDMS interface due to the low difference of the refractive indices, $n_{H_2O} = 1.33$ and $n_{PDMS} = 1.45$. The height of the pillars appears too short by a factor proportional to $1/n$. Therefore, artificial pillars are inserted in Figure 8.3d. The height of the pillars was taken from SEM images.

In the Wenzel state, the water drop follows the topography of the pillars, reflected in a periodic step profile (Figure 8.3c).[9] At the three-phase contact line the water–air interface sharply bends upwards. The tangent at the three-phase contact line to the drop gives the apparent contact angle. Independent of the topography, locally it is energetically favourable for a drop to assume its material contact angle, *i.e.* $\theta = 109°$ in case of PDMS. Furthermore, a drop tends to minimize its contact area with the surface to minimize its total surface energy. These competing restrictions cause a bending of the liquid surface close to the three-phase contact line. As a result, the apparent contact angle varies along the three-phase contact line (Figure 8.3d and e). The local maxima are due to the vertical slopes and the rim of the pillars' top surface. The minima correspond to the inter-pillar valleys and the flat area on the pillars' top surface, resulting in variations of the apparent contact angle by up to 20°.[9]

In the Cassie state the drop rests partially on air cushions and partially on the pillar's top face. The air cushion gives rise to the two well-separated horizontal reflections resulting from the PDMS–air (or SU8–air) and air–water interfaces (Figure 8.3f). At the rim of the drop the water–air interface bends upwards. The absence of reflection on top of the pillars, supported by the fluorescence images of water and PDMS (or SU8), proved that the pillars were completely wetted by water. The water–air interface curvature of the drop is governed by the Laplace equation ($P = 2\gamma_L/R$), which relates the pressure, P, inside the drop of radius R to its curvature, $1/R$. Flattening of the drop by gravity can be neglected for drops much smaller than the capillary length, $(\gamma_L/\rho g)^{1/2} = 2.7$ mm, where ρ is the density of water, $\gamma_L = 72$ mN m^{-1} and $g = 9.81$ m s^{-2}. The three-phase contact line is pinned at the edge of the pillars. As the curvature of the water–air interface is determined by the Laplace pressure, the minimal thickness of the air cushion decreases with increasing pillar spacing. The variation of the thickness of the air cushion along the main axis (black) and diagonal (red) for the drop shown in Figure 8.3g is given in Figure 8.3h.

8.1.3 Superhydrophobicity

Often the term "Cassie state" is used as a synonym for superhydrophobicity. However, a surface is defined as superhydrophobic if the apparent contact angle with water is above 150° and the roll-off angle is below 10°. Tilting a superhydrophobic surface by a few degrees is already sufficient for a drop to overcome adhesion, to roll off easily and wash away dirt particles encountered on its way (Figure 8.4). A drop in a Cassie state can have lower apparent

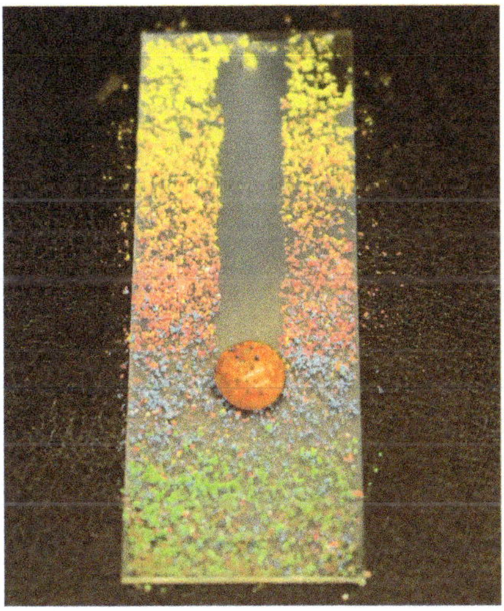

Figure 8.4 A water drop rolling-off a polluted superhydrophobic surface. The drop takes the dust along.

contact angles and higher roll-off angles.[10] Therefore, the Cassie state is not necessarily superhydrophobic.

This extreme water repellency was first observed in nature[11] and can be found on insect wings[12] and legs[13] (Figure 8.5a) and many plant leaves[14] like the lotus leaf[15] (Figure 8.5b). These surfaces possess a multiscale (hierarchic) surface structure in combination with a low surface tension material. The surface of a water strider leg, for example, consists of numerous needle-shaped setae (Figure 8.5c) with nanoscale grooves (Figure 8.5c, inset) that enable the water strider to stand and move on water.[13] The lotus leaf surface consists of micrometre-sized papillae covered with a layer of wax crystals (Figure 8.5d, inset).[16]

If the superhydrophobic state should be thermodynamically stable a material with low surface free energy is required ($\gamma_S < \gamma_{SL}$), as roughness amplifies the wetting properties. According to Young's equation it was expected that superhydrophobicity requires a material contact angle above 90°. For example, poly(tetrafluoroethylene) with a surface free energy $\gamma_S = 0.0185$ N m^{-1} has a Young contact angle with water of about 108°.[17] However, superhydrophobic properties can also be observed on intrinsically hydrophilic materials.[18,19] This was predicted by Herminghaus theoretically.[20] Notably, even the wax of the lotus leaf is weakly hydrophilic with a material contact angle θ of ~74°.[21] Cao *et al.* fabricated superhydrophobic surfaces with multiscale roughness by packing hydrophilic flower-like hematite particles (Figure 8.6a). The 1–2 μm-sized α-Fe_2O_3 particles possess a rough surface consisting of 40–60 nm thick distorted plates of interconnected crystals which remind one of the petals of a flower. These distorted plates present an overhanging structure, which prevents the water from penetrating the textures. Due to the overhangs the multilayer of α-Fe_2O_3 particles is superhydrophobic (Figure 8.6b) despite a material contact angle of $\theta = 45°$ for α-Fe_2O_3. The overhangs act as an energy barrier preventing a transition from the Cassie to the fully wetted state.

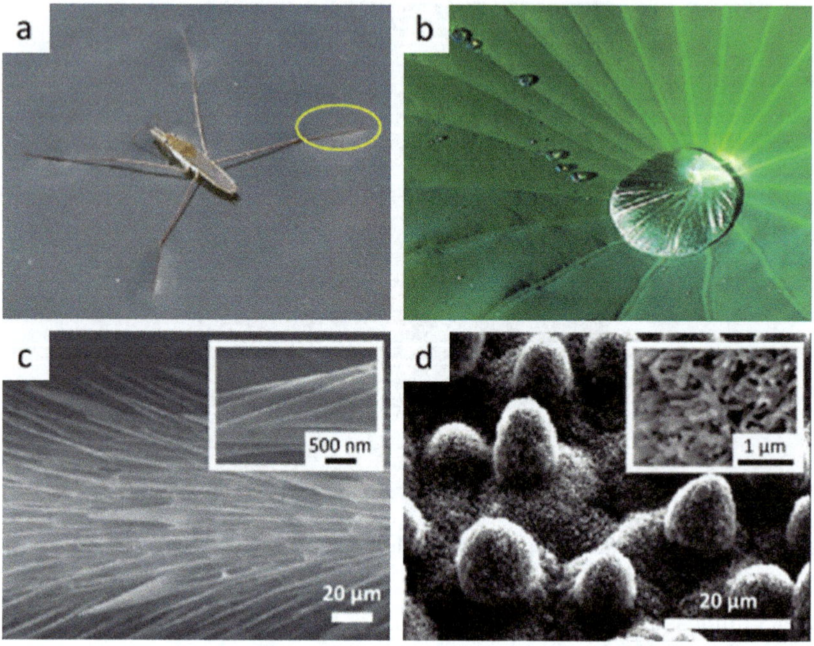

Figure 8.5 Examples for natural superhydrophobic surfaces, like the water strider leg (a) and the lotus leaf (b). Both surfaces possess a hierarchical structure on two length scales. The water strider leg surface consists of micrometer-sized setae (c) with nano grooves (c, inset) and the lotus leaf surface of micro bumps (d) covered with nanometre-sized wax crystals (d, inset). Reprinted with permission from Y. W. Su, *et al.*, Nature's Design of Hierarchical Superhydrophobic Surfaces of a Water Strider for Low Adhesion and Low-Energy Dissipation, *Langmuir*, 2010, **26**(24), 18926–18937. Copyright 2010 American Chemical Society.

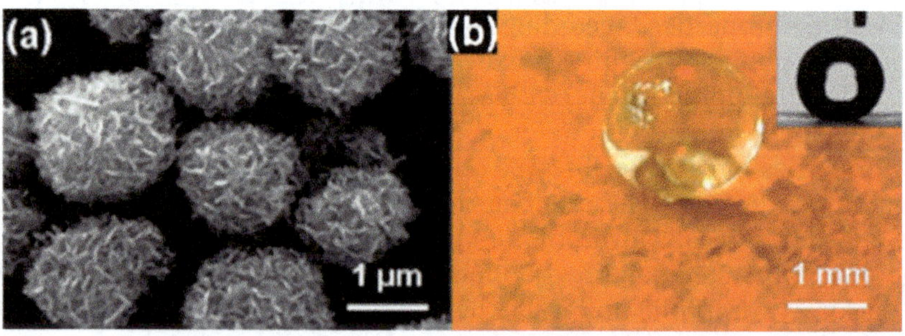

Figure 8.6 Superhydrophobic surface made of intrinsically hydrophilic hematite flower-like particles (a) showing an apparent contact angle with water of $\Theta = 159°$ (b). Reprinted with permission from A. M. Cao, L. L. Cao and D. Gao, Fabrication of nonaging superhydrophobic surfaces by packing flowerlike hematite particles, *Appl. Phys. Lett.*, 2007, **91**(3). Copyright 2007, AIP Publishing LLC.

8.1.4 Superamphiphobicity

Whereas innumerable plants and many insects are known for their superhydrophobicity, extremely oil and water repellent surfaces, in nature so-called superamphiphobic surfaces, are rare. Superamphiphobic surfaces show high apparent contact angles ($\Theta \geq 150°$) in conjunction with low roll-off angles ($\alpha \leq 5°$ to $10°$) with water and non-polar liquids.[22,23] To achieve an intrinsic contact angle higher than 90° for most alkanes with a surface tension γ_L = 20–30 mN m^{-1} the solid surface tension γ_S should be below ~6 mN m^{-1}.[24] Furthermore, γ_{SL} needs to be lower than γ_S; this has not been achieved yet. However, metastable superamphiphobicity can be induced to intrinsically oleophilic materials by creating local overhanging structures, acting as an energy barrier for the drop to impale into the coating.[23,25,26] Highly advanced design of the surface protrusions even permits fluorinated solvents to rest on hydrophilic nail-like silicon pillars with vertical overhangs in a Cassie state.[27] One example of an insect with superamphiphobic skin are the springtails, which are mainly soil-dwelling wingless arthropods (Figure 8.7a).[28] On their skin they can form a stable air layer, a so-called plastron. Upon immersion in water (Figure 8.7b) or oils (Figure 8.7c) the plastron protects the skin breathing animals against suffocation.

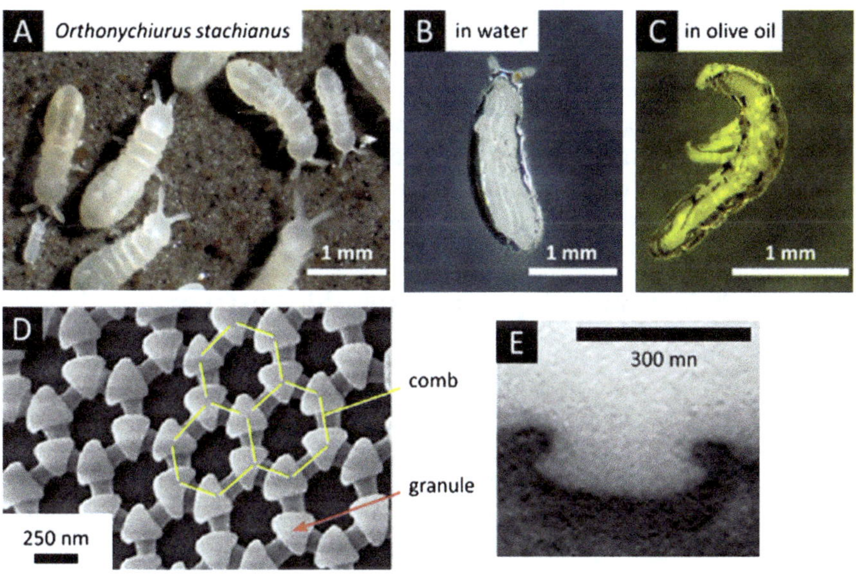

Figure 8.7 (a) Springtail colony of *Orthonychiurus stachianus*. (b and c) Plastron surrounding the entire animal upon immersion into (b) water and (c) olive oil. (d) Scanning electron microscopy image and (e) cross-sectional transmission electron microscopy image of the nanoscopic skin ornamentation. Adapted with permission from R. Hensel, *et al.*, Wetting Resistance at Its Topographical Limit the Benefit of Mushroom and Serif T Structures, *Langmuir*, 2013, **29**(4), 1100–1112. Copyright 2013 American Chemical Society.

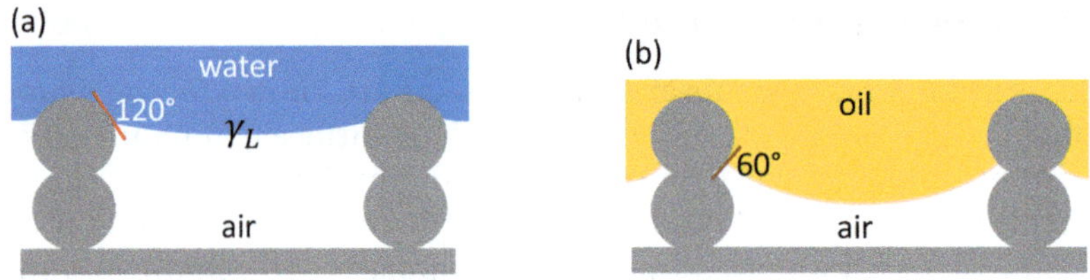

Figure 8.8 Sketch of a water drop deposited on a hydrophobic sphere. (a) The water drop rests at the top part of the sphere, to locally take its material's contact angle (here 120°). (b) The oil drop is pinned at the lower side of the sphere, as its material's contact is below 90° (here 60°). The oil drop is in a metastable state.

Scanning electron microscopy revealed the skin of springtails to be covered by a nanoscopic comb structure with hexagonally arranged cavities (Figure 8.7d).[28] Granules are located at the intersections of the combs. A cross-sectional profile of the skin, taken by transmission electron microscopy, illustrates that the cavities possess pronounced overhangs at their top edges (Figure 8.7e) similar to the petals of the flower-like particles.

Overhangs are required as on a submicrometre length scale the molecules of the drop want to minimize the interfacial energies. This results in a material contact angle below 90° for hydrophobic or non-polar liquids. As sketched in Figure 8.8, the contact line of a water drop is pinned at the upper side of a sphere (Figure 8.8a). Contrary, the material's contact angle for an oil drop is below 90°. Therefore, an oil drop is pinned at the underside of a sphere. With decreasing material's contact angle the position where the contact line is pinned moves towards the bottom of the sphere (Figure 8.8b). To prevent the drop touching the underlying sphere, the neck connecting both spheres should be narrow.

8.1.5 Fabrication of Superamphiphobic Surfaces

There are numerous techniques to construct superhydrophobic surfaces with hierarchical structures employing various materials and substrates.[29] The fabrication of superamphiphobic surfaces is more demanding due to the complex overhang structures required. The first superamphiphobic model surface was prepared in 2007, making use of lithography.[23] Different groups picked up the strategy and fabricated superamphiphobic structures consisting of regular arrays of inverse trapezes or nail-like morphologies.[9,30–33] Strategies include spontaneous bottom-up[26,34] like spray casting,[35] lithographic top-down[23,36] or combined approaches.[24,37] The bottom-up approach is usually simpler, cheaper and reaches smaller sub-100 nm features.[38] Also, superamphiphobic surfaces formed by fibrous structures have been developed.[39,40]

In 1997, Kaoru Tsujii *et al.* successfully fabricated a surface that showed an apparent contact angle of $\Theta = 151°$ for oil droplets. They modified the surface morphology of an aluminium plate by anodic oxidization (Figure 8.9). Subsequently they grafted fluorinated monoalkylphosphates to lower the surface energy.[41] The surface repels liquids which have a surface tension above 40 mN m^{-1}.

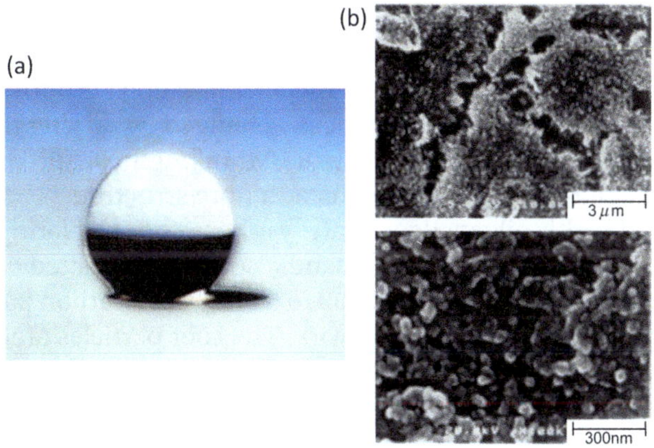

Figure 8.9 (a) A rapeseed oil droplet on a superamphiphobic surface. (b) SEM images of the surface taken at two different magnifications, showing an anodically oxidized aluminium plate surface. K. Tsujii, *et al.*, Super Oil-Repellent Surfaces, *Angew. Chem., Int. Ed. Engl.*, 1997, **36**(9), 1011–1012. Copyright Wiley-VCH Verlag GmbH & Co. KGaA. Reproduced with permission.

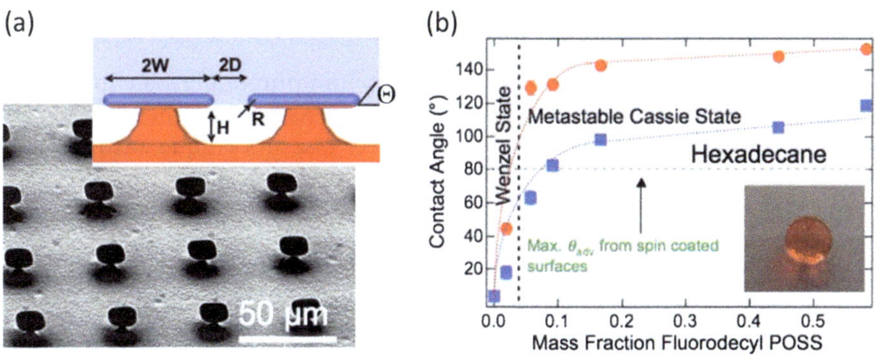

Figure 8.10 (a) Superamphiphobic surface prepared by lithography. Scanning electron microscopy images of a surface coated with "micro-nails" having flat caps. Inset: cartoon highlighting the formation of the composite interface with overhang structures. (b) Advancing apparent contact angle and receding apparent contact angle for hexadecane, respectively, on the electrospun surfaces. The inset shows a drop of hexadecane (dyed with Oil Red O) on a 44 weight % fluorodecyl POSS electrospun surface. From A. Tuteja, *et al.*, Designing superoleophobic surfaces, *Science*, 2007, **318**(5856), 1618–1622. Reprinted with permission from AAAS.

Tuteja *et al.* fabricated a superamphiphobic surface composed of so-called "micro-hoodoos" (Figure 8.10a). This model surface rendered it possible to calculate the impalement pressure independent of the size and spacing of the micro-hoodoos. The impalement pressure is the pressure that needs to be overcome so that a drop passes the transition from the Cassie to the Wenzel state. Furthermore, Tuteja *et al.* fabricated superamphiphobic meshes and showed their performance to separate water and oil. Therefore, they synthesized polyhedral oligomeric silsesquioxane (POSS) in which the rigid silsesquioxane cage was surrounded by perfluoro-alkyl groups. The surface energy of the matrix can be systematically changed by varying the mass fraction of fluoroPOSS molecules blend with PMMA (Figure 8.10b). After electrospinning, the fluoroPOSS molecule forms a fibre structure.

A simple and highly oil repellent strategy is based on templating soot (Figure 8.11).[26] First, several 10 μm layers of candle soot are deposited on a substrate (Figure 8.11a and b). Soot consists of 30 to 50 nm sized carbon particles which form a fractal-like highly porous network. The soot particles are covered with a 20 ± 5 nm thick silica shell (Figure 8.11b, d and e). After removing the carbon interior of the particles by calcination at 600 °C the layer is transparent (Figure 8.11j). Finally, the silica layer is hydrophobized by chemical vapour deposition with a semi-fluorinated silane to reduce the surface energy. Drops of water and hexadecane show apparent contact angles of 165 ± 1° and 156 ± 1° and tilting angles of 1 ± 1° and 5 ± 1°, respectively (Figure 8.11g and h).[26] The drops rest in a metastable state on the top layer of the coating.

8.1.6 Stability of the Cassie State

8.1.6.1 Stability Against Impalement

The stability of the superhydrophobic or superamphiphobic state is decisive for the use of the surfaces in real applications. As soon as the drop passes the transition to the Wenzel state, the surfaces loses its unique wetting properties such as self-cleaning and easy roll-off. In general the so-called Cassie-to-Wenzel transition is irreversible, in particular if the transition was caused by mechanical damage of the surface or an increase of the surface energy due to chemical decomposition or destruction of the coating by strong acids or bases. The Cassie-to-Wenzel transition can also be induced by applying an external pressure, evaporation of the drop, condensation of dew or a liquid in the coating.[6] In general, the transition is accompanied by a decrease of the apparent contact angle and an increase of the roll-off angle due to increased adhesion of the drop on the surface (Figure 8.12).[42]

To lift a drop from the Wenzel to the superhydrophobic Cassie state external energy is required to overcome the energy barrier.[43] A reversible switching between a wetted Wenzel and a superhydrophobic Cassie state could be realized by the application of electrical voltage and current (Figure 8.12).[44] A Pt wire was dipped into a water drop deposited on a superhydrophobic nanostructured surface (Figure 8.12a) that was in direct contact with a conductive substrate. Application of a potential difference between the water and the underlying substrate caused the Cassie-to-Wenzel transition (Figure 8.12b). After removal of the voltage the drop remained in the wetted state. To reverse the Cassie-to-Wenzel transition a short pulse of electrical current was transmitted through the substrate, causing a momentary increase of the surface temperature and the formation of a thin vapour layer which levitates the droplet back into the Cassie state (Figure 8.12c).

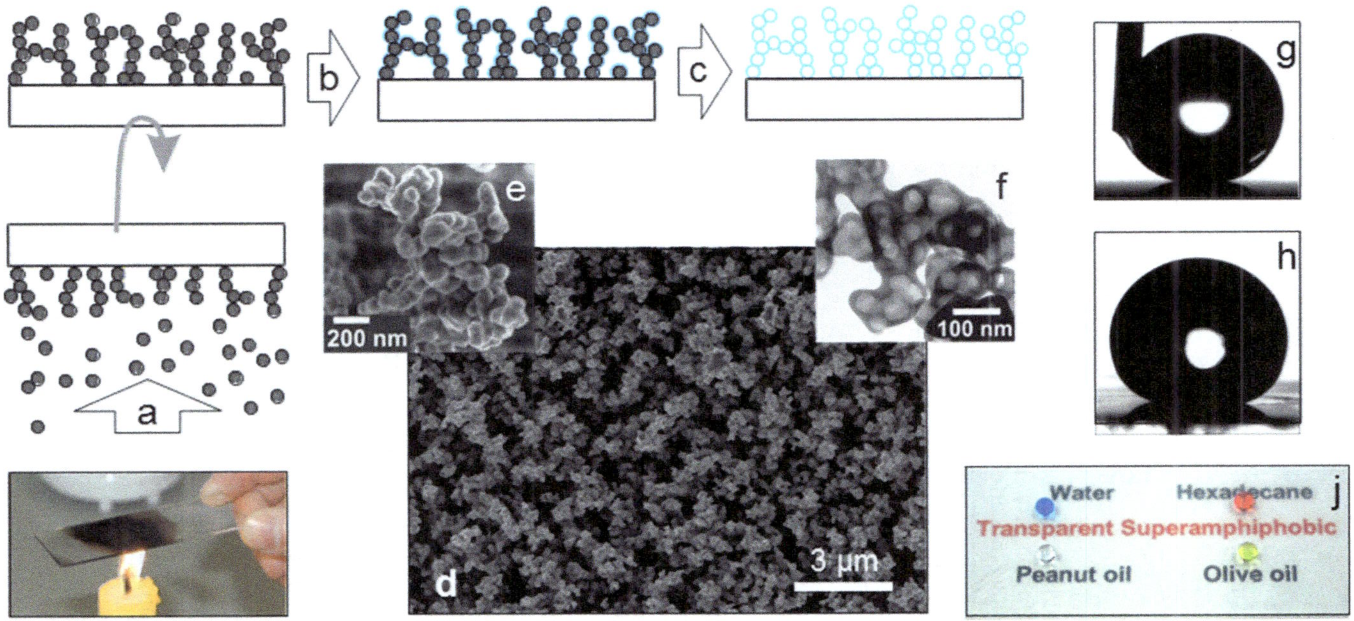

Figure 8.11 (a)–(c) Schematic to make superamphiphobic layers. (d) and (e) SEM image of the fractal-like structure of the layer before calcination. (f) TEM image of the structure after calcination. (g) Water drops form an apparent contact angle of 165°, (h) hexadecane forms an apparent contact angle of 156°. (j) Photograph of a drop of dyed water, peanut oil, olive oil, and hexadecane deposited on a superamphiphobic glass slide. The coated slide was placed on labelled paper to demonstrate transparency. Adapted from X. Deng, *et al.*, Candle Soot as a Template for a Transparent Robust Superamphiphobic Coating, *Science*, 2012, **335**(6064), 67–70. Reprinted with permission from AAAS.

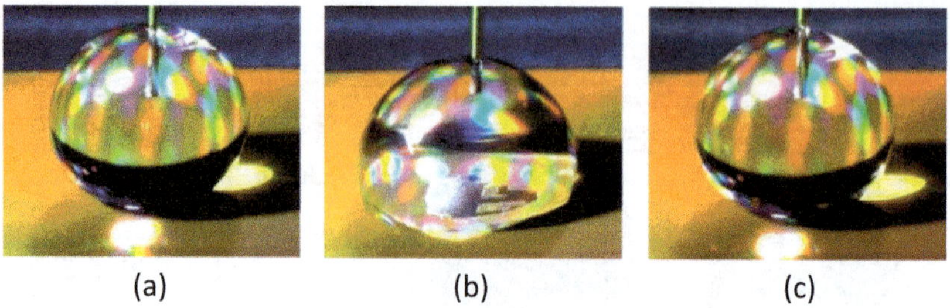

Figure 8.12 Electrically induced reversible transitions between different wetting states of a water drop on a nanostructured substrate. A voltage was applied between the drop and the substrate. The water drop was contacted by a Pt wire. (a) At $V = 0$ the drop was highly mobile. (b) At $V = 35$ the droplet transitioned to an immobile state. (c) After a short pulse of electrical current the drop returned to the original state. Adapted with permission from T. N. Krupenkin, *et al.*, Reversible wetting–dewetting transitions on electrically tuneable superhydrophobic nanostructured surfaces, *Langmuir*, 2007, **23**(18), 9128–9133. Copyright (2007) American Chemical Society.

The Cassie-to-Wenzel transition typically passes *via* the sag (Figure 8.13a) or the de-pinning (Figure 8.13c) mechanism[45–47] as illustrated by LSCM images. The transition was induced by an increase of the Laplace pressure as the water drop evaporated. The impalement of the drop into the layer can be monitored by recording the thickness of the air cushion (Figure 8.13b). In the sag mechanism the rim of the drop remained pinned at the top face of the pillars. The curvature of the water–air interface gradually increases while the drop evaporates (Figure 8.13b). The drop passes the transition to the Wenzel state as soon as the lowest point of the meniscus touches the bottom surface. The minimum thickness of the air cushion can be calculated if the maximum spacing of neighbouring protrusions and the radius of the drop, R, is known. In case of a micropillar array arranged on a square lattice the largest value of the mean curvature is obtained at the centre of the diagonal connecting neighbouring pillars. The thickness of the air cushion, h_{min}, follows from geometrical considerations and decreases with increasing pillar–pillar distance and decreasing radius of the water drop.

$$h_{min} = h - \left(R + \sqrt{R^2 - 0.5d^2}\right) \tag{8.5}$$

The liquid interface touches the bottom of the substrate for $h_{min} = 0$. The corresponding maximum Laplace pressure is given by:[38]

$$P = \frac{2\gamma}{R} = \frac{8\gamma h}{d^2 + 2h^2} \tag{8.6}$$

It is important to keep in mind that the largest spacing between pillars and the lowest height of the pillars determines the stability of the Cassie state. Mechanical damage of the surface can locally increase the spacing between

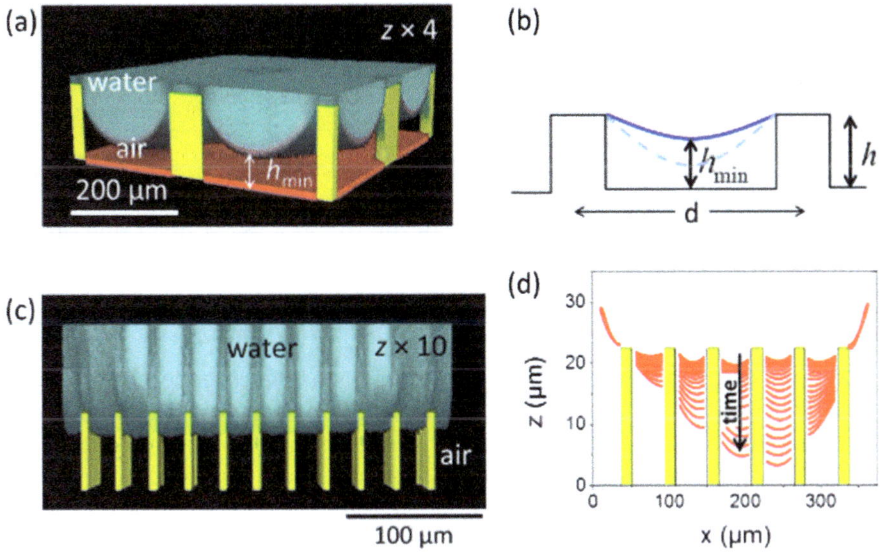

Figure 8.13 (a) LSCM images of the sagging mechanism of the Cassie-to-Wenzel transition. The array dimensions are $a = 50$ μm, $h = 23$ μm, $d = 200$ μm. For better visualization, the *z*-axis is 4 times magnified. (b) Variation of the thickness of the air cushion. h_{min} denotes the minimum thickness. (c) LSCM images of the de-pinning mechanism of the Cassie-to-Wenzel transition. Complete contact area view of a drop during the Cassie-to-Wenzel transition on an array with $a = 5$ μm, $h = 5$ μm, $d = 20$ μm. The *z*-axis is 10× magnified. (d) Dependence of the thickness of the air cushion along a diagonal passing through the centre of the drop. The red lines are fits to the drop profile. Profiles are taken at 3 s apart. The array dimensions are $a = 10$ μm, $h = 23$ μm, $d = 40$ μm. Adapted with permission from P. Papadopoulos, *et al.*, How superhydrophobicity breaks down, *Proc. Natl. Acad. Sci.*, 2013, **110**(9), 3254–3258.

neighbouring protrusions, causing the drop to touch its base. Sagging forms a problem for the design of transparent superhydrophobic surfaces. To reduce haze, the superamphiphobic layer should be as thin as possible, preferentially below half a micrometre. In that case, a few micrometres wide spacing between neighbouring protrusions can already be sufficient that the drop touches the base.

In the de-pinning mechanism the three-phase contact line unpins from the edge of the protrusions and slides down its wall (Figure 8.13c and d). De-pinning starts if the contact angle exceeds the advancing contact angle of the corresponding flat surface of identical material, measured from the vertical.[48] In reality, de-pinning does not proceed simultaneously from all pillars but depends on small differences of the smoothness and chemical composition of the pillar's edge (Figure 8.13d).

The pressure to induce de-pinning depends on the shape of the protrusions. For simplicity, we assume spherical protrusions. A water drop on a hydrophobic surface has a material contact angle above 90°. Therefore, a water drop is pinned at the upper side of the sphere. Deviations from the material contact angle are energetically unfavourable, *i.e.* the drop needs to overcome an energy barrier. In case of large spacing between the spheres, the maximum capillary force is given by:

$$F_c = 2\pi\gamma R \sin\beta \sin(\theta_A - \beta) \tag{8.7}$$

β is the angle describing the position of the three-phase contact line on the particle surface (Figure 8.14a).

$$F_c = 2\pi\gamma R \sin^2(\theta_A/2) \tag{8.8}$$

The maximal capillary force per protrusion can be exceeded by applying an external pressure. The critical pressure, the so-called impalement pressure, is given by the maximum capillary force per area covered by one pillar, $P = F_c/A$. If the spheres (or columns of spheres) are arranged on a square lattice ($A = d^2$), P is given by:[38]

$$P = \frac{2\pi\gamma R}{d^2 \sin^2(\theta_A/2)} \tag{8.9}$$

The impalement pressure decreases with increasing lattice spacing or large sphere size (Figure 8.14b). This derivation assumes that the necks connecting neighbouring spheres are small, *i.e.* the material advancing contact angle should not be affected by the sphere underneath. For a monolayer of spheres and low θ_A the drop may touch the top side of the sphere underneath or the base before de-pinning occurs.

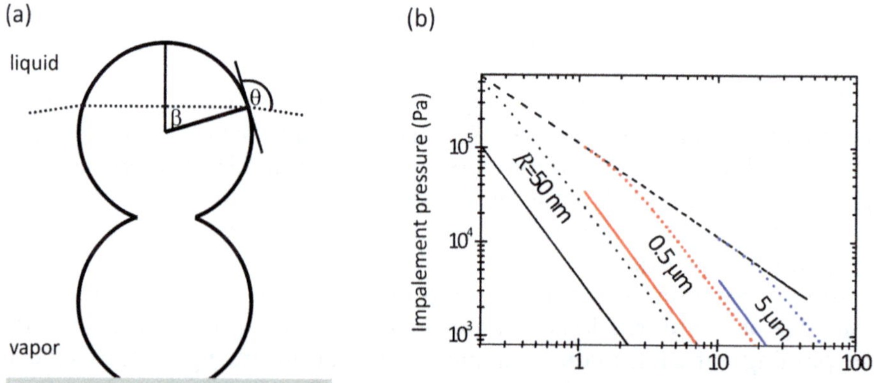

Figure 8.14 (a) Sketch of a drop on a hydrophobic sphere. (b) Impalement pressure for a liquid deposited on an array of spheres arranged in columns for different size and spacing between neighbouring columns. Adapted from H. J. Butt, *et al.*, Design principles for superamphiphobic surfaces, *Soft Matter*, 2013, **9**(2), 418–428 with permission from The Royal Society of Chemistry.

The impalement pressure does not depend on the hydrophilicity of the sphere's topmost part. In case of micropillar arrays this implies that the pillar's top face can be hydrophilic.[10] It is essential that the side walls are hydrophobic to ensure a large advancing contact angle. However, hydrophilic top faces reduce the apparent receding contact angle, resulting in an increase of hysteresis. The apparent receding contact angle follows from balancing the forces at the rim of a drop.[38]

$$\Theta_{\mathrm{r}} = \pi\left(1 - \frac{2R}{d}\cos^2\frac{\theta_{\mathrm{R}}}{2}\right) \tag{8.10}$$

For a given material receding contact angle θ_A, the apparent receding contact angle only depends on the ratio d/R, which should be as high as possible to achieve a high Θ_r. A decreasing apparent receding contact angle increases the solid–liquid adhesion force as:

$$F_{\mathrm{LA}} = \gamma bk(\cos\Theta_{\mathrm{r}} - \cos\Theta_{\mathrm{a}}) \tag{8.11}$$

Here, b is the effective width of the drop in contact with the solid. $k \approx 1.5$ is a constant which varies slightly depending on the specific shape of the contact area.[49] Therefore, surfaces with low Θ_r are rarely superhydrophobic or superamphiphobic. However, the drop still can be in the Cassie state.

8.1.6.2 Mechanical Stability

The mechanical stability of super liquid-repellent surfaces is essential for many real-life applications. The longevity of super liquid-repellent surfaces can be increased by two main approaches which are ideally combined.

The first approach focuses on reducing the consequences of surface damage and the second approach considers how to restore surface roughness/chemistry after damage.[50,51] The consequences of surface damage can be reduced by the combination of micro- and nano-rough structures.[52–54] Ideally, the nano-roughness should provide good liquid repellency and the micro-rough structures should ensure good mechanical stability (Figure 8.15a). Given that the damage only partially destroys the micro-roughness, the sheltered regions keep their nano-roughness. By this, liquid drops remain pinned at the nano-protrusions.

Often hydrophilic templates are used which are post-treated with a hydrophobic coating. Surface damage will cause the hydrophilic bulk material to be revealed (Figure 8.15b, bottom). These hydrophilic spots lead to an increase of the receding contact angle and therefore to an increase of solid–liquid adhesion. Therefore, another approach emphasizes the role of the bulk material used to prepare the super-liquid repellent surface. Upon surface damage bulk material can be exposed to the liquid. A motivation to use hydrophobic bulk materials is that the newly created interface will also be low in energy. Ideally, the surface chemistry is the same before and after the damage (Figure 8.15b, top).

Several methods have been applied to quantify mechanical stability. Assessing "how stable" surfaces are, is however not trivial. Depending on the

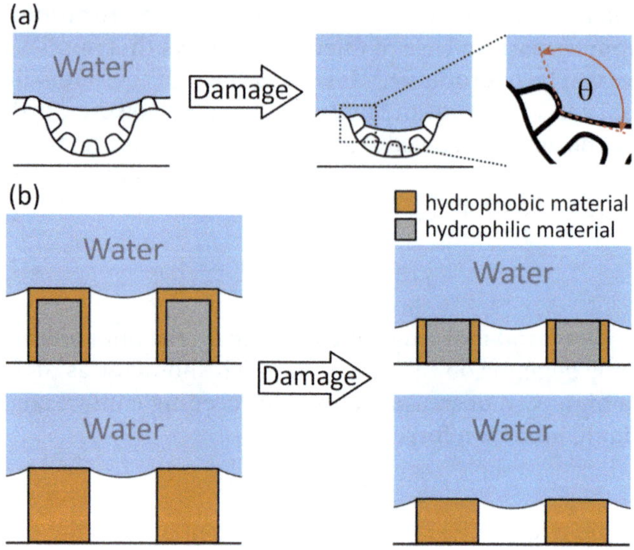

Figure 8.15 (a) A micro- and nano-rough surface is exposed to surface damage. The micro-roughness provides shelter for the nano-rough structures, which is required for a stable Cassie–Baxter state. Magnification: the three-phase contact line is suspended on the nano-rough structure details under local fulfilment of the material contact angle θ. (b, top) Damage of a hydrophobic coating can reveal hydrophilic bulk material (grey). The Cassie state can still be stable though drop adhesion is increased. (b, bottom) Damage of hydrophobic (orange) material will expose a new interface with same surface chemistry, ideally not changing repellency.

materials used, the surface thickness, the specific shape of the surface protrusions and the overall porosity, large differences in the mechanical stability of super liquid-repellent layers can be obtained. This strongly complicates the adoption of standard testing routines and so far no universal protocol exists. Accordingly, numerous protocols for testing of mechanical properties can be found in the literature, *e.g.* reviewed in ref. 55. The protocols differ in how and how much load/wear was applied to the surface and how the analysis of the experiment is conducted. For simplification, we suggest classification of tests in two categories.

Category 1: load/wear is applied to a macroscopic area (at least a few mm). Examples are tape test,[56,57] wiping of tissues,[58] sand impact[57] or laundry cycles[59] (Figure 8.16). After the test, the samples are analysed by image-based techniques, *e.g.* by optical or electron microscopy. Importantly, the apparent contact angles before and after surface treatments are measured. The advantages of this category are its good accessibility and simple analysis. Nevertheless, macroscopic tests are likely to fail resolving the difference of layers having similar mechanical properties.

Category 2: load/wear is applied to a microscopic area (several hundred μm too a few nm). In this case, the affected area is too small for reasonable contact

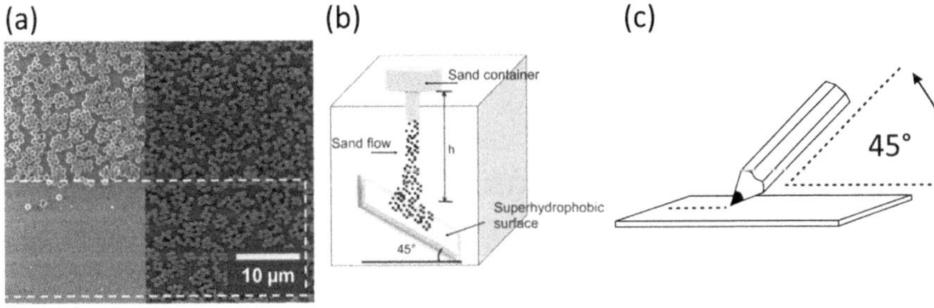

Figure 8.16 (a) A superhydrophobic surface made of hollow silica particles was partially exposed to double sided tape (white boxes indicate the exposed areas). (a, left) The particles only adhere by van der Waals force to the substrate and are easily removed by the tape test. (a, right) Particles are chemically fixed by chemical vapour deposition of silica. The particles were not removed from the substrate by the tape test. (b) Schematic setup to determine surface stability against impact of sand grains. The kinetic energy of the sand grains can be varied by changing the distance *h* between surface and container. (c) Scheme of the pencil hardness test. Pencils with different hardness (9H(hard) – 9B(soft)) are pushed/pulled over a surface at an angle of typically 45°. The resistance of the coating is given by the hardest pencil unable to scratch the surface. Adapted from X. Deng, *et al.*, Transparent, Thermally Stable and Mechanically Robust Superhydrophobic Surfaces Made from Porous Silica Capsules, *Adv. Mater.*, **23**(26), 2962. Copyright © 2011.

angles measurements before and after surface damage. One popular example is the pencil hardness test where pencils of different hardnesses are drawn over a surface until one pencil type fails to scratch the surface.[56] The International Organization of Standardization (ISO) and the American Society for Testing and Materials (ASTM) designed guidelines for surface testing, *e.g.* for the tape test and the pencil scratch test. These tests mainly aim at assessing the stability of thin film coatings and their adoption to fragile, porous surfaces can be problematic.[55] Nano-indentation or atomic force microscopy (AFM) based approaches are more sensitive and quantitative. By combining nano-indentation and AFM, a force range between a few nano-newton to several milli-newton can be covered. Both techniques are highly sensitive.[60] Nano-indentation and atomic force spectroscopy also gives the Young's modulus E and the hardness H.[58,61–63] This can provide precious information for optimization of individual systems. It also bears the potential to simplify comparison of surfaces between different groups. However, the measurements themselves can be challenging as the penetration depth should be smaller than 10% of the film thickness to neglect substrate effects. For individual super liquid-repellent surfaces this might cause a problem due to their poor mechanical stability or small film thickness or both. Visualization of the indentation area, often required for the elastic modulus and the hardness in nano-indentation, can be cumbersome for very porous surfaces. Last, but not least, due to the high surface roughness, the size ratio of the indenter to the surface features needs to be considered and taken care of for the interpretation of the results.

Improving and characterizing the mechanical stability of super liquid-repellency is and will remain one of the most challenging aspects towards their broad application in everyday life.

8.2 Potential Applications

Prospective applications of superhydrophobic and superamphiphobic surfaces range from self-cleaning,[14,64,65] drug reduction,[66–68] anti-biofouling,[69,70] sensors,[71–73] liquid repellent textiles,[74–76] gas contactor membranes,[77] to surfaces for particle synthesis.[78–80]

8.2.1 Polymeric Particles in the mm to μm Range

Polymeric particles grew more and more important over the last decades and became an important part of everyday life, *e.g.* in coatings, cosmetics and electronics.[81,82] Common polymerization techniques include suspension, dispersion, emulsion and mini-emulsion polymerization or microfluidics.[83,84] All methods have in common that particles are produced, functionalized and processed in solvents. The solvents are essential for heat transfer and to control the particle shape. Typically, minimization of free energy dictates a spherical shape of the initial monomer drops or growing particle and this geometry is maintained in the final product.

Superhydrophobic surfaces were used to prepare particles by solvent evaporation.[79,80,85,86] An aqueous dispersion was deposited on a superhydrophobic surface. After evaporation of water particles the dispersed nanoparticles formed spherical aggregates. The shape of the particles can be tuned by adjusting the concentration of the dispersion and the properties of the superhydrophobic or superamphiphobic surface.

Superamphiphobic surfaces renders further strategies possible as they are repellent towards many organic low surface tension liquids, including several monomers and functional liquids (Figure 8.17). Particles can be fabricated *via* physical (tuning temperature,[78] solvent evaporation[79,85]) or chemical polymerization[78] strategies. Compared to classical polymerization techniques, particle synthesis on superamphiphobic surfaces neither requires solvents nor additives.

8.2.2 Particle Synthesis *via* Tuning Temperature

This strategy makes use of the temperature dependence of the physical interactions of polymers. A polymer powder, a polymer blend, or a mixture of a polymer powder with nanoparticles is sprinkled onto the superamphiphobic layer and heated above its glass transition temperature, T_g (Figure 8.18a). While heating the powder, the viscosity gradually decreases and the agglomerates start to shrink and transform to spherical microspheres (Figure 8.18b). The spherical shape is preserved after cooling the particles

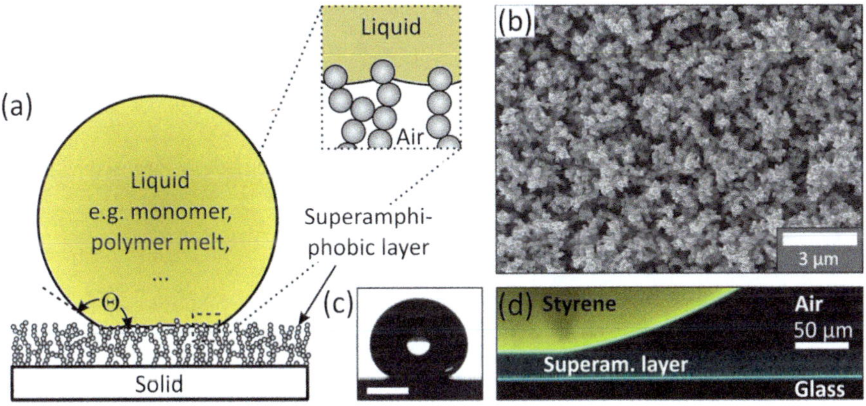

Figure 8.17 Liquid drop on a superamphiphobic layer. (a) Schematic of a liquid drop on a superamphiphobic layer and a magnified view of the interface between the liquid and the superamphiphobic layer. (b) Scanning electron microscope (SEM) image of a superamphiphobic layer. (c) Video image and (d) vertical section through a drop of styrene on a superamphiphobic layer imaged with a confocal microscope. Scale bar in (c) is 1 mm. The 6 µL drop was labelled with 0.04 mg mL^{-1} *N*-(2,6-diisopropylphenyl)perylene-3,4-dicarbonacidimide. It appears yellow. The original excitation light in blue caused by reflection at interfaces shows the top surface of the glass and the bottom surface of the styrene drop. In between, with some weak scattering at the silicon oxide nanostructures the superamphiphobic layer can be seen. Adapted from Deng, *et al.*, Solvent-Free Synthesis of Microparticles on Superamphiphobic Surfaces, *Angew. Chem., Int. Ed.*, **52**(43), 11286–11289. Copyright © 2013.

below T_g. The size of the final microspheres is given by the volume of the initial powder agglomerate and thus by the dispersion and size of the agglomerates. Janus microspheres can be produced by mixing two polymers, for example polystyrene and poly(methyl methacrylate) (Figure 8.18b). To distinguish the two polymers, the fluorescent dye rhodamine 6G was covalently attached to the polystyrene. The dyed blend was sprinkled on the superamphiphobic surface and annealed at 160 °C. This temperature is well above the glass transition temperatures of both dyed polystyrene (T_g = 91 °C) and poly(methyl methacrylate) (T_g = 120 °C). The different stages of the phase-separation process were monitored by video microscopy (Figure 8.18b). After 7 minutes a microparticle with two separate domains was obtained. Finally, the structure freezes when the temperature is lowered below T_g. This approach to produce Janus particles leads to particles with well-separated surface properties.

Magneto-responsive microspheres were prepared by mixing iron oxide nanoparticles with PS. The agglomerates were placed on the superamphiphobic surface and the surface was annealed at 165 °C. During annealing, a magnetic field was applied to orient the iron oxide nanoparticles (Figure 8.18c). The nanoparticles diffused through the PS matrix towards the magnet, leading to an anisotropic distribution inside the microsphere. After

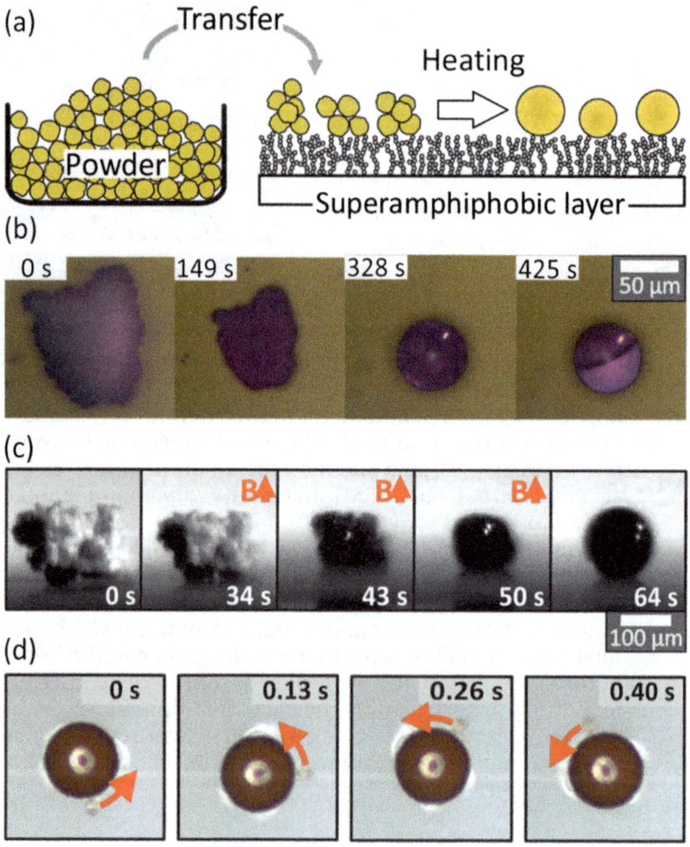

Figure 8.18 (a) Schematics of particle synthesis *via* tuning the temperature. A polymer powder/blend is placed on a superamphiphobic surface and annealed above T_g. (b) Sequence of video microscopy images showing annealing of an agglomerate of polystyrene dyed with rhodamine B (PS-dye, MW = 13 800 g mol^{-1}, T_g = 91 °C) and poly(methyl methacrylate) (PMMA, MW = 9500 g mol^{-1}, T_g = 120 °C, surface tension: γ = 43 mN m^{-1} at 120 °C) for 7 minutes. The polymer blend contains PS-dye/PMMA (1 : 1 w/w). (c) Sequence of optical microscope images starting with an iron oxide containing agglomerate of polystyrene (MW = 5800 g mol^{-1}, polydispersity, T_g = 78 °C, surface tension: γ = 32 mN m^{-1} at 120 °C). Microspheres formed after annealing for more than 1 minute at 100 °C. The polymer melt is exposed to a magnetic field during annealing for anisotropic arrangement of the iron oxide particles. (d) After cooling the magnetic microspheres are transferred to the air–water interface and exposed to a rotating magnetic field (1.3 ± 0.1 mT, 14 Hz). The magnetic microparticles rotated according to the external field. A small plastic bead was attached to the magnetic PS-particle to visualize the rotation (indicated by arrows). Adapted from Deng *et al.*, Solvent-Free Synthesis of Microparticles on Superamphiphobic Surfaces, *Angew. Chem., Int. Ed.*, **52**(43), 11286–11289. Copyright © 2013.

cooling, the nanoparticles are frozen in the matrix and the composite microspheres possess a permanent magnetic dipole moment. This was demonstrated by dispersing individual microspheres in water and exposing them to a weak rotating magnetic field. The microspheres rotate in a well-controlled manner with a rotation frequency corresponding to the external magnetic field (Figure 8.18d).

8.2.3 Particle Synthesis *via* Radical Polymerization

This strategy makes use of the high repellency of the superamphiphobic surface *versus* certain monomers, as shown in Table 8.1.

For demonstration, bi-functional methacrylate derivatives were radically polymerized since they have a broad application in industry and medicine. Triethylene glycol dimethacrylate (TEGDMA) and bisphenol A glycerolate dimethacrylate (Bis-GMA) mixtures, for example, were extensively studied as they form part of dental resins. For the convenience of the patient, these resins are cured photochemically on a short time scale (seconds to few minutes).[87]

A mixture of GMA, TEGDMA and initiator (phenylbis(2,4,6-trimethylbenzoyl)phosphine oxide, Irgacure 819) was prepared and a 20 μL sized drop was placed on a concave watch glass coated with a superamphiphobic layer. The polymerization was initiated by pulsed UV irradiation. The drop was moved continuously while the polymerization proceeded (Figure 8.19a). Drop motion reduced impalement of the mixture into the top-most part of the superamphiphobic layer and allowed the suppression of a flat contact area in the final particle. After 2 minutes reaction time, a sphere-like solid polymeric particle was synthesized (Figure 8.19b and c). However, depending on the mechanical stability of the superamphiphobic coating, the top layer can be peeled off by the moving particle in the course of the polymerization. As a result, particles can partially be coated with layer material. The size of the microsphere corresponds to the initial volume of the monomer drops. Depositing the monomer mixture with a pipette, the microparticles have a diameter up of ~2.5 mm. If the monomer mixture is deposited with a nanoplotter the particle size is in the order of a few tens of a micrometre (Figure 8.19d). There is a lower

Table 8.1 Apparent contact angle Θ, roll-off angle α, and surface tension γ of monomers on a superamphiphobic surface at 20 °C. Adapted from Deng, *et al.*, Solvent-Free Synthesis of Microparticles on Superamphiphobic Surfaces, *Angew. Chem., Int. Ed.*, 52(43), 11286–11289. Copyright © 2013.

Monomer	Θ	α	γ (mN m^{-1})
Styrene	158°	6°	34
Methyl methacrylate	156°	10°	28
Acrylic acid	154°	7°	29
Adipoyl chloride	152°	9°	38
Ethylenediamine	152°	16°	42

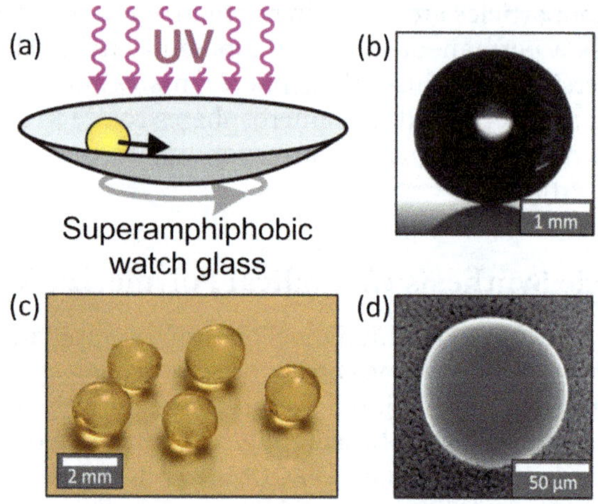

Figure 8.19 Synthesis of microspheres by radical polymerization. (a) Schematic of the set-up. (b) and (c) Particles synthesized from bis-GMA (15 wt%), TEGDMA (84 wt%) and photoinitiator (1 wt%). After mixing and sonication for 30 min, a drop (8–10 mL) was pipetted onto a concave watch glass coated with a superamphiphobic layer. The polymerization was initiated by pulsed UV irradiation for 1 min followed by continuous illumination for 4 min (LQ 400, UV-A: 200 mW cm^{-2} at the end of the glass fibre). (d) SEM image of a microsphere from 99 wt% TEGDMA with 1 wt% photoinitiator polymerized by UV exposure for 3 min. The mixture was deposited by an inkjet printer (Nano-Tip AJ 070-401) held at a distance of 4 cm. Adapted from Deng, *et al.*, Solvent-Free Synthesis of Microparticles on Superamphiphobic Surfaces, *Angew. Chem., Int. Ed.*, **52**(43), 11286–11289. Copyright © 2013.

fundamental limit for the particle size which is given by the average spacing between the protrusions. If the diameter of the liquid drop is smaller than about 10 times the average protrusion distance, adhesion starts to dominate and particles adhere strongly to the surface. The average protrusion distance for a candle soot-based superamphiphobic layer is 1–2 µm, which means the lower limit for particle size is about 20 µm in diameter.

Despite the success of the different strategies, particle synthesis on superamphiphobic surfaces is a two-dimensional process, and techniques for high throughput still need to be established.

8.2.4 Protein and Cell Adhesion on Superamphiphobic Layers

Adhesion of proteins, bacteria or cells on surfaces concerns a wide range of different domains, from food processing[88] to health care, and can cause serious health risks such as inflammation.[89] Once proteins, bacteria or cells found anchoring points on the surface, more and more of these will attach to the present ones and agglomerations will be formed.[90,91] One strategy to prevent the formation of biofilms focuses on coatings that release biocidal compounds, such as silver ions or antibiotics.[90,92–94] Another strategy is to suppress biofilm formation by designing coatings that prohibit the adsorption of cells, bacteria or proteins, for example using low-energy surfaces (*e.g.* PEG or Teflon)[95,96] or by inducing nano- or micro-roughness.[97–99]

Lately, superhydrophobic surfaces were tested for their ability to prevent attachment of blood,[100] platelets,[101] bacteria,[102] and proteins[69] under stationary or flow[70,101,102] conditions. Superhydrophobic Teflon tubes (PTFE) were implanted in pigs and rabbits to investigate their anti-adhesion performance for cells compared to classical Teflon tubes.[103] Reduced,[101,102] enhanced,[103] time[104] and cell type-[105] dependent adsorption was recorded. In general, laminar flow of the liquid slows down adhesion due to a reduced dwell time for anchoring, promoted desorption, and reduced collision frequency of the material with the surface due to a finite slip length.[70,101]

Blood easily impales a superhydrophobic surface, resulting in an increased contact area with the substrate. The reason for impalement is the low interfacial tension of blood, $\gamma = 47$ mN m^{-1}. Superamphiphobic coatings prevent impalement of blood into the coating. A drop of heparinized whole human blood deposited on a superamphiphobic surface showed high contact angle values of $162 \pm 1°$ and low tilting angles of $5 \pm 1°$ (Figure 8.20a). A sharp boundary between the superamphiphobic surface and the blood can be identified when imaging a drop of blood on a superamphiphobic surface by LSCM (Figure 8.20b).

Various protocols exist for quantitative analysis of proteins. The adhesion of proteins to superamphiphobic surfaces can be quantified after contact of a superamphiphobic substrate and blood for several hours. After removing the blood by tilting the wells no residuals were detected by eye (Figure 8.21).

The area of the wells which were in direct contact with the supernatant blood were analysed using the Pierce 660 nm Protein Assay Kit. As a result, one obtained the total amount of adhered protein per cm^2 of analysed mesh. Superamphiphobic meshes without any contact to blood were also

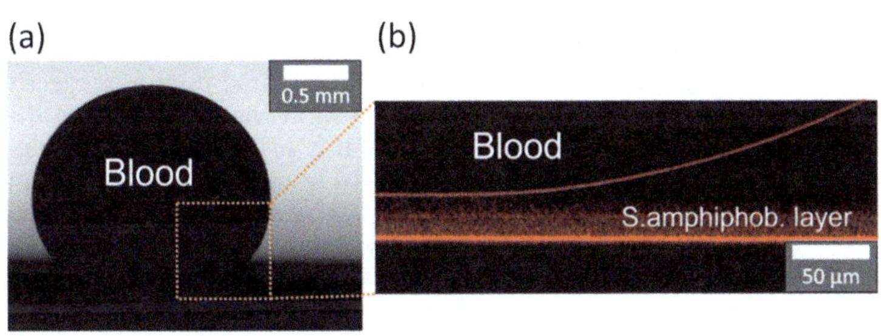

Figure 8.20 (a) A drop of heparinized whole human blood deposited on a superamphiphobic surface. The diameter of the drop of blood is *ca.* 3 mm. (b) Vertical cross-section of a blood drop on a superamphiphobic layer imaged by confocal microscopy in reflection mode. Macmillan Publishers Ltd., Nature Publishing Group: adapted from M. Paven, *et al.*, Super liquid-repellent gas membranes for carbon dioxide capture and heart–lung machines, *Nat. Commun.*, **4**, copyright 2013.

Figure 8.21 Heparinized blood was incubated for 6 h in a superamphiphobic well and removed afterwards by tilting. Macmillan Publishers Ltd., Nature Publishing Group: M. Paven, *et al.*, Super liquid-repellent gas membranes for carbon dioxide capture and heart-lung machines, *Nat. Commun.*, 2013, **4**, copyright 2013.

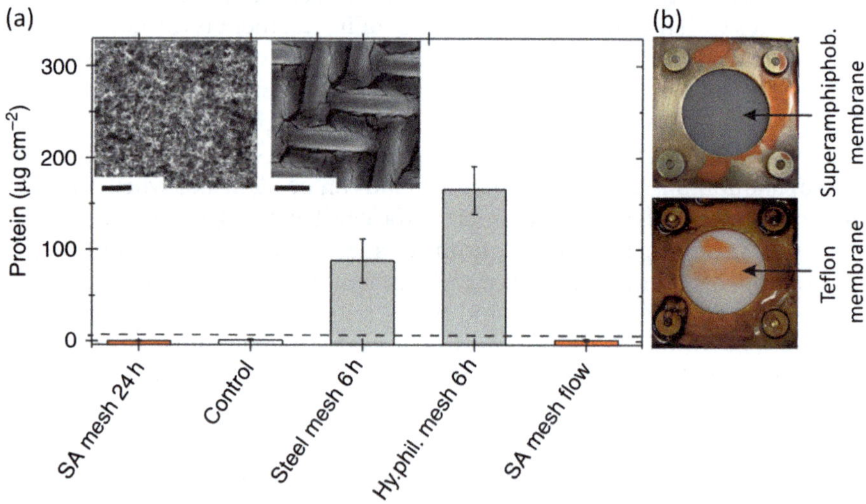

Figure 8.22 (a) Protein adhesion in µg cm^{-2} for metal, hydrophilic and superamphiphobic (SA) meshes at different time points, respectively. The indicated detection limit (dashed line) was derived from the known sensibility limit of the Pierce 660 nm Test. Error bars are based upon the root mean square deviation. (b) Images of flow cell equipped with superamphiphobic (top) and Teflon membrane (bottom) after being exposed to a flow of heparinized blood for 3 h. Macmillan Publishers Ltd., Nature Publishing Group: adapted from M. Paven, *et al.*, Super liquid-repellent gas membranes for carbon dioxide capture and heart–lung machines, *Nat. Commun.*, 2013, **4**, copyright 2013.

analysed and served as control samples. Also, a superamphiphobic surface was exposed to a circulating flow of whole human blood for 3 h. After the flow cell was opened, no blood was found on the membrane. In contrast, a Teflon membrane used as reference was covered by blood (Figure 8.22b). Figure 8.22a shows the average value of protein adsorption for each surface type.

In contrast to steel and hydrophilic meshes, superamphiphobic meshes incubated for 24 h, superamphiphobic meshes exposed to blood flow for 3 h and control meshes were below the detection limit of 6.25 μg cm^{-2}. Also, no cells or other blood components were found by electron microscopy on a superamphiphobic surface incubated for 48 h (Figure 8.22a, left inset). In contrast, Figure 8.22a (right inset) shows a steel mesh exposed to 2 h of human blood.

8.2.5 Superamphiphobic Membranes

So far, most superhydrophobic and superamphiphobic layers were prepared on solid surfaces. Few techniques exist to coat meshes or to fabricate superamphiphobic surfaces that are gas permeable. Superamphiphobic meshes can be used as highly efficient contactor membranes for gas exchange between a liquid and a gas phase, *i.e.* O_2 enrichment of haemoglobin, or for self-cleaning textiles (Figure 8.23).[75,106] Superamphiphobic membranes are less sensitive to membrane wetting and plugging than conventional materials.

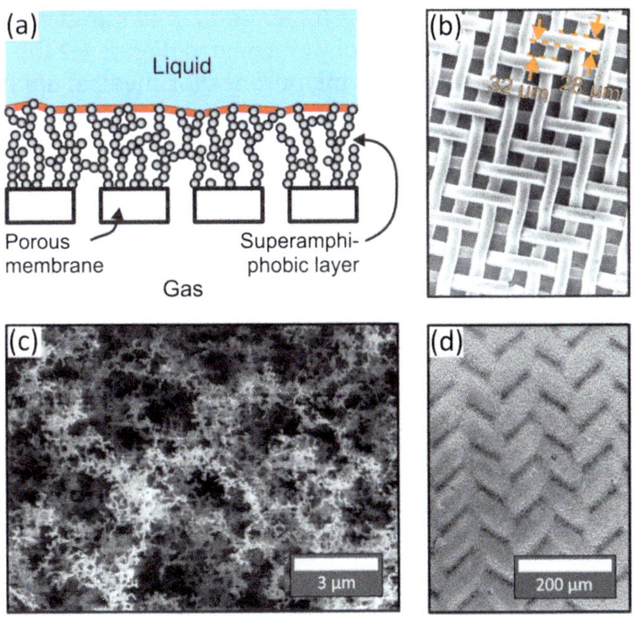

Figure 8.23 (a) Schematic drawing of a superamphiphobic membrane. (b) SEM of a blank stainless steel metal mesh. (c) Magnification of a stainless steel mesh coated with a superamphiphobic layer (scale bar: 3 μm). (d) Low magnification of a superamphiphobic membrane (scale bar: 200 μm). Macmillan Publishers Ltd., Nature Publishing Group: adapted from M. Paven, *et al.*, Super liquid-repellent gas membranes for carbon dioxide capture and heart–lung machines, *Nat. Commun.*, **4**, 2013, copyright 2013.

Typically, gas contactor membranes consist of non-porous, mostly polymeric membranes, supported liquid membranes or porous membranes.[107–109] Gas and liquid are separated by the contactor membrane. Gas molecules diffuse from the feed to the permeate side of the membrane, along a concentration gradient. The efficiency of the membrane is defined by the mass transfer coefficient K.[110,111] The mass transfer coefficient is defined as the amount of gas transferred per time $\dot{n}$ and per unit area A divided by the driving concentration difference Δc:

$$K = \dot{n}/(A\Delta c)$$

As one example, we demonstrated the oxygenation of blood. Therefore, first a 1.2% solution of whole human blood stabilized with lithium-heparin as anticoagulant in phosphate buffered saline (PBS, Dulbecco's PBS) was deoxygenized by bubbling a continuous stream of nitrogen through the solution for 30 min at 37 °C; blood was diluted to reach a high sensitivity in the UV/vis absorption spectrometer. Deoxygenated blood in PBS was transferred into the reservoir of the setup (Figure 8.24a) under nitrogen. We continuously recorded UV/vis absorption spectra, while diluted blood circulated for 16 min under nitrogen atmosphere until oxygen was introduced into the gas box at a flow rate of 0.7 L min^{-1}. Spectra measured at t = 0 min and t = 45 min demonstrate that pumped blood in PBS was successfully oxygenated (Figure 8.24b).

When using superamphiphobic membranes for medical applications, as for the oxygenation of blood, it needs to be ensured that the coating preserves its stability for a long time. In particular, breaking off of parts of the

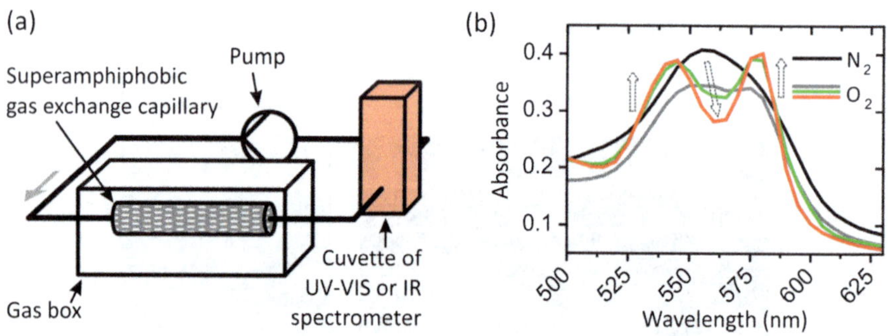

Figure 8.24 (a) Sketch of a setup used to measure gas exchange. A gas box contained a flow cell equipped with superamphiphobic membranes. Sodium hydroxide solution or deoxygenized blood was pumped through the flow cell and UV-vis spectra were subsequently recorded to monitor the gas exchange. (b) Absorbance spectra of 1.2% deoxygenated blood before (black) and its subsequent change while being exposed to oxygen (grey to red). The mono-peak of deoxygenated blood (black, N_2) transformed into the characteristic double-peak of oxygenated blood at 560 nm (grey to red, O_2), proving successful exchange. Macmillan Publishers Ltd., Nature Publishing Group: adapted from M. Paven, *et al.*, Super liquid-repellent gas membranes for carbon dioxide capture and heart–lung machines, *Nat. Commun.*, 2013, **4**, copyright 2013.

topmost region of the coating needs to be prevented as otherwise fluorinated nanoparticles might enter the bloodstream. This still remains a challenge.

8.2.6 Fog Harvesting

Water supply in arid regions is commonly performed by desalination. More than 15 000 desalination plants are installed and operated worldwide.[112] The installed capacity around the Mediterranean is about $12 \times 10^6\ m^3$ per day,[113] of which 70% is used for irrigation. Water desalination suffers from high operational energy consumption. Depending on the input water salinity, the specific energy requirement is 2–3 kWh m^{-3} H_2O, corresponding to a fingerprint of 1.4 to 1.8 kg CO_2 per cubic meter of produced water.[114] Lately, the possibilities to collect water directly from humid atmospheric air or fog have been investigated. Fog forms when humid air is cooled below its dew point, *e.g.* when it moves upward. Foggy air contains 0.05–0.5 g H_2O m^{-3}. Thus, a 40 m^2 surface could ideally collect up to 1700 L per day at a wind speed of 1 m s^{-1}. "Fog harvesting" is meant to be a process in which water is collected from supersaturated humid air, mist or fog using "traps" made of engineered materials (Figure 8.25).[115,116] Tiny water droplets are taken along by the wind until they hit the traps, adhere on the surface and grow by coalescence or condensation until they reach a certain size and roll-off. This water is collected.

Figure 8.25 (a) Water drops on a spider's web. Reproduced from http://www.rsc.org/images/Spider_Web_250_tcm18-209682.jpg. (b) The charitable organization FogQuest installed one of the first fog harvesting devices in Chungungo, Chile in the Atacama Desert in 1992. Six years later with the help of 80 fog collectors, village residents were enjoying an average of 10 000–15 000 litres of fresh drinking water each day, which amounts to an average of 35 litres per person. Reproduced from http://www.climateprep.org. Copyright by Anne Lummerich © at the moment the meshes are made of simple polymer fibres. Reproduced from http://www.fogquest.org/.

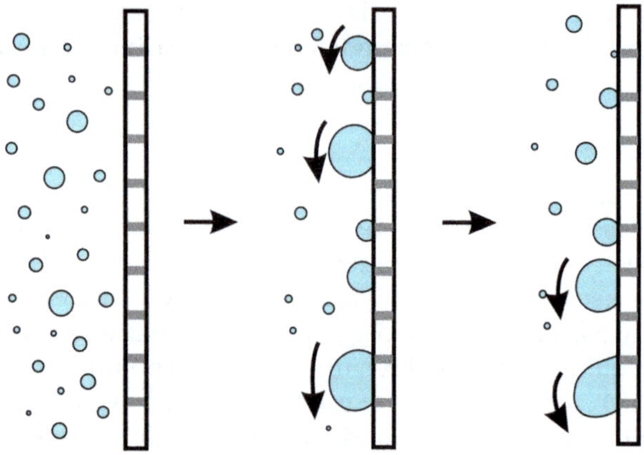

Figure 8.26 Sketch of adhesion and roll-off of water drops on a surface possessing hydrophilic (grey) and superhydrophobic regions.

The idea of collecting water from air in arid areas originates from the ability of beetles (*Stenocara gracilipes*) in the Namib Desert on the west coast of Africa to drink water from the morning dew.[117] The beetle's surface is composed of alternating hydrophilic and hydrophobic areas. The hydrophilic parts collect the fog and the hydrophobic areas repel the condensed drops as soon as they reach a certain size.[118]

Currently meshes collect only a small fraction of the air humidity (Figure 8.25a and b). By facilitating the mechanisms of water condensation and drop collection on solid surfaces and in particular the subsequent droplet roll-off, the efficiency of fog harvesting will be increased. McKinley *et al.* have intensively investigated the optimum design of the mesh with respect of their hydrophobicity, wire thickness and spacing.[119] To maximize the efficiency for fog harvesting efficient condensation of water molecules or water droplets on the surface of the mesh is needed.[119–121] Simultaneously, easy roll-off of the drops is required to prevent evaporation of drops that stick to the mesh. Therefore, surfaces and fabrics that imitate the structure of the *Stenocara gracilipes* beetle have been synthesized.[122] Hydrophilicity favours condensation; however, it inhibits droplet roll-off (Figure 8.26, grey spots). On the contrary, superhydrophobicity inhibits condensation, while favouring roll off. Still it is unclear which material and which topography combine hydrophilicity with superhydrophobicity in a way that optimizes water collection performance. However, no cheap procedure of sufficient mechanical stability has been designed yet, able to provide large-scale production. Furthermore, the optimized meshes need to be cheap, mechanically stable and UV resistant.

8.3 Challenges

For industrial applications, the methods to fabricate superhydrophobic or superamphiphobic surfaces need to be reproducible, cheap and scalable. Compared to superhydrophobic surfaces, it is much more demanding to fabricate the overhang structures that are required to achieve superamphiphobicity. Precise control of the spacing, size and geometry of surface protrusions is crucial to establish a stable Cassie state and to achieve high liquid repellency. The lower the surface tension of the liquid becomes the more careful these parameters need to be chosen. The durability of the superhydrophobic or superamphiphobic coatings should be high. The coating should resist UV exposure, acidic and basic solutions and common solvents. This depends mostly on the chemical nature of the coating's materials. The most challenging factor for industrial application is the long-term mechanical stability of the surface. In general, overhang structures show weaker mechanical resistance towards shearing than pillars or pyramidal shapes. During abrasion, not only will the structure change its geometry, but also the surface chemistry may change from hydrophobic to hydrophilic. This increases adhesion and therefore increases contact angle hysteresis.

Acknowledgements

Financial support from ERC for the advanced grant 340391-SuPro and COST1106 is gratefully acknowledged.

References

1. T. Young, *Philos. Trans. R. Soc. London*, 1805, **95**, 65–87.
2. C. W. Extrand, *Langmuir*, 2006, **22**, 1711–1714.
3. T. Nishino, M. Meguro, K. Nakamae, M. Matsushita and Y. Ueda, *Langmuir*, 1999, **15**, 4321–4323.
4. H. J. Butt, K. Graf and M. Kappl, *Physics and Chemistry of Interfaces*, WILEY-VCH Verlag GmbH & Co. KGaA, Weinheim, 2006.
5. R. N. Wenzel, *Ind. Eng. Chem.*, 1936, **28**, 988–994.
6. A. Lafuma and D. Quere, *Nat. Mater.*, 2003, **2**, 457–460.
7. A. B. D. Cassie and S. Baxter, *Trans. Faraday Soc.*, 1944, **40**, 0546–0550.
8. E. L. Decker and S. Garoff, *Langmuir*, 1997, **13**, 6321–6332.
9. P. Papadopoulos, X. Deng, L. Mammen, D. M. Drotlef, G. Battagliarin, C. Li, K. Müllen, K. Landfester, A. del Campo, H.-J. Butt and D. Vollmer, *Langmuir*, 2012, **28**, 8392–8398.
10. L. Mammen, K. Bley, P. Papadopoulos, F. Schellenberger, N. Encinas, H.-J. Butt, C. K. Weiss and D. Vollmer, *Soft Matter*, 2015, **11**, 506–515.
11. C. V. Boys, *Soap bubbles*, Society for Promoting Christian Knowledge, London, 1902.
12. T. L. Sun, L. Feng, X. F. Gao and L. Jiang, *Acc. Chem. Res.*, 2005, **38**, 644–652.
13. X. F. Gao and L. Jiang, *Nature*, 2004, **432**, 36.
14. C. Neinhuis and W. Barthlott, *Ann. Bot.*, 1997, **79**, 667–677.
15. W. Barthlott and C. Neinhuis, *Planta*, 1997, **202**, 1–8.
16. L. C. Gao and T. J. McCarthy, *Langmuir*, 2006, **22**, 2966–2967.
17. D. K. Owens and R. C. Wendt, *J. Appl. Polym. Sci.*, 1969, **13**, 1741–1747.

18. A. M. Cao, L. L. Cao and D. Gao, *Appl. Phys. Lett.*, 2007, **91**, 034102.
19. L. L. Cao, H. H. Hu and D. Gao, *Langmuir*, 2007, **23**, 4310–4314.
20. S. Herminghaus, *Europhys. Lett.*, 2000, **52**, 165–170.
21. Y. T. Cheng and D. E. Rodak, *Appl. Phys. Lett.*, 2005, **86**, 144101.
22. W. Chen, A. Y. Fadeev, M. C. Hsieh, D. Oner, J. Youngblood and T. J. McCarthy, *Langmuir*, 1999, **15**, 3395–3399.
23. A. Tuteja, W. Choi, M. L. Ma, J. M. Mabry, S. A. Mazzella, G. C. Rutledge, G. H. McKinley and R. E. Cohen, *Science*, 2007, **318**, 1618–1622.
24. L. L. Cao, T. P. Price, M. Weiss and D. Gao, *Langmuir*, 2008, **24**, 1640–1643.
25. A. Ahuja, J. A. Taylor, V. Lifton, A. A. Sidorenko, T. R. Salamon, E. J. Lobaton, P. Kolodner and T. N. Krupenkin, *Langmuir*, 2008, **24**, 9–14.
26. X. Deng, L. Mammen, H. J. Butt and D. Vollmer, *Science*, 2012, **335**, 67–70.
27. T. L. Liu and C.-J. C. J. Kim, *Science*, 2014, **346**, 1096–1100.
28. R. Hensel, R. Helbig, S. Aland, H. G. Braun, A. Voigt, C. Neinhuis and C. Werner, *Langmuir*, 2013, **29**, 1100–1112.
29. Z. G. Guo, W. M. Liu and B. L. Su, *J. Colloid Interface Sci.*, 2011, **353**, 335–355.
30. J. Bico, C. Marzolin and D. Quéré, *Europhys. Lett.*, 1999, **47**, 220–226.
31. D. Bartolo, F. Bouamrirene, E. Verneuil, A. Buguin, P. Silberzan and S. Moulinet, *Europhys. Lett.*, 2006, **74**, 299–305.
32. C. Dorrer and J. Rühe, *Soft Matter*, 2009, **5**, 51–61.
33. C. Priest, T. W. J. Albrecht, R. Sedev and J. Ralston, *Langmuir*, 2009, **25**, 5655–5660.
34. D. Xiong, G. J. Liu, L. Z. Hong and E. J. S. Duncan, *Chem. Mater.*, 2011, **23**, 4357–4366.
35. A. Steele, I. Bayer and E. Loth, *Nano Lett.*, 2009, **9**, 501–505.
36. R. Dufour, M. Harnois, V. Thomy, R. Boukherroub and V. Senez, *Soft Matter*, 2011, 7, 9380–9387.
37. T. Darmanin, F. Guittard, S. Amigoni, E. T. de Givenchy, X. Noblin, R. Kofman and F. Celestini, *Soft Matter*, 2011, 7, 1053–1057.
38. H. J. Butt, C. Semprebon, P. Papadopoulos, D. Vollmer, M. Brinkmann and M. Ciccotti, *Soft Matter*, 2013, **9**, 418–428.
39. M. R. Flynn and J. W. M. Bush, *J. Fluid Mech.*, 2008, **608**, 275–296.
40. M. L. Blow and J. M. Yeomans, *Langmuir*, 2010, **26**, 16071–16083.
41. K. Tsujii, T. Yamamoto, T. Onda and S. Shibuichi, *Angew. Chem., Int. Ed. Engl.*, 1997, **36**, 1011–1012.
42. B. He, N. A. Patankar and J. Lee, *Langmuir*, 2003, **19**, 4999–5003.
43. E. Bormashenko, R. Pogreb, G. Whyman and M. Erlich, *Langmuir*, 2007, **23**, 6501–6503.
44. T. N. Krupenkin, J. A. Taylor, E. N. Wang, P. Kolodner, M. Hodes and T. R. Salamon, *Langmuir*, 2007, **23**, 9128–9133.
45. P. Papadopoulos, L. Mammen, X. Deng, D. Vollmer and H. J. Butt, *Proc. Natl. Acad. Sci.*, 2013, **110**, 3254–3258.
46. M. Reyssat, J. M. Yeomans and D. Quere, *Europhys. Lett.*, 2008, **81**, 26006.
47. D. Bartolo, F. Bouamrirene, E. Verneuil, A. Buguin, P. Silberzan and S. Moulinet, *Europhys. Lett.*, 2006, **74**, 299–305.

48. J. F. Oliver, C. Huh and S. G. Mason, *J. Colloid Interface Sci.*, 1977, **59**, 568–581.
49. A. ElSherbini and A. Jacobi, *J. Colloid Interface Sci.*, 2006, **299**, 841–849.
50. T. Verho, C. Bower, P. Andrew, S. Franssila, O. Ikkala and R. H. A. Ras, *Adv. Mater.*, 2011, **23**, 673–678.
51. C.-H. Xue and J.-Z. Ma, *J. Mater. Chem. A*, 2013, **1**, 4146–4161.
52. D. Ebert and B. Bhushan, *J. Colloid Interface Sci.*, 2012, **368**, 584–591.
53. J. Groten and J. Rühe, *Langmuir*, 2013, **29**, 3765–3772.
54. Y. C. Jung and B. Bhushan, *ACS Nano*, 2009, **3**, 4155–4163.
55. B. P. Dyett, A. H. Wu and R. N. Lamb, *ACS Appl. Mater. Interfaces*, 2014, **6**, 18380–18394.
56. E. J. Lee, J. J. Kim and S. O. Cho, *Langmuir*, 2010, **26**, 3024–3030.
57. X. Deng, L. Mammen, Y. F. Zhao, P. Lellig, K. Mullen, C. Li, H. J. Butt and D. Vollmer, *Adv. Mater.*, 2011, **23**, 2962–2965.
58. Y. Wu, M. Bekke, Y. Inoue, H. Sugimura, H. Kitaguchi, C. Liu and O. Takai, *Thin Solid Films*, 2004, **457**, 122–127.
59. H. Wang, Y. Xue, J. Ding, L. Feng, X. Wang and T. Lin, *Angew. Chem., Int. Ed.*, 2011, **50**, 11433–11436.
60. P. S. Brown and B. Bhushan, *Sci. Rep.*, 2015, **5**, 8701.
61. X. Shi, T. A. Nguyen, Z. Suo, J. Wu, J. Gong and R. Avci, *Surf. Coat. Technol.*, 2012, **206**, 3700–3713.
62. M. L. B. Palacio and B. Bhushan, *Mater. Charact.*, 2013, **78**, 1–20.
63. B. Bhushan, *Wear*, 1999, **225–229**, **Part 1**, 465–492.
64. X. Feng and L. Jiang, *Adv. Mater.*, 2006, **18**, 3063–3078.
65. B. Bhushan and Y. C. Jung, *Prog. Mater. Sci.*, 2011, **56**, 1–108.
66. G. McHale, M. I. Newton and N. J. Shirtcliffe, *Soft Matter*, 2010, **6**, 714–719.
67. J. P. Rothstein, *Annual Review of Fluid Mechanics*, Annual Reviews, Palo Alto, 2010, vol. 42, pp. 89–109.
68. P. Joseph, C. Cottin-Bizonne, J. M. Benoit, C. Ybert, C. Journet, P. Tabeling and L. Bocquet, *Phys. Rev. Lett.*, 2006, **97**, 156104.
69. R. B. Pernites, C. M. Santos, M. Maldonado, R. R. Ponnapati, D. F. Rodrigues and R. C. Advincula, *Chem. Mater.*, 2012, **24**, 870–880.
70. Y. Koc, A. J. de Mello, G. McHale, M. I. Newton, P. Roach and N. J. Shirtcliffe, *Lab Chip*, 2008, **8**, 582–586.
71. C. Pang, G. Y. Lee, T. I. Kim, S. M. Kim, H. N. Kim, S. H. Ahn and K. Y. Suh, *Nat. Mater.*, 2012, **11**, 795–801.
72. N. Andreeva, T. Ishizaki, P. Baroch and N. Saito, *Sens. Actuators, B*, 2012, **164**, 15–21.
73. Y. Gao, T. Chen, S. Yamamoto, T. Miyashita and M. Mitsuishi, *ACS Appl. Mater. Interfaces*, 2015, **7**, 3468–3472.
74. H. F. Hoefnagels, D. Wu, G. de With and W. Ming, *Langmuir*, 2007, **23**, 13158–13163.
75. G. R. J. Artus, J. Zimmermann, F. A. Reifler, S. A. Brewer and S. Seeger, *Appl. Surf. Sci.*, 2012, **258**, 3835–3840.
76. M. Yu, G. Gu, W.-D. Meng and F.-L. Qing, *Appl. Surf. Sci.*, 2007, **253**, 3669–3673.

77. M. Paven, P. Papadopoulos, S. Schoettler, X. Deng, V. Mailaender, D. Vollmer and H.-J. Butt, *Nat. Commun.*, 2013, **4**, 2512.
78. X. Deng, M. Paven, P. Papadopoulos, M. Ye, S. Wu, T. Schuster, M. Klapper, D. Vollmer and H.-J. Butt, *Angew. Chem., Int. Ed.*, 2013, **52**, 11286–11289.
79. V. Rastogi, S. Melle, O. G. Calderon, A. A. Garcia, M. Marquez and O. D. Velev, *Adv. Mater.*, 2008, **20**, 4263–4268.
80. V. Rastogi, A. A. Garcia, M. Marquez and O. D. Velev, *Macromol. Rapid Commun.*, 2010, **31**, 190–195.
81. B. Comiskey, J. D. Albert, H. Yoshizawa and J. Jacobson, *Nature*, 1998, **394**, 253–255.
82. J. Du and R. K. O'Reilly, *Chem. Soc. Rev.*, 2011, **40**, 2402–2416.
83. K. Matyjaszewski and T. P. Davis, *Handbook of radical polymerization*, Wiley Online Library, 2002.
84. D. Dendukuri and P. S. Doyle, *Adv. Mater.*, 2009, **21**, 4071–4086.
85. M. Sperling, O. D. Velev and M. Gradzielski, *Angew. Chem., Int. Ed.*, 2014, **53**, 586–590.
86. A. G. Marin, H. Gelderblom, A. Susarrey-Arce, A. van Houselt, L. Lefferts, J. G. E. Gardeniers, D. Lohse and J. H. Snoeijer, *Proc. Natl. Acad. Sci. U. S. A.*, 2012, **109**, 16455–16458.
87. K. Ikemura and T. Endo, *Dent. Mater. J.*, 2010, **29**, 481–501.
88. R. A. N. Chmielewski and J. F. Frank, *J. Food Prot.*, 2004, **67**, 2712–2718.
89. R. M. Klevens, J. R. Edwards, C. L. Richards Jr., T. C. Horan, R. P. Gaynes, D. A. Pollock and D. M. Cardo, *Public Health Rep.*, 2007, **122**, 160–166.
90. J. W. Costerton, P. S. Stewart and E. P. Greenberg, *Science*, 1999, **284**, 1318–1322.
91. J. N. Wilking, T. E. Angelini, A. Seminara, M. P. Brenner and D. A. Weitz, *MRS Bull.*, 2011, **36**, 385–391.
92. X. Chen and H. J. Schluesener, *Toxicol. Lett.*, 2008, **176**, 1–12.
93. P. S. Stewart and J. W. Costerton, *Lancet*, 2001, **358**, 135–138.
94. I. Banerjee, R. C. Pangule and R. S. Kane, *Adv. Mater.*, 2011, **23**, 690–718.
95. K. D. Park, Y. S. Kim, D. K. Han, Y. H. Kim, E. H. B. Lee, H. Suh and K. S. Choi, *Biomaterials*, 1998, **19**, 851–859.
96. S. Mazumder, J. O. Falkinham, A. M. Dietrich and I. K. Puri, *Biofouling*, 2010, **26**, 333–339.
97. D. C. Leslie, A. Waterhouse, J. B. Berthet, T. M. Valentin, A. L. Watters, A. Jain, P. Kim, B. D. Hatton, A. Nedder, K. Donovan, E. H. Super, C. Howell, C. P. Johnson, T. L. Vu, D. E. Bolgen, S. Rifai, A. R. Hansen, M. Aizenberg, M. Super, J. Aizenberg and D. E. Ingber, *Nat. Biotech.*, 2014, **32**, 1134–1140.
98. R. J. Crawford, H. K. Webb, T. Vi Khanh, J. Hasan and E. P. Ivanova, *Adv. Colloid Interface Sci.*, 2012, **179**, 142–149.
99. J. Genzer and K. Efimenko, *Biofouling*, 2006, **22**, 339–360.
100. X. Hou, X. Wang, Q. Zhu, J. Bao, C. Mao, L. Jiang and J. Shen, *Colloids Surf., B*, 2010, **80**, 247–250.

101. H. Fan, P. Chen, R. Qi, J. Zhai, J. Wang, L. Chen, L. Chen, Q. Sun, Y. Song, D. Han and L. Jiang, *Small*, 2009, **5**, 2144–2148.
102. B. J. Privett, J. Youn, S. A. Hong, J. Lee, J. Han, J. H. Shin and M. H. Schoenfisch, *Langmuir*, 2011, **27**, 9597–9601.
103. G. J. Toes, K. W. van Muiswinkel, W. van Oeveren, A. J. H. Suurmeijer, W. Timens, I. Stokroos and J. van den Dungen, *Biomaterials*, 2002, **23**, 255–262.
104. T. Ishizaki, N. Saito and O. Takai, *Langmuir*, 2010, **26**, 8147–8154.
105. S. M. Oliveira, W. Song, N. M. Alves and J. F. Mano, *Soft Matter*, 2011, **7**, 8932–8941.
106. B. Leng, Z. Shao, G. de With and W. Ming, *Langmuir*, 2009, **25**, 2456–2460.
107. M. C. Yang and E. L. Cussler, *AIChE J.*, 1986, **32**, 1910–1916.
108. R. Prasad and K. K. Sirkar, *AIChE J.*, 1988, **34**, 177–188.
109. A. Gabelman and S. T. Hwang, *J. Membr. Sci.*, 1999, **159**, 61–106.
110. W. D. Zhang, J. A. Li, G. Chen, W. You, Y. Jiang and W. Sun, *Ind. Eng. Chem. Res.*, 2010, **49**, 6641–6648.
111. C. S. Feng, R. Wang, H. Y. Zhang and L. Shi, *J. Appl. Polym. Sci.*, 2011, **119**, 1259–1267.
112. L. F. Greenlee, D. F. Lawler, B. D. Freeman, B. Marrot and P. Moulin, *Water Res.*, 2009, **43**, 2317–2348.
113. J. C. Cuenca, *Report on the Water Desalination Status in the Mediterranean Countries*, Instituto Murciano de Investigación y Desarollo Agroalimentario Murcia, Spain, 2012.
114. M. Elimelech and W. A. Phillip, *Science*, 2011, **333**, 712–717.
115. J. Olivier, *Water Sa*, 2002, **28**, 349–360.
116. S. A. Abdul-Wahab, H. Al-Hinai, K. A. Al-Najar and M. S. Al-Kalbani, *Environ. Eng. Sci.*, 2007, **24**, 446–456.
117. T. Norgaard and M. Dacke, *Front. Zool.*, 2010, **7**, 23.
118. A. R. Parker and C. R. Lawrence, *Nature*, 2001, **414**, 33–34.
119. K.-C. Park, S. S. Chhatre, S. Srinivasan, R. E. Cohen and G. H. McKinley, *Langmuir*, 2013, **29**, 13269–13277.
120. M. A. K. Azad, D. Ellerbrok, W. Barthlott and K. Koch, *Bioinspiration Biomimetics*, 2015, **10**, 016004.
121. L. Zhang, J. Wu, M. N. Hedhili, X. Yang and P. Wang, *J. Mater. Chem. A*, 2015, **3**, 2844–2852.
122. C. Dorrer and J. Ruhe, *Langmuir*, 2008, **24**, 6154–6158.

Subject Index